Lecture Notes in Artificial Intelligence 13212

Subseries of Lecture Notes in Computer Science

Series Editors

Randy Goebel
University of Alberta, Edmonton, Canada

Wolfgang Wahlster
DFKI, Berlin, Germany

Zhi-Hua Zhou
Nanjing University, Nanjing, China

Founding Editor

Jörg Siekmann
DFKI and Saarland University, Saarbrücken, Germany

More information about this subseries at https://link.springer.com/bookseries/1244

Zygmunt Vetulani · Patrick Paroubek ·
Marek Kubis (Eds.)

Human Language Technology

Challenges for Computer Science and Linguistics

9th Language and Technology Conference, LTC 2019
Poznan, Poland, May 17–19, 2019
Revised Selected Papers

 Springer

Editors
Zygmunt Vetulani 🆔
Adam Mickiewicz University
Poznań, Poland

Patrick Paroubek 🆔
LIMSI-CNRS
Orsay, France

Marek Kubis 🆔
Adam Mickiewicz University
Poznań, Poland

ISSN 0302-9743   ISSN 1611-3349 (electronic)
Lecture Notes in Artificial Intelligence
ISBN 978-3-031-05327-6   ISBN 978-3-031-05328-3 (eBook)
https://doi.org/10.1007/978-3-031-05328-3

LNCS Sublibrary: SL7 – Artificial Intelligence

Preface

Sometimes the sudden occurrence of dire events like natural disasters, pandemic outbreaks, or war force us to face the question of our motivations for doing science and of the pertinence of our actions as time accelerates and uncertainty prevails. Considering the answer to such questions may seem a daunting task and a vain endeavour, but not doing anything would be much worse than facing the risk of failure. Resilience. It is a word that we seem to encounter more and more nowadays. It is definitely not the full answer, but we think it is good candidate for a start. To pursue the quest of new knowledge gained by the joint forces of everybody in an open and friendly collaboration and making it widely available is what we strive for and will continue to do. This is the answer we chose. Humbly, we offer you this volume which contains a selection of 24 revised and, in most cases, substantially updated and extended versions of papers presented at the 9th Language and Technology Conference (LTC 2019). The LTC reviewing process was completed by an international jury composed of the Program Committee members or experts nominated by them. The selection was made among 61 contributions presented at the conference and essentially represents the preferences of the reviewers. The 72 authors of the selected contributions represent the research achievements of more than 30 institutions from Australia, France, Georgia, Germany, Japan, Kazakhstan, Nigeria, Poland, Russia, Spain, Thailand, Uzbekistan, and Vanuatu.

Papers selected for this volume contribute to various fields of human language technologies and illustrate the large thematic coverage of the LTC conferences. We have grouped them into 7 chapters

1. Speech Processing (5 papers)
2. Language Resources and Tools (5 papers)
3. Computational Semantics (2 papers)
4. Emotions, Decisions and Opinions (4 papers)
5. Digital Humanities (3 papers)
6. Evaluation (4 papers)
7. Legal Aspects (1 paper)

The clustering of these papers into chapters is approximate as many papers address several thematic areas. Unlike the previous LTC post-conference monographs published in the LNAI series, we chose not to single out contributions on less-resourced languages, although many of them referred directly to this language group. Within chapters, we have ordered the contributions alphabetically according to the surname of the first author. A summary of the content of this volume is provided below[1].

---

[1] In the review, the editors decided to synthetically and accurately indicate the subject of each paper, so it should come as no surprise that they often use the expressions of the authors themselves, which are easy to find in the text.

Following the above mentioned thematic order, we start this volume with the Speech Processing chapter containing five contributions. In the paper "A set of tools for extraction and manipulation of speech fundamental frequency" Jolanta Bachan describes a set of tools created and used for fundamental frequency extraction and manipulation, prosody and speech perception analysis, and speech synthesis. The aim of the paper "The automatic search for sounding segments of SPPAS: application to Cheese! corpus" by Brigitte Bigi and Béatrice Priego-Valverde is to describe a method to search for interpausal units in recorded speech corpora. The third contribution concerning speech, "The Phonetic Grounding of Prosody: Analysis and Visualisation Tools" by Dafydd Gibbon, provides a detailed description of a suite of related online and offline analysis and visualisation tools for training students of phonetics in the acoustics of prosody. The paper "Hybridised Deep Ensemble Learning for Tone Pattern Recognition", by Udoinyang G. Inyang and Moses E. Ekpenyong, proposes a hybridised ensemble of three heterogeneous classifiers operating on a single speech dataset for tone languages (e.g., Ibibio and Yoruba from Nigeria). The research on speech segmentation and recognition is the main topic of the final paper of this chapter, "ANNPRO: a desktop module for automatic segmentation and transcription" by Katarzyna Klessa, Danijel Koržinek, Brygida Sawicka-Stępińska, and Hanna Kasperek.

The Language Resources and Tools chapter also contains five papers. The paper "Analysis and processing of the Uzbek language on the multi-language modelled computer translator technology", by Mersaid Aripov, Muftakh Khakimov, Sanatbek Matlatipov, and Ziyoviddin Sirojiddinov, presents the construction of logical-linguistic models for words and sentences of the Uzbek language. The next paper, "Computer Application of Georgian Words" by Irakli Kardava, Nana Gulua, Jemal Antidze, Beka Toklikishvili, and Tamta Kvaratskhelia, proposes a set of software tools for generating Georgian words. The article "Community-led documentation of Nafsan (Erakor, Vanuatu)", by Ana Krajinović, Rosey Billington, Lionel Emil, Gray Kaltap̃au, and Nick Thieberger, presents outcomes of collaboration between local community members and visiting linguists in Erakor, Vanuatu, aiming to build the capacity of community-based researchers to undertake and sustain documentation of Nafsan, the local indigenous language. The fourth contribution, "Design and Development of Pipeline of Preprocessing Tools for Kazakh Language Texts" by Madina Mansurova, Vladimir B. Barakhnin, Gulmira Madiyeva, Nurgali Kadyrbek, and Bekzhan Dossanov, is about the design and development of pre-processing tools for a media-corpus of the Kazakh language. This chapter ends with the paper "Thai Named Entity Corpus Annotation Scheme and Self Verification by BiLSTM-CNN-CRF", written by Virach Sornlertlamvanich, Kitiya Suriyachay, and Thatsanee Charoenporn, where the authors propose a Thai language oriented approach to clean up the existing named entity corpus for the Thai language for purposes of further research.

Two papers constitute the Computational Semantics chapter. Yves Lepage, the author of the first contribution, "Analogies Between Short Sentences: a Semantico-Formal Approach", proposes a method to solve analogies between sentences by combining existing techniques to solve formal analogies between strings and semantic analogies between words. In the second paper, "Effective development and deployment of domain and application conceptualization in wordnet-based ontologies using Ontology Repository Tool", Jacek Marciniak presents the process of building and developing the PMAH wordnet-based ontology using the Ontology Repository Tool software, which is created in parallel with the process of indexing content in two information systems: the E-archaeology Content Repository and the Hatch system storing multimodal data.

The area of Emotions, Decisions and Opinions is represented by four papers. The paper "Speech Prosody Extraction for Ibibio Emotions Analysis and Classification", by Moses E. Ekpenyong, Aniekan J. Ananga, Ememobong O. Udoh, and Nnamso M. Umoh, reports on basic prosodic features of speech for the analysis and classification of Ibibio (New Benue Congo, Nigeria) emotions, at the suprasegmental level. In the second paper in this chapter, "Multilingual and language-agnostic recognition of emotions, valence and arousal in large-scale multi-domain text reviews", Jan Kocoń, Piotr Miłkowski, Małgorzata Wierzba, Barbara Konat, Katarzyna Klessa, Arkadiusz Janz, Monika Riegel, Konrad Juszczyk, Damian Grimling, Artur Marchewka, and Maciej Piasecki present their results in emotion recognition in Polish texts with the help of machine learning and text engineering techniques used to automatically detect emotions expressed in natural language. The next paper, "Construction and evaluation of sentiment datasets for low-resource languages: the case of Uzbek" by Elmurod Kuriyozov, Sanatbek Matlatipov, Miguel A. Alonso, and Carlos Gómez-Rodíguez, presents the results of research undertaken in order to provide the first annotated corpora for Uzbek language polarity classification. The final paper of this chapter, "Using Book Dialogues to Extract Emotions from Texts" by Paweł Skórzewski, describes two methods of training a text-based emotion detection model using a corpus of annotated book dialogues as a training set.

Digital Humanities is an area of research that started with the advent of the computer era after World War II, but its name appeared only at the beginning of the 21st century, and the field itself is experiencing a period of dynamic growth, perfectly fitting into the scope of the LTC conference, particularly given its subtitle of "Human Language Technologies as a Challenge for Computer Science and Linguistics". This field is represented in this volume by three contributions. In the first paper, "NLP tools for lexical structure studies of the literary output of a writer. Case study: literary works of Tadeusz Boy-Żeleński and Julia Hartwig, Zygmunt Vetulani, Marta Witkowska, and Marek Kubis present the use of NLP tools to study the lexical structure of a representative part of the literary output of two outstanding Polish writers of the 20th century. The next contribution, "Neural nets in detecting word level metaphors in Polish" by Aleksander Wawer, Małgorzata Marciniak, and Agnieszka Mykowiecka, presents an experiment in detecting metaphorical usage of adjectives and nouns in Polish data, describes two methods for literal/metaphorical sense classification, and provides test results for the two architectures for identification of metaphorical use of noun and adjective pairs in Polish texts. The last paper of this chapter, "Frame-based annotation in the corpus of synesthetic metaphors" by Magdalena Zawisławska, describes the project aimed at creating a semantically and grammatically annotated corpus of Polish synesthetic metaphors.

The domain of digital language resources quality evaluation and processing, initiated by Antonio Zampolli in 1990s, countinues to gain more and more importance in the languages industry. The first of the four contributions in the Evaluation chapter is the paper "PolEval 2019 — the next chapter in evaluating Natural Language Processing tools for Polish", by Łukasz Kobyliński, Maciej Ogrodniczuk, Jan Kocoń, Michał Marcińczuk, Aleksander Smywiński-Pohl, Krzysztof Wołk, Danijel Koržinek, Michał Ptaszynski, Agata Pieciukiewicz, and Paweł Dybała, which presents a SemEval-inspired evaluation campaign for natural language processing tools for Polish – PolEval (first organized in 2017). In the second paper, "Open Challenge for Correcting Errors of Speech Recognition Systems", Marek Kubis, Zygmunt Vetulani, Mikołaj Wypych, and Tomasz Ziętkiewicz present a newly defined (2019) long-term challenge for improving automatic speech recognition by investigating methods of correcting the recognition results on the basis of previously made errors by the speech processing system. The next paper, "Evaluation of basic modules for isolated spelling error correction in Polish texts" by Szymon Rutkowski, presents evaluation experiments for isolated spelling errors in Polish texts. The last of the four papers, "Assessment of document similarity visualisation methods" by Mateusz Gniewkowski and Tomasz Walkowiak, concerns the problem of assessing the similarity of document visualization.

The last chapter is devoted to Legal Aspects and contains just one paper. This subject, of particular importance to producers and users of language technology, is unfortunately rarely presented at conferences addressed to this field, such as LTC or LREC. The reader will find the paper "Legal Regime of the Language Resources in the Context of the European Language Technology Development" by Ilya Ilin, in which the author examines language resources from two perspectives: firstly, with language resources being considered as a database covered by the protection regulations, and secondly, as the object of legal analysis from the point of view of the materials used for their creation.

We wish you all an interesting read.

March 2022

Zygmunt Vetulani
Patrick Paroubek
Marek Kubis

Organization

Organizing Committee Chair

Zygmunt Vetulani
Adam Mickiewicz University, Poland

Organizing Committee

Jolanta Bachan	Adam Mickiewicz University, Poland
Marek Kubis	Adam Mickiewicz University, Poland
Jacek Marciniak	Adam Mickiewicz University, Poland
Tomasz Obrębski	Adam Mickiewicz University, Poland
Tamara Greinert Czekalska	Adam Mickiewicz University Foundation, Poland
Marta Witkowska	Adam Mickiewicz University, Poland
Mateusz Witkowski	Adam Mickiewicz University, Poland

Program Committee Chairs

Zygmunt Vetulani	Adam Mickiewicz University, Poland
Patrick Paroubek	LIMSI-CNRS, France

Program Committee

Victoria Arranz	ELRA, France
Jolanta Bachan	Adam Mickiewicz University, Poland
Núria Bel	Universitat Pompeu Fabra, Spain
Krzysztof Bogacki	University of Warsaw, Poland
Christian Boitet	Université Grenoble Alpe, France
Gerhard Budin	University of Vienna, Austria
Nicoletta Calzolari	ILC-CNR, Italy
Nick Campbell	Trinity College Dublin, Ireland
Khalid Choukri	ELRA, France
Christopher Cieri	Linguistic Data Consortium, USA
Paweł Dybała	Jagiellonian University, Poland
Katarzyna Dziubalska-Kołaczyk	Adam Mickiewicz University, Poland
Moses Ekpenyong	Uyo University, Nigeria
Cedrick Fairon	University of Louvain, Belgium
Piotr Fuglewicz	TIP Sp. z o.o., Poland
Maria Gavrilidou	ILSP, Greece

Dafydd Gibbon	University of Bielefeld, Germany
Marko Grobelnik	Jožef Stefan Institute, Slovenia
Eva Hajičová	Charles University, Czech Republic
Krzysztof Jassem	Adam Mickiewicz University, Poland
Girish Nath Jha	Jawaharlal Nehru University, India
Katarzyna Klessa	Adam Mickiewicz University, Poland
Cvetana Krstev	University of Belgrade, Serbia
Yves Lepage	Waseda University, Japan
Gerard Ligozat	LIMSI-CNRS, France
Natalia Loukachevitch	Moscow State University, Russia
Wiesław Lubaszewski	AGH UST, Poland
Bente Maegaard	Centre for Language Technology, Denmark
Bernardo Magnini	FBK-ICT Irst, Italy
Jacek Marciniak	Adam Mickiewicz University, Poland
Joseph Mariani	LIMSI-CNRS, France
Jacek Martinek	Poznań University of Technology, Poland
Gayrat Matlatipov	Urgench State University, Uzbekistan
Keith J. Miller	MITRE, USA
Asunción Moreno	UPC, Spain
Agnieszka Mykowiecka	IPI PAN, Poland
Jan Odijk	Utrecht University, The Netherlands
Maciej Ogrodniczuk	IPI PAN, Poland
Karel Pala	Masaryk University, Czech Republic
Pavel S. Pankov	National Academy of Sciences, Kyrgyzstan
Maciej Piasecki	Wrocław University of Technology, Poland
Stelios Piperidis	ILSP, Greece
Gabor Proszeky	Morphologic, Hungary
Georg Rehm	DFKI, Germany
Michał Ptaszyński	University of Hokkaido, Japan
Rafał Rzepka	University of Hokkaido, Japan
Kepa Sarasola Gabiola	Universidad del Pas Vasco, Spain
Sanja Seljan	University of Zagreb, Croatia
Claudia Soria	ILC-CNR, Italy
Janusz Taborek	Adam Mickiewicz University, Poland
Ryszard Tadeusiewicz	AGH UST, Poland
Marko Tadić	University of Zagreb, Croatia
Dan Tufiş	RCAI, Romania
Tamás Váradi	RIL, Hungarian Academy of Sciences, Hungary
Andrejs Vasiljevs	Tilde, Latvia
Cristina Vertan	University of Hamburg, Germany
Dusko Vitas	University of Belgrade, Serbia
Piek Vossen	Vrije Universiteit Amsterdam, The Netherlands

Jan Węglarz   Poznań University of Technology, Poland
Bartosz Ziółko   AGH UST, Poland
Mariusz Ziółko   AGH UST, Poland
Andrzej Zydroń   XTM International, UK

Reviewers

Jolanta Bachan   Michael Maxwell
Delphine Bernhard   Michal Mazur
Brigitte Bigi   Alice Millour
Nergis Biray   Dominika Narożna
Krzysztof Bogacki   Girish Nath Jha
Nicoletta Calzolari   Tomasz Obrębski
Paweł Dybała   Jan Odijk
Moses Ekpenyong   Maciej Ogrodniczuk
Jeremy Evas   Patrick Paroubek
Mikel Forcada   Laurette Pretorius
Dimitrios Galanis   Gábor Prószéky
Maria Gavrilidou   Piotr Przybyła
Dafydd Gibbon   Michał Ptaszyński
Eva Hajicova   Georg Rehm
Annette Herkenrath   Mike Rosner
Muhammad Imran   Rafał Rzepka
Krzysztof Jassem   Kevin Scannell
Rafał Jaworski   Sanja Seljan
László Károly   Marcin Sikora
Maciej Karpinski   Paweł Skórzewski
Katarzyna Klessa   Claudia Soria
Steven Krauwer   Janusz Taborek
Cvetana Krstev   Ryszard Tadeusiewicz
Marek Kubis   Marko Tadic
Yves Lepage   Dan Tufis
Anne-Laure Ligozat   Zygmunt Vetulani
Belinda Maia   Dusko Vitas
Jacek Marciniak   Leo Wanner
Małgorzata Marciniak   Marcin Włodarczak
Joseph Mariani   Bartosz Ziolko
Jacek Martinek

Contents

Computational Semantics

Emotions, Decisions and Opinions

Digital Humanities

Evaluation

Legal Aspects

Speech Processing

A Set of Tools for Extraction and Manipulation of Speech Fundamental Frequency

Jolanta Bachan$^{(\boxtimes)}$ (iD)

Adam Mickiewicz University, ul. Wieniawskiego 1, 61-712 Poznań, Poland
jbachan@amu.edu.pl

Abstract. The present paper describes a set of tools created and used for fundamental frequency extraction and manipulation, prosody and speech perception analysis and speech synthesis. The tools were implemented as Praat and Python scripts and were created for different purposes in projects over the last few years. The paper presents the functionality and possible usage of the tools in phonetic research.

Keywords: Fundamental frequency (F_0) · Prosody analysis · F_0 extraction · Speech synthesis · Speech annotation

1 Introduction

The present paper describes a set of tools created for fundamental frequency (F_0) extraction and manipulation, prosody and speech perception analysis and speech synthesis, for use as direct empirical models rather than mediated through a Tilt, Fujisaki, Hirst or other model. The tools were implemented as Praat (Boersma 2001) and Python scripts and were created for different purposes and projects over the last few years. The paper presents the functionality and possible usage of the tools in phonetic research and teaching.

2 Automatic Close Copy Speech Synthesis

Automatic close copy speech (ACCS) synthesis tool extracts the sequence of phones, their durations and fundamental frequency from the original recording and annotations, to create a file in the format required by the speech synthesiser. In the ACCS synthesis the original voice is replaced by the synthetic voice, while keeping the original speech prosody parameters (Dutoit 1997). Details of the ACCS synthesis method were already described in (Bachan 2007) and (Gibbon and Bachan 2008), but the currently presented version is a new standalone script written in Praat. It takes the phone labels and their durations from the TextGrid annotation file and extracts the mean F_0 for each phone interval. Such data are printed to an output file in the PHO file format which is required by the MBROLA speech synthesiser (Dutoit 2006). MRBOLA uses an older diphone concatenation strategy, but is preferred for this purpose because of the clear linguistic interface: each phone is associated explicitly with a duration and a fundamental frequency vector. The general architecture of the ACCS synthesis is presented in Fig. 1.

© Springer Nature Switzerland AG 2022
Z. Vetulani et al. (Eds.): LTC 2019, LNAI 13212, pp. 3–15, 2022.
https://doi.org/10.1007/978-3-031-05328-3_1

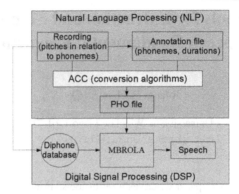

Fig. 1. The architecture of ACCS synthesis system.

The format of the PHO file in columns is as follows:

1. phone labels (usually SAMPA for a given language, Wells 1997),
2. duration of the phones in milliseconds,
3. pitch pairs:

    a. pitch position (by default equals 50, meaning 50% of the phone duration, i.e. the middle of the phone),
    b. pitch values in Hz.

An excerpt of the PHO file is presented in Fig. 2. The unvoiced phones do not have to have the pitch pairs, they may only have the durations.

Z	98	50	110
e	42	50	118
b	51	50	117
I	39	50	121
k	56		
t	50	50	139
o	46	50	124
s'	75	50	108
n	53	50	122
a	60	50	113
t	70	50	154
I	40	50	134
x	60	50	131
m	46	50	138
j	25	50	133
a	29	50	126
s	80	50	123
t	65		

Fig. 2. An excerpt of the automatically generated PHO file in ACCS synthesis.

If the phone labels in a TextGrid annotation file are the same as the ones used by the MBROLA voice, the PHO file can be input to MBROLA and synthesised without any additional phone-to-phone mapping rules (i.e. when there is a mismatch between the phone labels used in the annotation file and their counterparts in the MBROLA voice inventory). An exemplar speech file is presented in Fig. 3, uttered by a male speaker. On the top, there is the original speech sound (oscillogram and spectrogram) and its annotation, with the solid line on the spectrogram corresponding to the F_0. On the bottom there is the same utterance, but synthesised in MBROLA using the pl2 male voice (Bachan 2011).

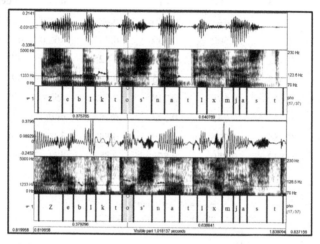

Fig. 3. The original male recording (top) and its ACCS synthesised counterpart (bottom).

The ACCS synthesis tool may be used for creating stimuli for speech perception tests as well as checking the quality of time-aligned annotations of big speech corpora.

3 F_0 Manipulation and Speech Resynthesis

3.1   Pitch Contour Replacement

The F_0 manipulation tool is used for the replacement of the F_0 contour between two the same utterances of different prosody. For example, on the neutral utterances the F_0 values extracted from the emotional speech are imposed. The durations stay the same for the neutral speech, but the F_0 contour is replaced with the emotional one (Oleśkowicz-Popiel and Bachan 2017).

The F_0 manipulation tool is composed of a set of Praat and Python scripts. First, the F_0 is extracted from the target speech, for example from the emotional recording, using a Praat script. The script takes information from the TextGrid annotation file about the phones and their durations on the phone annotation tier. Then each phone duration is divided into three intervals: 0%–20%, 20%–80% and 80%–100% of phone duration and the mean pitch value is extracted for each of the intervals from the corresponding WAV file. The data from this step are saved in a text file with .F0 extension for each of the file in a directory. The extracted data and the .F0 format are presented in Table 1.

Table 1. Exemplar data from the .F0 file format, female voice, emotional recording; time is in seconds, the three last columns show the F_0 values in Hz, *undef* stands for undefined (voiceless phones).

Phone	Duration	Start time	Time 20%	Time 80%	End time	0%–20%	20%–80%	80%–100%
s	0.089	0.343	0.360	0.414	0.432	undef	undef	undef
k	0.055	0.432	0.443	0.476	0.487	undef	429.36	451.66
o	0.064	0.487	0.500	0.539	0.552	460.01	450.03	489.57
Z	0.073	0.552	0.567	0.610	0.625	436.69	373.13	429.97
y	0.036	0.625	0.632	0.655	0.662	466.49	493.85	480.03
s	0.070	0.662	0.676	0.718	0.732	459.49	480.92	undef
t	0.051	0.732	0.742	0.773	0.784	undef	418.60	416.97
a	0.058	0.784	0.795	0.830	0.842	381.68	368.52	407.55
w	0.041	0.842	0.850	0.875	0.883	473.97	441.40	425.48

```
File type = "ooTextFile"
Object class = "PitchTier"

xmin = 0
xmax = 2.235929
points: size = 62
points [1]:
        number = 0.4520983
        value = 429.3607438711908
points [2]:
        number = 0.47262966
        value = 451.66478192073487
points [3]:
        number = 0.48331423
        value = 460.0143901582947
points [4]:
        number = 0.50552115
        value = 450.03339849324544
points [5]:
        number = 0.52772807
        value = 489.577976832166
```

Fig. 4. Praat PitchTier format, with combined data: "neutral" durations (seconds), "emotional" F_0 values (Hz).

The following step is the creation of PitchTier files in the Praat format using a Python script. The PitchTier format takes pairs of the time point in seconds of the F_0 measure – called number, and the F_0 value – called value. Depending on what F_0

contour replacement is to be done, the Python script, for example, extracts information about the phone duration from a "neutral" file and combines it with F_0 values from the corresponding "emotional" file. The "emotional" F_0 values are inserted in 10%, 50% and 90% of phone duration in "neutral" file/template. An excerpt of a PitchTier file is presented in Fig. 4. The duration is taken from a "neutral" file and the F_0 values are extracted from the "emotional" recording (cf. Table 1).

The final step in the F_0 manipulation process is to replace the neutral pitch tier with the newly created pitch tier with the emotional F_0 values and to re-synthesise the neutral recording using the overlap-add synthesis in another Praat script. An exemplar utterance set used for the F_0 manipulation process is presented in Fig. 5. On the top there is the original "neutral" recording, in the middle the original "emotional" recording and at the bottom the re-synthesised "neutral" recording using overlap-add synthesis with the imposed "emotional" F_0 contour and duration for the "neutral" recording. The solid lines on the spectrograms correspond to the F_0 contour.

The F_0 manipulation tool may be used in studies of perception of emotions and how it is affected by different prosody patterns, such as durations and F_0 contour. In some studies, F_0 manipulation is performed manually using the Praat built-in tool, but the advantage of direct, automatic extraction from an authentic utterance is naturalness, avoiding filtering by local human decisions.

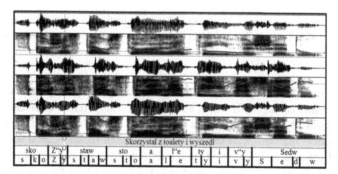

Fig. 5. F_0 manipulation data – from the top: "neutral", "emotional" and re-synthesised "neutral" recording with imposed "emotional" F_0 values.

3.2 Pitch Contour Shift Using Praat Built-in Function

As mentioned in the previous section, Praat offers a built-in function for manipulating the F_0 contour. The function is called "Shift pitch frequencies". For some smaller tasks, a researcher may want to use the built-in function for one file, instead of running a Praat script on a collection of files which can be found online (Rivas 2018). The steps to generate a resynthesised voice in Praat with the shifted F_0 contour are presented below:

1. Read in sound file (WAV) to Praat.
2. Mark with the cursor the Sound file from the Object list and choose from the right menu: *Manipulate → To Manipulation.*

8   J. Bachan

3. Select with the cursor the Manipulation file and go to *View & Edit*.
4. On the edited Manipulation file mark the whole contour with the cursor and from the top menu select: *Pitch → Shift pitch frequencies*.
5. In the pop-up menu write the number of frequency shift (the default unit is Hz).
6. Close the window with the edited Manipulation file
7. Select the Manipulation file from the Object list and choose: *Get resynthesis (overlap-add)* from the right menu
8. Save the new Sound file, for example by giving it the _synth extension to differentiate it from the original Sound file.

Figure 6 shows the original F_0 contour (light grey) and the shifted contour (bold dark green) by +40%.

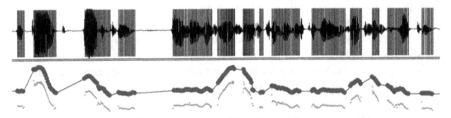

Fig. 6. The original F_0 contour (light grey) and its manipulation by +40% (bold dark green) of an utterance in Polish *"O, tak! Bardzo! Ale najważniejszą ocenę wystawiają zawsze słuchacze"*. (Color figure online)

Speech stimuli created by this method were used in speech perception test described in detail in Bachan et al. (2020) which aimed at assessing alignment to natural speech and resynthesised speech. The factor evaluated was pitch, and as the test speech stimuli repeated sentences in Polish were used. Two recordings of female natural voice and two recordings of male natural voice with small absolute mean F_0 difference to the model female or male voices, respectively, were selected as a base for resynthesis of female and male stimuli for speech perception test. The values of F_0 shift are presented in Table 2. Additionally, for the test natural female and male voices were used with high absolute mean F_0 difference to the model voice, 18 Hz and 42 Hz for the female voices, and 12 Hz and 37 Hz for the male voices. Altogether, there were 48 utterance–pairs used in total in the speech perception test in 6 categories:

1. similar – natural female and male voices with similar F_0 (8 utterance–pairs: 4 female–female pairs, 4 male–male pairs);
2. different – natural female and male voices with different F_0 (8 utterance–pairs: 4 female–female pairs, 4 male–male pairs);
3. 25_minus – resynthesised female and male voices with F_0 originally similar to the model voice, but after modification F_0 was shifted down by 25% in the test stimulus (8 utterance–pairs: 4 female–female pairs, 4 male–male pairs, original recording paired with the resynthesised stimulus);

4. 25_plus – resynthesised female and male voices with F_0 originally similar to the model voice, but after modification F_0 was shifted up by 25% in the test stimulus (8 utterance–pairs: 4 female–female pairs, 4 male–male pairs, original recording paired with the resynthesised stimulus);
5. 40_minus – resynthesised female and male voices with F_0 originally similar, but after modification F_0 was shifted down by 40% (8 utterance–pairs: 4 female–female pairs, 4 male–male pairs, original recording paired with the resynthesised stimulus);
6. 40_plus – resynthesised female and male voices with F_0 originally similar, but after modification F_0 was shifted up by 40% (8 utterance–pairs: 4 female–female pairs, 4 male–male pairs, original recording paired with the resynthesised stimulus).

Table 2. The values of F_0 shift in Hertz in resynthesised female and male stimuli.

Voice	Filename	Mean F_0	Absolute mean F_0 difference to model voice	25% of F_0	40% of F_0
female	N05_Z03_O1_01	223.17	1.42	55.79	89.27
female	N09_Z03_O1_01	231.05	9.30	57.76	92.42
male	N12_Z01_O2_01	109.73	2.58	27.43	43.89
male	N06_Z01_O2_01	113.17	6.02	28.29	42.27

There were 57 participants of the speech perception test, who assessed their hearing on the MOS scale as 3 – 5 (4.39 average). The youngest participant was 19 and the eldest was 62. The average age was 23.91 years. There were 35 women, 19 men and 3 people did not disclose their gender. There were 2736 answers given, 456 per category.

In the test the subjects were instructed to assess the similarity of voices on the scale 1 – 5, where 5 was the highest grade. There were 2 recordings per one question and the subjects could play them as many times as they needed to assess the similarity of pitch height of the voices.

The utterance–pairs from the 6 categories were mingled and presented to the subjects at random. The subjects did not know about the categorisation. They based their assessment only on their hearing.

The summary of the grades given per category is shown in Fig. 7. The recordings of natural voices with the smallest F_0 difference received the highest grades. The second best scores were assigned to the natural voices with the biggest F_0 difference. Resynthesised stimuli which were paired with the natural model voices received worse results. High F_0 manipulations were sensed by the subjects and therefore resynthesised stimuli with plus 40% pitch shift were graded the lowest. The similarity of pitch was graded highest with many scores of 3 points. The more manipulation added to the speech audio file, the lower grades were given. Independently from the movement up or down, resynthesised stimuli with 40% pitch shift received the highest number of scores of 1 point.

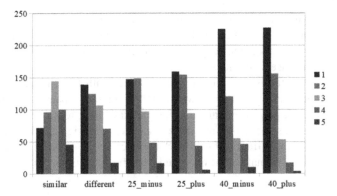

Fig. 7. Grades per category in speech perception test. The colours represent the MOS values.

4 F$_0$ Extraction

The F$_0$ extraction tool extracts F$_0$ for all speech files in a directory and calculates simple statistics. The tool uses the Praat autocorrelation algorithm for F$_0$ estimation.

The F$_0$ extraction package may work with mono- and stereo-files. If the files are stereo (two-channel), then first a Praat script must be run to extract the separate channels which is a part of this F$_0$ extraction tool. Then the automatic F$_0$ extraction may be run on the mono-recordings using another Praat script. The script was dedicated for dialogue recordings (around 5 min, maximum length 7 min) to study the phonetic convergence between speakers in different stages of conversation, therefore the script divides the recording into 4 sections: initial (I, 0–25% of time duration), initial-medial (IM, 25–50%), medial-final (MF, 50–75%) and final (F, 75–100%) and extracts the F$_0$ for these intervals (Demenko and Bachan 2018). The Praat script commands and value parameters for the male voice are:

```
To Pitch... 0.001 60 300
tmin = startTime + (step - 1) * 0.01
tmax = tmin + 0.01
mean = Get mean: tmin, tmax, "Hertz"
minimum = Get minimum: tmin, tmax, "Hertz", "Parabolic"
maximum = Get maximum: tmin, tmax, "Hertz", "Parabolic"
stdev = Get standard deviation: tmin, tmax, "Hertz"
```

In the script excerpt above, first, the F$_0$ range is declared (between min and max values). For males, the F$_0$ range is 60–300 Hz and for females it is 110–500 Hz. The measure window is 10 ms (step = 0.01 s) and for each of the windows, F$_0$ mean, F$_0$ min, F$_0$ max, and standard deviation are calculated using the Praat built-in functions. All the data are printed out to 4 separate CSV files (0–25% of time duration, 25–50%, 50–75%, 75–100%). Additionally, if no F$_0$ value was extracted for a given part of the recording (because the speaker did not say anything), this information is printed to a report file.

The final part of the F$_0$ extraction tool calculates the descriptive statistics for each of the file in a Python script, using the NumPy library functions: mean, median, max, min, stdv and some other information. The F_0 values are rounded for this analysis in Hz. The whole list of values is described below and exemplar data is presented in Table 3 – the list numbers correspond to the table columns:

1. filename
2. F_0_mean
3. F_0_median
4. F_0_most_common – the F_0 value which appeared most often
5. F_0_most_common_count – how often the most common value appeared
6. len_F_0_mean_list – how often the mean value appeared
7. F_0_max
8. F_0_min
9. F_0_std – standard deviations
10. len_F_0_values – the number of F_0 values used in the statistics

The F_0 extraction tool is useful in research of F_0 changes in a longer speech recordings.

Table 3. The exemplar F_0 statistics for a female voice, total duration: 309.51 s, expressive speech, column headings described in the text above.

1	2	3	4	5	6	7	8	9	10
N1_0_25	296	288	294	22	10	495	111	65	1330
N1_25_50	244	237	232	52	28	489	111	61	3080
N1_50_75	262	252	237	46	22	492	110	68	2350
N1_75_100	254	244	223	36	22	499	110	76	2457

5 Extraction of Prosodic Information

The tool for extraction of prosodic information from annotation is designed to work on 5 annotation tiers in Praat TextGrid format. The tool was dedicated to extract data from *Polish Rhythmic Database* (Wagner et al. 2016) in which speech is annotated on 5 tiers (Wagner and Bachan 2017):

1. phonemes (IntervalTier)
2. syllables (IntervalTier)
3. words (IntervalTier)
4. INT – intonational features and prominence (PointTier)
5. BI (from Break Index (Beckman et al. 2005)) – prosodic structure (PointTier)

The tool implemented as a Python script analyses annotations on all 5 tiers and generates a table with 26 columns for all TextGrid files in a directory. An excerpt of the annotation for a Polish text *"północny wiatr i słońce"*, pronunciation: /puwnot͡sny vjatr i swon't͡se/ (Eng. *"the north wind and the sun"*) is shown in Fig. 8.

Initially, the tool extracts information about start and end times and durations of the events (syllables) and pauses in the preceding and following context (and their durations). In addition, the script extracts prosodic information from the INT and BI tiers concerning the degree of stress, break index indicating the prosodic constituency, presence of pitch accent or prominence. Further, a number of positional/structural features are calculated, such as syllable position in clitic group (initial, medial or final), or phonological phrase position in the intonational phrase, and length of the prosodic constituents, e.g., intonational phrase length measured in the number of phonological phrases and clitic groups. These data in table format can be directly used as an input to rhythm (and generally, prosody) analysis, or can be further processed to obtain any additional prosodic information that may be required for a particular study.

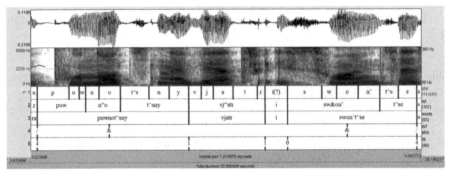

Fig. 8. An excerpt of annotation on 5 tiers of a Polish text *"północny wiatr i słońce"* from the Polish Rhythmic Database.

6   Drawing Waveform, Pitch and Annotation for Two-Channel Sounds

The last tool is used for drawing a waveform, annotation tiers and two F_0 contours for a two-channel (stereo) sound in Praat. The built-in functions in Praat allow to display a waveform, a spectrogram, a pitch contour and annotation tiers for a mono-sound, but when stereo-sound is read in Praat, the pitch contour for both channels is displayed as one line. The presented script allows to draw pitch contours for each channel on separate pictures. The script works on short audio files, up to 10 s (the standard length for which the spectrogram is shown in a Praat window). To generate the drawing as shown in Fig. 9 for each of the stereo channels the pitch contour must be extracted using the built-in function: *Pitch → Extract visible pitch contour*. To get this data the stereo sound should be extracted as mono-channels, for example using the function: *Convert → Extract all channels*. As a result, the script for drawing waveforms for a stereo-sound, two pitch contours and the annotation takes as input 4 files:

- WAV: stereo-sound up to 10 s
- TextGrid: annotation
- Pitch: channel 1
- Pitch: channel 2

The Praat drawing shown in Fig. 9 was generated using the Praat script which allowed to draw pitch contours for each speaker of a dialogue separately. The script is to be found in Bachan (2011).[1] To make it work, all 4 objects must be marked (selected) from the Praat Objects list and then the script may be run.

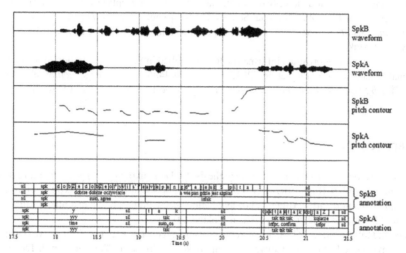

Fig. 9. Speaker's A and Speaker's B waveforms, pitch contours and annotation tiers of a dialogue excerpt for a two-channel sound.

7 Summary and Conclusion

A set of tools for automatic extraction and manipulation of speech fundamental frequency was presented. The automatic close copy speech synthesis take empirical models directly from authentic utterances, rather than filtering them through abstract models or human manipulation and synthesises them using diphone concatenation speech synthesis in MBROLA. The tools for F_0 contour replacement and shift use overlap-add synthesis in Praat. Creating the synthetic stimuli based on natural human speech recordings make it possible to test the hypotheses for which these stimuli were created in speech perception tests with humans. An extractor of different levels and types of prosodic information from annotation was described, which can be further analysed by advanced statistical

---

[1] The author is grateful to David Weenink for help in creating the Praat script at Interspeech 2009 for the mono-sounds, which was later modified by the author to analyse two-channel audio files.

methods. Last but not least, a tool for drawing F_0 contours for two-channel audio files, often used in dialogue corpora, was shown.

The presented tools are easy to use and have saved a lot of time and effort in procedures which in many previous studies have been performed manually by linguists. The large amount of data extracted automatically may be used for further statistical analyses.

Acknowledgements. The presented tools were created within the following projects:
• The F_0 manipulation tool described in Sect. 3.1 was supported by the Polish National Science Centre, project no.: 2013/09/N/HS2/02358, *"Vocal schemes of verbal emotion communication in linguistic perspective"*, Principal Investigator: Magdalena Oleśkowicz-Popiel.
• The F_0 extraction tool described in Sect. 4 and the speech perception test described in Sect. 3.2 were supported by the Polish National Science Centre, project no.: 2014/14/M/HS2/00631, *"Automatic analysis of phonetic convergence in speech technology systems"*, Principal Investigator: Grażyna Demenko.
• The extraction of prosodic information tool described in Sect. 5 was supported by the Polish National Science Centre, project no.: 2013/11/D/HS2/04486, *"Rhythmic structure of utterances in the Polish language: A corpus analysis"*, Principal Investigator: Agnieszka Wagner.

References

Bachan, J. Automatic close copy speech synthesis. In: Demenko, G. (ed.) Speech and Language Technology, vol. 9/10, pp. 107–121. Polish Phonetic Association, Poznań (2007) (2006/2007)

Bachan, J.: Communicative Alignment of Synthetic Speech. Ph.D. Thesis. Institute of Linguistics. Adam Mickiewicz University in Poznań, Poland (2011)

Bachan, J., Jankowska, K., Pikus, S.: Perception of convergence in native and not native speech. In: Demenko, G. (ed.) Phonetic Convergence in Spoken Dialogues in View of Speech Technology Applications, pp. 101–136. Akademicka Oficyna Wydawnicza Exit (2020). ISBN 978-83-7837-093-2

Beckman, M.E., Hirschberg, J., Shattuck-Hufnagel, S.: The original ToBI system and the evolution of the ToBI framework. In: Jun, S.-A. (ed.) Prosodic Typology -- The Phonology of Intonation and Phrasing (2005)

Boersma, P.: Praat, a system for doing phonetics by computer. Glot Int. **5**(9/10), 341–345 (2001)

Demenko, G., Bachan, J.: Convergence of speaking fundamental frequency in dialogues. In: 17th Speech Science and Technology Conference (SST 2018), 4–7 December 2018, Sydney, Australia (2018)

Dutoit, T.: An Introduction To Text-To-Speech Synthesis. Kluwer Academic Publishers, Dordrecht (1997)

Dutoit, T.: The MBROLA project (2006). http://tcts.fpms.ac.be/synthesis/mbrola.html. Accessed 30 Apr 2019

Gibbon, D., Bachan, J.: An automatic close copy speech synthesis tool for large-scale speech corpus evaluation. In: Choukri, K. (ed.) Proceedings of the Sixth International Language Resources and Evaluation Conference (LREC 2008), 28–30 May 2008, Marrakech, Morocco, pp. 902–907. ELDA, Paris (2008)

Oleśkowicz-Popiel, M., Bachan, J.: Manipulations of F_0 contours in affective speech analysis. In: Trouvain, J., Steiner, I., Möbius, B. (eds.) Proceedings of 28th Conference on Electronic Speech Signal Processing (ESSV), 15–17 March 2017, Saarbrücken, Germany, pp. 9–16 (2017)

Python (2019). https://www.python.org/. Accessed 30 Apr 2019

Rivas, A.: Is Pitch Shifting accurate in Praat? https://stackoverflow.com/questions/55886164/is-pitch-shifting-accurate-in-praat. Accessed 15 Nov 2021

Wagner, A., Kless, K., Bachan, J.: Polish rhythmic database – new resources for speech timing and rhythm analysis. In: Calzolari, N. et al. (eds.) Proceedings of the International Conference on Language Resources and Evaluation (LREC 2016), pp. 4678–4683. Portorož, Slovenia (2016)

Wagner, A., Bachan, J.: Speaking rate variation and the interaction between hierarchical rhythmic levels. In: Trouvain, J., Steiner, I., Möbius, B. (eds.) Proceedings of 28th Conference on Electronic Speech Signal Processing (ESSV), 15–17 March 2017, Saarbrücken, Germany, pp. 308–315 (2017)

Wells, J.C.: SAMPA computer readable phonetic alphabet. In: Gibbon, D., Moore, R., Winski, R. (eds.) Handbook of Standards and Resources for Spoken Language Systems. Mouton de Gruyter, Berlin and New York. Part IV, Section B (1997)

The Automatic Search for Sounding Segments of SPPAS: Application to Cheese! Corpus

Brigitte Bigi[(✉)] and Béatrice Priego-Valverde

LPL, CNRS, Aix-Marseille Université,
5, Avenue Pasteur, 13100 Aix-en-Provence, France
brigitte.bigi@lpl-aix.fr, beatrice.priego-valverde@univ-amu.fr

Abstract. The development of corpora inevitably involves the need for segmentation. For most of the corpora, the first segmentation to operate consist in determining silences vs Inter-Pausal Units - IPUs, i.e. sounding segments. This paper presents the "Search for IPUs" feature included in SPPAS - the automatic annotation and analysis of speech software tool distributed under the terms of public licenses. Particularly, this paper is focusing on its evaluation on Cheese! corpus, a corpus of reading then conversational speech between two participants. The paper reports the number of manual actions which was performed manually by the annotators in order to obtain the expected segmentation: add new IPUs, ignore irrelevant ones, split an IPU, merge two consecutive ones and move boundaries. The evaluation shows that the proposed fully automatic method is relevant.

Keywords: Speech · IPUs · Segmentation · Silence · Corpus · Conversation

1 Introduction

Corpus Linguistics is a Computational Linguistics field which aims to study the language as expressed in corpora. Nowadays, Annotation, Abstraction and Analysis (the 3A from [9]) is a common perspective in this field. Annotation consists of the application of a scheme to recordings (text, audio, video, ...). Abstraction is the mapping of terms in the scheme to terms in a theoretically motivated model or dataset. Analysis consists of statistically probing, manipulating and generalizing from the dataset. Within these definitions, this paper focuses on *annotation* which "can be defined as the practice of adding interpretative, linguistic information to an electronic corpus of spoken and/or written language data. 'Annotation' can also refer to the end-product of this process" [6]. Annotating corpora is of crucial importance in Corpus Linguistics. More and more annotated corpora are now available, and so are tools to annotate automatically and/or manually. Large multimodal corpora are now annotated with detailed

© Springer Nature Switzerland AG 2022
Z. Vetulani et al. (Eds.): LTC 2019, LNAI 13212, pp. 16–27, 2022.
https://doi.org/10.1007/978-3-031-05328-3_2

information at various linguistic levels. The temporal information makes it possible to describe behaviour or actions of different subjects that happen at the same time.

An orthographic transcription is often the minimum requirement for a speech corpus. It is often at the top of the annotation procedure, and it is the entry point for most of the other annotations and analysis. However, in the specific annotation context of a multimodal corpus, the time synchronization of the transcription is of crucial importance.

In recent years, SPPAS [2] has been developed by the first author to automatically produce time-aligned annotations and to analyze annotated data. The SPPAS software tool is multi-platform (Linux, MacOS and Windows) and open source issued under the terms of the GNU General Public License. It is specifically designed to be used directly by linguists. As a main functionality, SPPAS allows to perform all the automatic annotations that are required to obtain the speech segmentation at the word and phoneme level of a recorded speech audio and its orthographic transcription [3].

Figure 1 describes the full process of this method in order to annotate a multimodal corpus and to get time-synchronized annotations, including the speech segmentation. The audio signal is analyzed at the top-level of this procedure in order to search for the Inter-Pausal Units. IPUs are defined as sounding segments surrounded by silent pauses of more than X ms. They are time-aligned on the speech signal. IPUs are widely used for large corpora in order to facilitate speech alignments and for the analyses of speech, like prosody in (Peshkov et al., 2012). The orthographic transcription is performed manually at the second stage and is done inside the IPUs.

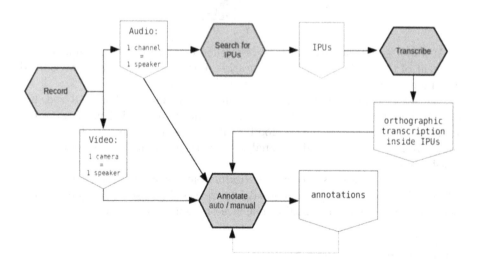

Fig. 1. Annotation process method

We applied this annotation method to the conversational French multimodal corpus 'Cheese!' [7,8]. "Cheese!" is made of 11 face-to-face dyadic interactions, lasting around 15 min each.

This paper presents the automatic annotation "Search for IPUs" of the SPPAS software tool. Given a speech recording of Cheese!, the goal was to generate an annotation file in which the sounding segments between silences are marked. This paper describes the method we propose for a fully automatic search for IPUs. The second section of this paper briefly presents Cheese!. We then propose a user-oriented evaluation method based on the amounts of manual interventions which were required to obtain the final IPUs annotation. Finally, the results are presented as appropriate for the intended use of the task and accompanied by a qualitative discussion about the errors.

2   The Method to Search Automatically for IPUs

2.1   Algorithm and Settings

The search for IPUs is performed on a recorded audio file made of only one channel (mono) and lossless.

Evaluation of a Threshold Value:
The method is using the Root-Mean-Square (rms), a measure of the power in an audio signal. The rms can be evaluated on the whole audio channel or on a fragment of n of its samples. It is estimated as follow:

$$rms = \sqrt{\frac{\sum_i^n (S_i^2)}{n}} \tag{1}$$

At a first stage, the search for IPUs method estimates the rms value of each fragment of the audio channel. The duration of these fragment windows is fixed by default to 20 ms and can be configured by the user.

The statistical distribution of the obtained rms values is then analyzed. Let min be the minimum rms value, μ the mean and σ the coefficient of variation. Actually, if the audio is not as good as it is expected, the detected outliers values are replaced by μ and the analysis is performed on this new normalized distribution.

A threshold value Θ *is fixed automatically* from the obtained statistical distribution. The estimation of Θ is of great importance: it is used to decide whether each window of the audio signal is a silence or a sounding segment:

– silence: $rms < \Theta$;
– sounding segment: $rms \geq \Theta$

We fixed the value of Θ as follow:

$$\Theta = min + \mu - \delta \tag{2}$$

The δ value of Eq. 2 was fixed to 1.5σ. All these parameters were empirically fixed by the author of SPPAS from her past experience on several corpora and from the feedback of the users.

It has to be noticed that Θ strongly depends on the quality of the recording. The value fixed automatically may not be appropriate on some recordings, particularly if they are of low-quality. By default it is estimated automatically but it can optionally be turned off and the user can fix it manually.

Get Silence vs Sounding Fragment Intervals:
The rms of each fragment window is compared to Θ and the windows below and above the threshold are identified respectively as silence and sounding. The neighboring silent and neighboring sounding windows are grouped into intervals. At this stage, we then have identified intervals of silences and intervals with "sounds". However, their duration can be very short and a filtering/grouping system must be applied to get intervals of a significant duration.

Because the focus is on the sounding segments, the resulting silent intervals with a too small duration are removed first (see the discussion section below). The minimum duration is fixed to 200 ms by default. This is relevant for French, however it should be changed to 250 ms for English language. This difference is mainly due to the English voiceless velar plosive /k/ in which the silence before the plosion could be longest than the duration fixed by default.

Construction of the IPUs:
The next step of the algorithm starts by re-grouping neighboring sounding intervals that resulted because of the removal of the too short silences. At this stage, the new resulting sounding intervals with a too small duration are removed. This minimum duration is fixed to 300 ms by default. This value has to be adapted to the recording conditions and the speech style: in read speech of isolated words, it has to be lowered (200 ms for example), in read speech of sentences it could be higher but it's not necessary to increase it too much. In spontaneous speech like in conversational speech, it has to be lowered mainly because of some isolated feedback items, often mono-syllabic, like 'mh' or 'ah'.

The algorithm finally re-groups neighboring silent intervals that resulted because of the removal of the too short sounding ones. It then make the Inter-Pausal Units it searched for. Silent intervals are marked with the symbol '#' and IPUs are marked with 'ipus_' followed by its number.

The Algorithm and Its Settings in a Nutshell:

1. fix a window length to estimate rms (default is 20 ms);
2. estimate rms values on the windows and their statistical distribution;
3. fix automatically a threshold value to mark windows as sounding or silent - this value can be fixed manually if necessary;
4. fix a minimum duration for silences and remove too short silent intervals (default is 200 ms);

5. fix a minimum duration for IPUs and remove too short sounding intervals (default is 300 ms);
6. tag the resulting intervals with # or *ipu_i*.

2.2 Optional Settings

From our past experience of distributing this tool, we received users' feedback. They allowed us to improve the values to be fixed by default this paper mentioned in the previous section. These feedbacks also resulted in adding the following two options:

– move systematically the boundary of the begin of all IPUs (default is 20 ms);
– move systematically the boundary of the end of all IPUs (default is 20 ms).

A duration must be fixed to each of the two options: a positive value implies to increase the duration of the IPUs and a negative to reduce them. The motivation behind these options comes from the need to never miss aso unding part. To illustrate how this might work, one of the users fixed the first value to 100 ms because his study focused on the plosives at the beginning of isolated words.

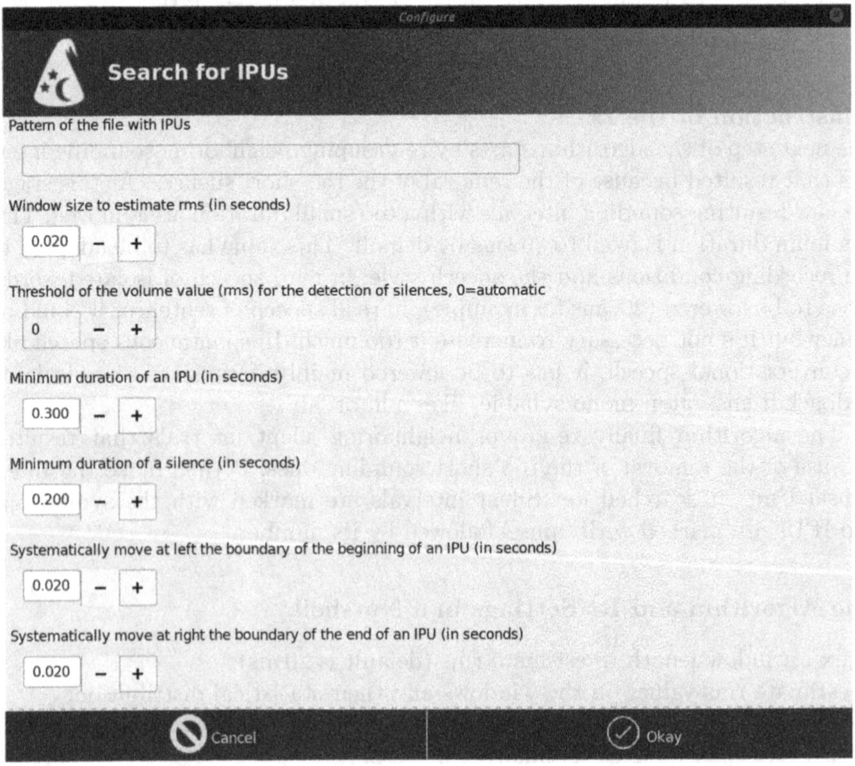

Fig. 2. Configuration with the Graphical User Interface

Figure 2 shows the full list of required parameters and optional settings when using the Graphical User Interface. The same parameters have to be fixed when using the Command-Line User Interface named `searchipus.py`.

2.3 Discussion

If the search for IPUs algorithm is as generic as possible, some of its parameters have to be verified by the user. It was attempted to fix the default values as relevant as possible. However, most of them highly depend on the recordings. They also depend on the language and the speech-style. It is strongly recommended to the users to check these values: special care and attention should be given to each of them.

Another issue that has to be addressed in this paper concerns the fact that the algorithm removes silence intervals first then sounding ones instead of doing it in the other way around. This choice is to be explained by the concern to identify IPUs as a priority: the problem we are facing with is to search for sounding segments between silences, but not the contrary. Removing short intensity bursts first instead of short silences results in possibly removing some sounding segments with for example a low intensity, or an isolated plosive, or the beginning of an isolated truncated word, i.e. any kind of short sounding event that we are interested in. However, removing short silences first like we do results in possibly assigning a sounding interval to a silent segment.

It has to be noticed that implementing a "Search for silences" would be very easy-and-fast but at this time none of the users of SPPAS asked for.

3 Cheese! Corpus

3.1 Description

"Cheese!" is a conversational corpus recorded in 2016 at the LPL - Laboratoire Parole et Langage, Aix-en-Provence, France. The primary goal of such data was a cross-cultural comparison on speaker-hearer smiling behavior in humorous and non-humorous segments of conversations in American English and French. For this reason, "Cheese!" has been recorded in respect with the American protocol [1], as far as possible.

"Cheese!" is made of 11 face-to-face dyadic interactions, lasting around 15 min each. It has been audio and video recorded in the anechoic room of the LPL. The participants were recorded with two headset microphones (AKG-C520) connected by XLR to the RME Fireface UC, which is connected with a USB cable to a PC using Audacity software. Two cameras were placed behind each of them in such a way each participant was shown from the front. A video editing software was used to merge the two videos into a single one (Fig. 3) and to embed the high quality sound of the microphones.

The 22 participants were students in Linguistics at Aix-Marseille University. The participants of each pair knew each other because they were in the same class. All were French native students, and all signed a written consent form before the recordings. None of them knew the scope of the recordings.

Two tasks were delivered to the participants: they were asked to read each other a canned joke chosen by the researchers, before conversing as freely as they wished for the rest of the interaction. Consequently, although the setting played a role on some occasions, the participants regularly forgot that they were being recorded, to the extent that sometimes they reminded each other that they were being recorded when one of the participants started talking about quite an intimate topic.

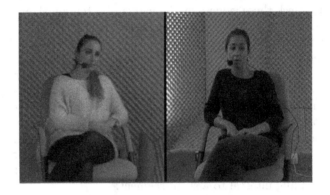

Fig. 3. Experimental design of Cheese!

In a previous study based on 4 dialogues of Cheese! [3], we observed a larger amount of laughter compared to other corpora: 3.32% of the IPUs of the read part contain laughter and 12.45% of IPUs of the conversation part. The laughter is the 5th most frequent token.

3.2   The IPUs of Cheese!

All the 11 dialogues were annotated by using the audio files re-sampled at 16,000 Hz. For each of the speakers, the "Search for IPUs" automatic annotation of SPPAS was performed automatically with the following settings:

- minimum silence duration: 200 ms because it's French language;
- minimum IPUs duration: 100 ms because it's conversational speech;
- shift begin: 20 ms;
- shift end: 20 ms.

The IPUs were manually verified with Praat [5]. Five dialogs were verified by 2 annotators and 6 by only one.

Table 1 reports the minimum (min), mean (μ), median, σ of the statistical distribution of the rms values. The second last column indicates the resulting automatically estimated Θ. The last column indicates if the rms values were normalized. This table shows that even with the same recording conditions, the recorded rms values are ranging from very different values depending on the speaker. It confirms the need to fix a specific threshold value for each recorded file in order to get the appropriate segmentation. Fixing automatically the Θ value is then important and a great advantage for users of the software tool.

It results in the following files:

- 22 files with the IPUs SPPAS 4.1 created fully automatically;
- 22 files with the manually corrected IPUs.

4   Evaluation Method

There are numerous methods and metrics to evaluate a segmentation task in the field of Computational Linguistics. Most of the methods are very useful to compare several systems and so to improve the quality of a system while developing it but their numerical result is often difficult to interpret.

Table 1. Distribution of the rms and the threshold value Θ fixed automatically

spk	min	μ	median	σ	Θ	Norm.	spk	min	μ	median	σ	Θ	Norm.
AA	12	548	58	177	295		OR	5	1009	19	160	773	
AC	6	818	371	209	510		MZ	5	1313	38	175	1056	
AW	7	595	96	164	355		CG	2	492	90	147	273	
CM	12	876	257	141	675		MCC	4	397	44	203	96	
ER	3	502	30	159	266		AG	8	328	38	168	84	
CB	4	753	63	174	495		FB	17	1058	89	194	783	
CL	3	1151	2918	15	256	x	JS	3	564	23	230	221	
LP	8	659	92	164	420		MA	3	672	123	152	446	
MA	3	608	24	178	343		PC	2	373	28	202	71	
AD	3	1680	3744	138	855	x	MD	10	844	164	199	555	
EM	4	1162	181	157	929		PR	12	427	61	188	156	

In this paper, we developed an evaluation method and a script that is distributed into the SPPAS package. It evaluates the number of manual "actions" the users had to operate in order to get the expected IPUs. We divided these manual actions into several categories described below. For a user who is going to read this paper, it will be easy to know what to expect while using this software on a conversational corpus, and to get an idea of the amount of work to do to get the expected IPUs segmentation.

In the following, the manually corrected IPUs segmentation is called "reference" and the automatic one is considered the "hypothesis". The evaluation reports the number of IPUs in the reference and in the hypothesis and the following "actions" to perform manually to transform the hypothesis into the reference:

add: number of times an IPU of the reference does not match any IPU of the hypothesis. The user had to *add* the missing IPUs;

merge: number of times an IPU of the reference matches with several IPUs of the hypothesis. The user had to *merge* two or more consecutive IPUs;

split: number of times an IPU of the hypothesis matches with several IPUs of the reference. The user had to *split* an IPU into several ones;

ignore: number of times an IPU of the hypothesis doesn't match any IPU of the reference. The user had to *ignore* a silence which was assigned to an IPU;

move_b: number of times the begin of an IPU must be adjusted;

move_e: number of times the end of an IPU must be adjusted.

The *add* action is probably the most important result to take into account. In fact, if *add* is too high it means the system failed to find some IPUs. It is critical because it means the user have to listen the whole content of the audio file to add such missing IPUs which is time consuming. If none of the IPUs is missed by the system, the user had only to listen the IPUs the system found and to check them by merging, splitting or ignoring them and by adjusting the boundaries.

In order to be exhaustive, this paper presents the *ignore* action. However, from our past experience in checking IPUs, we don't really consider this result an action to do. In practice, the user is checking IPUs at the same time of the orthographic transcription. If there's nothing interesting to transcribe, the interval is ignored: there's nothing specific to do. Moreover, we developed a plugin to SPPAS which deletes automatically these un-transcribed IPUs.

5   Results

5.1   Quantitative Evaluation

We applied the evaluation method presented in the previous section on the 11 dialogs of Cheese! corpus. The evaluated actions are reported into a percentage according to the 6922 IPUs in the reference for add, merge, move_b, move_e or according to the 7343 IPUs in the hypothesis for split and ignore:

- **add:** 54 (0.79% of the IPUs in the reference)
- **merge:** 104 (1.51% of the IPUs in the reference)
- **split:** 273 (3.72% of the IPUs in the hypothesis)
- **ignore:** 724 (9.86% of the IPUs in the hypothesis are false positives)
- **move_b:** 497 (7.23% of the IPUs in reference)
- **move_e:** 788 (11.46% of the IPUs in reference)

Reducing the number of missed IPUs was one of the main objective while developing the algorithm and we can see in the evaluation results that the number of IPUs to *add* is very small: it represents only 0.78% of the IPUs of the reference. It means that the user can be confident with the tool: the sounding segments are found.

The same holds true for the *merge* action: only very few IPUs are concerned which is very good because it's relatively time-consuming to do it manually. However, the number of IPUs to *split* is relatively high which is also relatively time-consuming to do manually.

Finally, the highest number of actions to perform is to move boundaries of the IPUs but this action is done very easily and fastly with Praat.

5.2 Qualitative Evaluation

We observed the durations of the added and ignored IPUs. It is interesting to mention that the duration of the IPUs to *add* and the IPUs to *ignore* are less than the average. Actually, the duration of the IPUs of the reference is 1.46 s in average but the 39 IPUs we added are only 0.93 s in average. This difference is even more important for the IPUs we ignored: their duration is 0.315 s in average.

Another interesting aspect is related to the speech style of the corpus: 14.11% of the IPUs contain a laughter or a sequence of speech while laughing. These events have a major consequence on the results of the system. Most of the actions to do contain a high proportion of IPUs with a laughter or a laughing sequence:

- 11 of the 39 IPUs to *add* (28.21%);
- 86 of the 171 IPUs to *merge* (50.29%);
- 5 of the 7 IPUs to *split* (71.42%).

This analysis clearly indicates that laugher, or laughing while speaking, is responsible for a lot of the errors of the system, particularly for the actions to *split* and to *merge*. Figure 4 illustrates this problem: the first tier is the manually corrected one - the reference, and the second tier is the system output - the hypothesis.

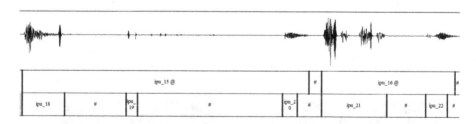

Fig. 4. Example of merged IPUs: laughter items are often problematic

In this scope of analyzing the errors, we also used the filtering system of SPPAS [4] that allowed us to create various tiers with: (1) the first phoneme of each IPU of the reference; (2) the first phoneme of each IPU of the reference for which a "move begin" action was performed; (3) the last phoneme of each IPU of the reference; (2) the last phoneme of each IPU of the reference for which a "move end" action was performed. We observed a high proportion of the fricatives /s/, /S/, /Z/ and the plosives /t/ at the beginning of the moved beginnings. On the contrary, the phonemes /w/ and /m/ which are the 2 most frequent ones in the reference are relatively less frequent in the "move begin" action. The following percentages indicate the proportion of the most frequent phonemes in the IPUs requiring the *move_b* action:

- /s/ is starting 7.96% of the IPUs of the reference but it concerns 20.08% of the IPUs of the *move_b* errors;
- /t/ is starting 4.31% IPUs of the reference but 7.19% of the *move_b* ones;
- /m/ is starting 8.66% IPUs of the reference but 5.92% of the *move_b* ones;
- /w/ is starting 11.76% IPUs of the reference but 5.70% of the *move_b* ones;
- /S/ is starting 2.83% IPUs of the reference but 4.65% of the *move_b* ones;
- /Z/ is starting 2.72% IPUs of the reference but 3.81% of the *move_b* ones;

Moreover, we observed that 10.6% of the *move_b* actions concern a laughter item.

We finally have done the same analysis for the last phoneme of the IPUs of the reference versus the last phoneme of IPUs with the *move_e* actions:

- /s/ is ending 2.67% IPUs in the reference but 7.11% in the *move_e* ones;
- /E/ is ending 9.08% IPUs in the reference but 6.85% in the *move_e* ones;
- /a/ is ending 10.42% IPUs in the reference but 6.72% in the *move_e* ones;
- /R/ is ending 5.72% IPUs in the reference but 6.32% in the *move_e* ones;
- /t/ is ending 3.88% IPUs in the reference but 6.19% in the *move_e* ones;

Like for the beginning, the /s/ and /t/ are relatively more frequent in the "move end" action than in the reference. And we observed that 17.26% of the *move_e* actions concern a laughter item which makes it the most frequently required "move end" action; but it's also the most frequent one to end an IPU with 11.11% in the reference.

Figure 5 illustrates the two actions *move_b* and *move_e* on the same IPU even if this situation is quite rare. In this example, the first phoneme is /s/ and the last one is /k/.

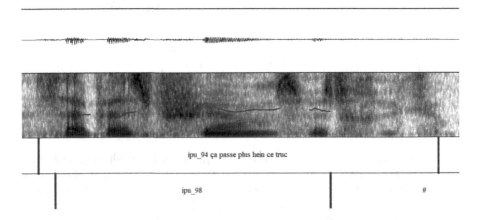

Fig. 5. Example of the *move_b* and *move_e* actions on an IPU with a low rms

6   Conclusion

This paper described a method to search for inter-pausal units. This program is part of SPPAS software tool. The program has been evaluated on the 11 dialogues of about 15 min each of Cheese! corpus, a corpus made of both read speech (1 min) and spontaneous speech.

We observed that the program allowed to find properly the IPUs, even on this particularly difficult corpus of conversations. To check the output of this automatic system, we had to perform the following actions on the IPUs the system found: to add new ones, to merge, to split, to ignore; and to perform the following actions on their boundaries: to move the beginning, to move the end. The analysis of the results showed that laughter are responsible for a large share of the errors. This is mainly because a laughter is a linguistic unit but acoustically it's often an outcome of alternate sounding and silence segments (Fig. 4).

Acknowledgments. We address special thanks to the Centre d'Expérimentation de la Parole (CEP), the shared experimental platform for the collection and analysis of data, at LPL.

References

1. Attardo, S., Pickering, L., Baker, A.: Prosodic and multimodal markers of humor in conversation. Pragmatics Cogn. **19**(2), 224–247 (2011)
2. Bigi, B.: SPPAS - multi-lingual approaches to the automatic annotation of speech. Phonetician **111–112**, 54–69 (2015)
3. Bigi, B., Meunier, C.: Automatic segmentation of spontaneous speech. Rev. Estud. Ling. (Int. Thematic Issue Speech Segm.) **26**(4), 1489–1530 (2018)
4. Bigi, B., Saubesty, J.: Searching and retrieving multi-levels annotated data, 4th edn. In: Proceedings of Gesture and Speech in Interactionin, Nantes, France, pp. 31–36 (2015)
5. Boersma, P., Weenink, D.: Praat: doing phonetics by computer (Computer Software). Department of Language and Literature, University of Amsterdam, Amsterdam (2011). http://www.praat.org/
6. Leech, G.: Introducing corpus annotation. In: Corpus Annotation: Linguistic Information from Computer Text Corpora, pp. 1–18. Longman, London (1997)
7. Priego-Valverde, B., Bigi, B., Attardo, S., Pickering, L., Gironzetti, E.: Is smiling during humor so obvious? A cross-cultural comparison of smiling behavior in humorous sequences in American English and French interactions. Intercult. Pragmat. **15**(4), 563–591 (2018)
8. Priego-Valverde, B., Bigi, B., Amoyal, M.: "Cheese!": a corpus of face-to-face French interactions. a case study for analyzing smiling and conversational humor. In: Proceedings of the 12th Language Resources and Evaluation Conference, pp. 467–475. European Language Resources Association, Marseille, France (2020)
9. Wallis, S., Nelson, G.: Knowledge discovery in grammatically analysed corpora. Data Min. Knowl. Disc. **5**, 307–340 (2001)

The Phonetic Grounding of Prosody: Analysis and Visualisation Tools

Dafydd Gibbon$^{(\boxtimes)}$ (iD)

Bielefeld University, Universitätsstraße 25, 33619 Bielefeld, Germany
gibbon@uni-bielefeld.de

Abstract. A suite of related online and offline analysis and visualisation tools for training students of phonetics in the acoustics of prosody is described in detail. Prosody is informally understood as the rhythms and melodies of speech, whether relating to words, sentences, or longer stretches of discourse, including dialogue. The aim is to contribute towards bridging the epistemological gap between phonological analysis, based on the linguist's intuition together with structural models, on the one hand, and, on the other hand, phonetic analysis based on measurements and physical models of the production, transmission (acoustic) and perception phases of the speech chain. The toolkit described in the present contribution applies to the acoustic domain, with analysis of the low frequency (LF) amplitude modulation (AM) and frequency modulation (FM) of speech, with spectral analyses of the demodulated amplitude and frequency envelopes, in each case as LF spectrum and LF spectrogram. Clustering functions permit comparison of utterances.

Keywords: Speech rhythm · F0 estimation · Frequency modulation · Amplitude modulation · Prosody visualisation

1 Introduction

A suite of related online and standalone analysis and visualisation tools for training students of phonetics in the acoustics of prosody is described. Prosody is informally understood as the rhythms and melodies of speech, whether relating to words, sentences, or longer stretches of discourse, including dialogue. The aim is to contribute towards the epistemological gap between phonological analysis of prosody, based on the linguist's qualitative intuition, hermeneutic methods and structural models on the one hand, and, on the other, the phonetic analysis of prosody based on quantitative measurements, statistical methods and causal physical models of the production, transmission (acoustic) and perception phases of the speech chain. The phonetician's methodology also starts with intuitions, even if only to distinguish speech from other sounds, or more specifically to provide categorial explicanda for quantitatively classifying the temporal events of speech. Nevertheless, the two disciplines rapidly diverge as the domains and methods become more complex, and issues of the empirical grounding of phonological categories beyond hermeneutic intuition arise. The present contribution addresses two of these issues:

© Springer Nature Switzerland AG 2022
Z. Vetulani et al. (Eds.): LTC 2019, LNAI 13212, pp. 28–45, 2022.
https://doi.org/10.1007/978-3-031-05328-3_3

1. models of rhythm as structural patterns which correspond to intuitions of stronger and weaker syllables and words in sequence;
2. intuitive perception of globally rising and falling pitch contours.

The present account focusses exclusively on the acoustic phonetics of speech transmission, not on production or perception, with a tutorial method of data visualisation in two main prosodic domains:

1. LF (low frequency) amplitude modulation (AM) of speech as phonetic correlates of sonority curves covering the time-varying prominences of syllables, phrases, words and larger units of discourse as contributors to speech rhythms;
2. LF frequency modulation (FM) of speech as the main contributor to tones, pitch accents and stress-pitch accents, to intonations at phrasal and higher ranks of discourse patterning, and also as a contributor to speech rhythms.

In Sect. 2, components of the online tool are described, centring on the demodulation of LF AM and FM speech properties. In Sect. 3 an open source extended offline toolkit for Rhythm Formant Analysis (RFA) procedure is described, followed by a demonstration of the RFA tool in a comparison of readings of translations of a narrative into the two languages of a bilingual speaker. Finally, conclusions are discussed in Sect. 4.

2 An Online Tool: *CRAFT*

2.1 Motivation

The motivation for the *CRAFT (Creation and Recovery of Amplitude and Frequency Tracks)* online speech analysis tutorial tool and its underlying principles are described, together with some applications, in [9] and [12]. Well-known tools such as *Praat* [3], *WinPitch* [19], *AnnotationPro* [18], *ProsodyPro* [29] and *WaveSurfer* [22] are essentially dedicated offline research tools. The online CRAFT visualisation application is a tutorial supplement to such tools, based on the need to develop a critical and informed initial understanding of strengths and weaknesses of different algorithms for acoustic prosody analysis. The main functional specifications for this online tool are: accessibility, version-consistency, ease of maintenance, suitability for distance tutoring, face-to-face teaching and individual study, and also interoperability using browsers on laptops, tablets and (with size restrictions) smartphones.

CRAFT is implemented in functional programming style using Python3 and the libraries *NumPy*, *SciPy* and *MatPlotLib*, with input via a script-free HTML page with frames (sometimes deprecated, but useful in this context), server-side CGI processing and HTML output. The graphical user interface (GUI) has one output frame and four frames for different input types[1] (Fig. 1):

1. study of selected published F0 estimators (*Praat, RAPT, Reaper, SWIPE, YAAPT, Yin*) and F0 estimators custom-designed for *CRAFT* (*AMDF* or *Average Magnitude Difference Function*, and *SOFT, Simple F0 Tracker*);

---

[1] http://wwwhomes.uni-bielefeld.de/gibbon/CRAFT/; code accessible on GitHub.

2. visualisation of amplitude and frequency modulation and demodulation operations;
3. display of alternative time and frequency domain based FM demodulation (F0 estimation, 'pitch' tracking);
4. visualisation of low-pass, high-pass and band-pass filters, Hann and Hamming windows, Fourier and Hilbert transformations;
5. visualisation of the output generated by the selected input frame.

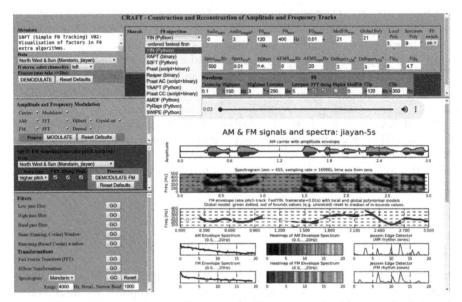

Fig. 1. CRAFT GUI: parameter input frame (top) for 9 F0 extractor algorithms; amplitude demodulation (left upper mid); F0 estimators (left lower mid); filters, transforms, spectrogram (left bottom); output frame (lower right).

In the following subsections, LF AM and LF FM visualisations are discussed, followed by descriptions of AM and FM demodulation, operations and transforms and long-term spectral analysis.

2.2 Amplitude Modulation (AM) and Frequency Modulation (FM)

The mindset behind the *CRAFT* tool is *modulation theory* (cf. also [28]): in the transmission of speech signals a *carrier signal* is modulated by an information-bearing lower frequency *modulation signal*, and in perception the *modulated* signal is *demodulated* to extract the information-bearing signal; cf. Fig. 2. Two main types of modulation are provided in the speech signal: *amplitude modulation* and *frequency modulation*. Both these concepts are familiar from the audio modulation of HF and VHF broadcast radio with amplitude modulation between about 100 kHz and 30 MHz (AM radio), and frequency modulation between about 100 MHz and 110 MHz (FM radio). The top frame in Fig. 1 provides inputs and parameters for the two core *CRAFT* tasks, exploring two prosodic subdomains: properties of F0 estimation algorithms and long term spectral analysis of demodulated amplitude and frequency envelopes. Corpus snippets are provided, including *The North Wind and the Sun* read aloud in English and in Mandarin.

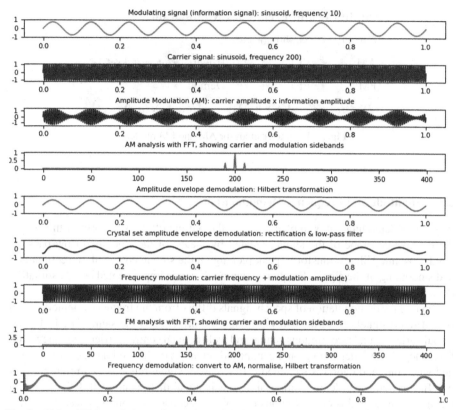

Fig. 2. AM and FM panels (top to bottom): modulation signal, carrier signal, amplitude modulated carrier, amplitude modulation spectrum, demodulated AM signal (rectification, peak-picking or Hilbert); frequency modulated carrier, frequency modulation spectrum, demodulated FM signal.

Alternating sequences of consonant clusters (lower amplitude) and vowels (higher amplitude) provide a low frequency (LF) AM sonority cycle of syllable sequences. The syllable lengths and corresponding modulation frequencies are around 100 ms (10 Hz) and 250 ms (4 Hz), and there are longer term, lower frequency amplitude modulations corresponding roughly to phrases, sentences and longer units of speech, constituting the LF formants of speech rhythms. Consonant-vowel sequences also involve a more complex kind of amplitude modulation: variable filtering of the amplitude of the high frequency (HF) harmonics of the fundamental frequency (and consonant noise filtering), creating the HF formants which distinguish speech sounds.

The *CRAFT* tool is designed for analysis of the LF rhythm formants which correlate with long-term LF AM sequences of syllables and of larger linguistic units, not for the analysis of HF formants. Global falling, rising or complex intonation patterns and local lexical tones, pitch accents or stress-pitch accents underlie the FM patterns of speech. The *CRAFT* input form for visualisation of amplitude and frequency modulation and demodulation procedures is shown in Fig. 3.

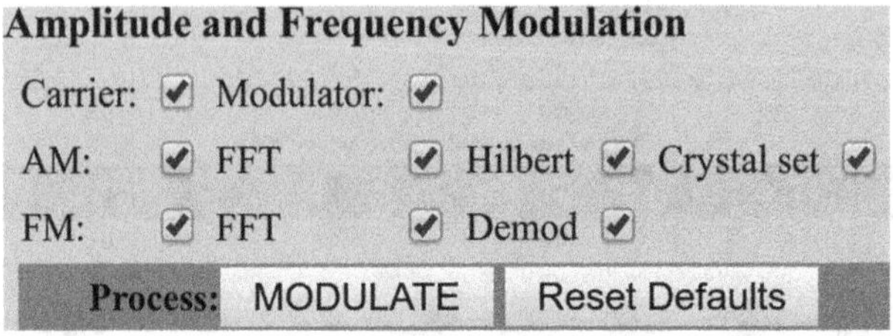

Amplitude and Frequency Modulation

Carrier: ☑ Modulator: ☑

AM: ☑ FFT   ☑ Hilbert ☑ Crystal set ☑

FM: ☑ FFT   ☑ Demod ☑

Process:	MODULATE	Reset Defaults

Fig. 3. Parameter input for AM and FM module.

2.3   Amplitude and Frequency Demodulation

Amplitude demodulation is implemented as the outline of the waveform of the signal, the positive envelope of the signal, created by means of the smoothed (low-pass filtered) absolute Hilbert transform, peak picking on the smoothed signal, or rectification and smoothing of the signal. Frequency demodulation is implemented as F0 estimation ('pitch' tracking) of the voiced segments of the speech signal.

Frequency demodulation of speech signals differs from the frequency demodulation of FM radio signals in various ways, though the principle is the same. The FM radio signal varies around a continuous, stable and well-defined central carrier frequency, with frequency changes depending on the amplitude changes of the modulating audio signal, and when the modulation is switched off the carrier signal remains as a reference signal. But in frequency demodulation of a speech signal there is no well-defined central frequency, the signal is discontinuous (in voiceless consonants and in pauses), the frequency cycles are uneven, since the vocal cords are soft and moist (and not mechanically or mathematically precisely defined oscillators), and when the modulation disappears, so does the fundamental frequency carrier signal – there is no unmodulated carrier; cf. [28]. For these reasons, in order to demodulate the speech fundamental frequency, the signal has to undergo a range of transformations.

Several comparisons and analyses of techniques for frequency demodulation (F0 estimation) of the speech signal are discussed in [1, 15, 17, 20] are facilitated. The following ten panels show visualisations of F0 estimates by a selection of algorithms, each selected and generated separately using *CRAFT*, with the same data. The algorithms produce similar results, though there are small local and global differences. The superimposed polynomial functions illustrate some non-obvious differences between the estimates: a local polynomial model for voiced signal segments and a global model for the entire contour with dotted lines interpolating across voiceless segments are provided.

1. *Average Magnitude Difference Function (AMDF*; custom implementation)

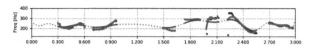

2. Praat (autocorrelation) [3]

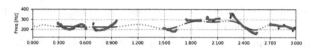

3. *Praat* (cross-correlation) [3]

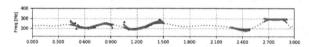

4. *RAPT* [23]

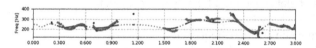

5. *PyRAPT* (Python emulation) [8]

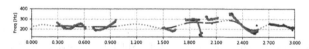

6. *Reaper* [24]

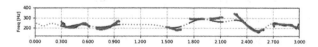

7. *SOFT* (custom implementation)

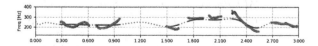

8. *SWIPE (Python emulation)* [4, 7]

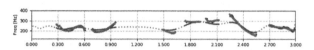

9.   *YAAPT* (Python emulation) [21, 30]

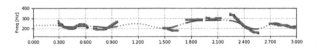

10. *YIN (Python emulation)* [6, 13]

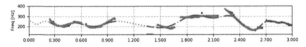

The F0 estimations in Panels 1 to 10 are applications of the different algorithms to the same data with default settings. Two polynomial models are superimposed in each case. The colour coding is: F0: blue, local polynomial: orange, global polynomial: red plus dotted green interpolation. The algorithms achieve quite similar results and correlate well (cf. Table 1), and are very useful for informal 'eyeballing'. Most of the algorithms use time domain autocorrelation or cross-correlation, while the others use frequency domain spectrum analysis or combinations of these techniques. *CRAFT* provides basic parametrisation for all the algorithms as follows:

- start and length of signal (in seconds),
- F0min and F0max for frequency analysis and display,
- length of frame for F0 analysis,
- length of median F0 smoothing filter,
- orders of global and local F0 polynomial model,
- on/off switch for F0 display,
- spectrum min and max frequencies,
- display min and max for envelope spectrum,
- power value for AM and FM difference spectra,
- display width and height.

Table 1. Selected F0 estimator correlations for S0FT, RAPT, Python RAPT and Praat on a single data sample.

Correlation	Pearson's r	p
S0FT:RAPT	0.897	<0.01
S0FT:PyRAPT	0.807	<0.01
S0FT:Praat-autocorr	0.843	<0.01
RAPT:PyRAPT	0.883	<0.01
RAPT:Praat-autocorr	0.868	<0.01
PyRAPT:Praat-autocorr	0.791	<0.01

AMDF and *SOFT* are minimalistic implementations which were designed specifically for the *CRAFT* tool. Except for a couple of sub-octave errors in the default configuration shown, *AMDF* compares favourably with *Praat* autocorrelation, which it resembles, except for the use of subtraction and not multiplication, resulting in a faster algorithm. Absolute speed is dependent on the implementation environment, of course, with C or C++ being in principle faster than Python. Interestingly, the Python YIN implementation is the fastest of all.

The *SOFT* F0 estimation algorithm has a different purpose from the others: parameters for 'tweaking' of the analysis are provided in order to find an optimal agreement with aural-visual inspection and accepted standard algorithms. In addition to the general parameters for adjustment by the user, which are available to all algorithms, *SOFT* also provides specific parameters:

- initial choice:

  - *voice type* (higher, middle, lower pitch), or
  - *custom*: levels of centre-clipping, high and low pass,

- if initial choice is *custom*:

  - filter frequency and order,
  - algorithm (FFT, zero-crossing, peak-picking),
  - length of F0 median filter,
  - min and max y-axis display clipping.

An example of *SOFT* output is shown in Panel 7 above. With the parameter defaults provided, the results can be very close to the output of standard algorithms such as the autocorrelation algorithm of *Praat*.

The quantitative measurements which underlie the visualisations are also available for further use, as in Table 1, which shows correlations between the algorithms under the same conditions and for a single data item (bearing in mind that the goal here is not to report an experiment but to outline the potential of this online tool).

2.4 Operations and Transforms

CRAFT also includes a panel for illustrating low-pass, high-pass and band-pass filters, Hann (cosine; also, correctly: von Hann, and incorrectly: 'Hanning'), and Hamming (raised cosine) windows, as well as Fourier and Hilbert Transforms and a parametrised spectrogram display. The following three panels show low-pass, high-pass and band-pass filters and the von Hann window.

1. High-pass and low-pass filters.

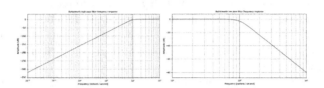

2. Band-pass filter with illustration of application.

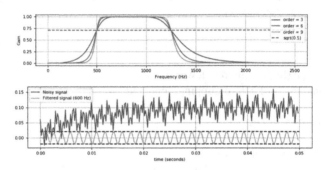

3. von Hann window with illustration of application.

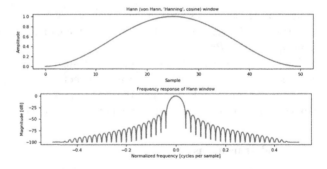

Figure 4 shows Fast Fourier Transforms (FFT) of six frequency estimations. Clearly the F0 spectra of these algorithms differ considerably in spite of the rather high correlations noted previously, because of differences in frequency vs. time domain processing, in window lengths and window skip distances, as well as in internal filtering. *SOFT*, *AMDF* and *PyRAPT* are rather similar, while *YAAPT*, *Praat* (cross-correlation) and *SWIPE* are very different. The implication is that when demodulated F0 is further processed quantitatively, these differences between the algorithms may need to be taken into account, in spite of the relatively high correlations between them.

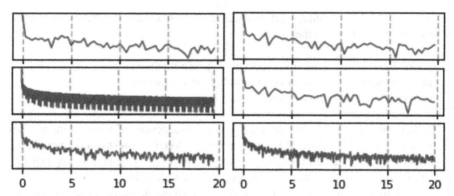

Fig. 4. Spectral analysis 0...20 Hz of frequency envelopes of selected F0 estimation algorithms for the same signal: S0FT, YAAPT, Praat (cross-correlation), AMDF, PyRAPT, SWIPE (top to bottom, left column before right).

3 The RFA Offline Extension Toolkit

3.1 Functional Specification

The main aim of the RFA (Rhythm Formant Analysis) toolkit extension is detailed investigation of the contributions of LF AM and LF FM to the analysis of speech rhythms, based on the concept of *rhythm formant*, that is, a region of high magnitudes in the low frequency spectrum and spectrogram of the speech signal, relating to the rhythms of words and syllables, and, in long utterances, to slower rhythms of phrases and longer discourse units. Theoretical foundations of the underlying Rhythm Formant Theory (RFT) and applications of the RFA toolkit are described in [12]. The code is available on GitHub (https://github.com/dafyddg/RFA).

Intuitively, rhythm is understood to be a real-time sequence of regular acoustic beats, usually between one and four per seconds (1...4 Hz). These beats are often said to be associated with stressed syllables in foot-based stress-pitch accent languages like English, and with each syllable in syllable-based languages like Chinese.

There are many approaches to the analysis of speech rhythms in linguistics and phonetics. Phonological approaches use numerical values, or tree structures with nodes labelled 'strong' and 'weak', or bar-chart like 'grids' to visualise a qualitative abstract notion of intuitively identified rhythm. Descriptive phonetic approaches annotate speech recordings with boundaries of intervals associated with phonological categories such as vocalic, consonantal, syllable or word segments and use strategies to form averages of interval duration differences and to apply the averages as indices for characterising language types.

A major issue with the annotation based duration average approaches is the lack of phonetic grounding in the reality of speech signals beyond crude segmentation. The reality of rhythm is that it involves more than just duration averages: it involves oscillations in real time of beats and waves with approximately equal intervals – relative or 'fuzzy' isochrony. The restrictive duration average methods have failed to find isochrony, for a number of simple reasons which have been discussed on many occasions; cf. the summary in [12]. In particular, the duration average methods do not actually capture the rhythms of speech, because they…

1. ignore the 'beat' or oscillation property of speech rhythms;
2. assume constant duration patterning throughout utterances;
3. assume a single duration average for each language.

In fact, rhythms may hold over quite short subsequences of three or more beats and then change in frequency, also on occasion in a longer term rhythmic pattern, the 'rhythms of rhythm' [12]. Further, rhythms vary not only from language to language and dialect to dialect, but also with different pragmatic speech styles.

Parallel to and in stark contrast with the annotation based duration average approaches are the signal processing approaches which start with the assumption of rhythm as oscillating signal modulations, and work with spectral analysis and related transformations to discover speech rhythms of different frequencies below about 10 Hz; cf. overviews in [9, 12] and Sect. 2.3, For example, a syllable speech rate of 5 syll/s corresponds to a low oscillation frequency of 5 Hz with an average syllable length of 0.2 s; a foot speech rate of 1.5 ft/s corresponds to an oscillation frequency of 1.5 Hz and average foot length of 0.6 s. The prediction is that with an appropriate spectral analysis, these and other rhythm frequencies can be detected inductively from the speech signal. Correspondingly, the intuitive understanding of 'rhythm' is explicated 'bottom-up', unlike top-down phonological approaches, starting with the intuition of rhythm as oscillation and then analysing physical properties of the speech signal, based on the modulation theoretic perspective of signal processing (Gibbon 2021:3):

> *Speech rhythms are fairly regular oscillations below about 10 Hz which modulate the speech source carrier signal and are detectable in spectral analysis as magnitude peaks in the LF spectrum of both the amplitude modulation (AM) envelope of the speech signal, related to the syllable sonority outline of the waveform, and the frequency modulation (FM) envelope of the signal, related to perceived pitch contours.*

The central requirement for a tool to be used for identifying rhythm frequencies by demodulation of the low frequency oscillating modulations of the speech signal is thus the identification of temporally regular oscillations with specific frequencies or, more realistically, frequency ranges. The frequency zones define multiple fuzzy-edged *rhythm formants* (cf. [9–12]), which can be associated with signal modulations by syllables, phrases and other categories [2, 5, 16]. In other words: the tool must include a method for *demodulating* the rhythmically modulated signal.

For this purpose, long-term spectrogram analysis of the positive signal amplitude envelope is introduced, both to analyse rhythms quantitatively, and to model the perception of varying rhythms in the LF AM and FM oscillations of speech signals (cf. [14, 25–27]).

The full procedure of demodulation and spectral analysis is shown with a stylised example in Fig. 5. In the example, the signal length is 1 s, the sampling frequency is

44.1 kHz, the modulation frequency is 10 Hz and the modulation index (modulation depth) is 0.75.

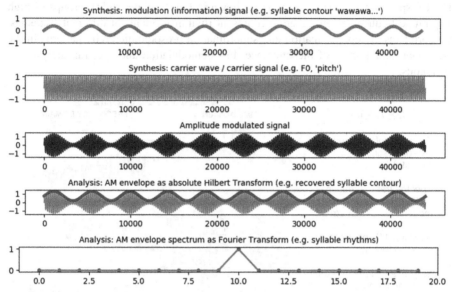

Fig. 5. Panels showing simplified aspects of amplitude modulation, demodulation and rhythm detection: 1. sine modulation wave, 2. sine carrier wave, 3. modulated carrier, 4. demodulated amplitude envelope, 5. amplitude envelope spectrum.

3.2 Standalone Offline Tools

Online tools are useful for teaching demonstrations in teaching situations where the user is a software consumer rather than developer, but have the disadvantage that they are constrained by the designer's goals and the further disadvantages of software and data integrity and possibly also unwanted logging of interactive activities.

For more flexibility, though at the cost of ease of use, a companion set of standalone offline tools was developed, also using Python3 and the libraries *NumPy*, *SciPy*, *MatPlotLib*, plus *GraphViz*. The toolset provides:

1. analysis and visualisation of AM and FM demodulation;
2. low frequency spectral analysis of AM and FM;
3. both a global spectrum and a 2 s or 3 s long windowed spectrogram for entire utterances;
4. trajectories of highest magnitude frequencies through the spectrogram frame series;
5. comparison of utterances using these criteria.

The goal here is not to produce off-the-shelf point-and-click consumer software, but to produce a suite of basic 'alpha standard' command-line tool prototypes which can be further developed by the interested user.

Figure 6 (upper left) shows the waveform (grey) and the rectified and low pass filtered amplitude envelope (red). The low pass filtered long-term amplitude envelope is taken to be the acoustic phonetic correlate of the 'sonority curve' of phonological analyses. The LF spectrum from 0 Hz to 5 Hz is shown in Fig. 6 (upper right); groups of high magnitude frequencies are taken to represent rhythm formants at different frequcies, the correlates of superordination and subordination prosodic hierarchy patterns in the locutionary component of the utterance. FM demodulation and spectral analysis are analogous.

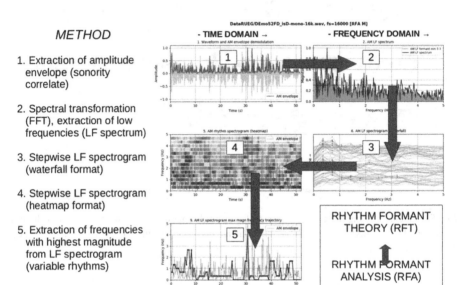

Fig. 6. Rhythm frequency identification procedure.

In the mid panels of Fig. 6, spectrograms of the utterance are shown, extracted in 2 ms overlapping FFT windows, first visualised as a waterfall spectrum, from top to bottom, consisting of a vertical sequence of spectra in the frequency domain (mid right), and also as a more conventional heatmap spectrogram in the time domain (mid left), with higher magnitudes shown as darker colours.

The main innovations in the offline toolkit are:

1. the LF spectrogram, which permits the observation and further analysis of changes in rhythm patterns through the utterance;
2. the extraction of a trajectory through the spectrogram, in which at every FFT analysis window the frequency with the highest magnitude is selected;
3. the custom design of the AMDF FM demodulation algorithm, in which frame duration and correlation domain are adjusted automatically in terms of minimum and maximum limit parameters for the frequency search space;
4. clustering procedures for comparing sets of utterances:

- prosodic *k-means* clustering;
- prosodic *distance mapping* of utterances with a selection of distance metrics;
- prosodic *hierarchical clustering* with selections of distance metric and clustering condition combinations.

In order to be able to combine these functions in different ways for different purposes, for example only the waveform, or with the AM envelope and the FM track (F0 estimation track), with the low frequency AM and FM spectra, or only the waveform, F0 estimation, AM LF spectrogram and FM LF spectrogram, a set of libraries was developed for specific analysis tasks:

1. waveform and LF amplitude demodulation (in the main application);
2. `module_fm_demodulation.py` (LF FM, LF F0 estimation, using a custom variant of AMDF, the Average Magnitude Difference Function);
3. `module_drawdendrogram.py` (spectral frequency grouping of magnitude peaks interpreted as rhythm formants);
4. `module_spectrogram.py` (LF spectrogram of utterance, typically with 2s or 3s LF FFT window);
5. `module_kmeans.py` (classification of sets of utterances using Euclidean distance-based *k-means* clustering);
6. `module_distancenetworks.py` (distance-based linking of utterances according to time domain and frequency domain spectrum and spectrogram data vectors, using a selection of distance metrics: *Canberra, Chebyshev, Cosine, Euclidean, Manhattan*);
7. `module_hierarchicalclustering.py` (hierarchical clustering based on a selection of distance metrics and clustering conditions).

3.3 Prosodic Comparison of Narrative Readings

For demonstration purposes an analysis of spoken narrative data was conducted on a small data set of readings aloud of the IPA benchmark narrative *The North Wind and the Sun* in English and German by a female bilingual speaker. The readings in each language are numbered in order of production; the German readings were produced before the English readings. While rhythms of spontaneous speech and dialogue may appear more interesting at first glance, reading aloud is a cultural technique with independent inherent value in report presentation, news-reading or reading to children and the sight-afflicted. Moreover, spontaneous more complex and it is advisable to introduce a new method with simpler clear cases.

As the first basic step, *k-means* analysis was chosen. The analysis is intended to demonstrate the value of both time and frequency domain parameters in prosodic typology. The prediction is that the readings in English and German can be distinguished by means of selected prosodic parameters, even though the readings are by the same speaker.

There are many prosodic properties which can be addressed. The present analysis uses variance in two time domain vectors:

1. *x*: the variance of the trajectory of the highest magnitude frequencies in the LF spectrogram;
2. *y*: the variance of the FM (F0) track.

The two measures ignore the facts that (a) the data set is tiny and (b) the parameters concerned are locally and globally varying time functions, not static populations, thus not being ideal candidates for variance analysis. However, with durations of approximately 60 s the utterances are much longer than the domains of rhythmic variation such as syllable, word and phrase, so that 'the end justifies the means' in this case. The readings in English cluster in the upper right quadrant, while the readings in German cluster in the lower left quadrant; cf. Fig. 7. The result shows that the bilingual speaker makes a clear distinction between her readings in English and her readings in German In Fig. 7, data positions are marked with filled circles coloured by cluster; centroid positions are marked with "C" in a square in the cluster colour.

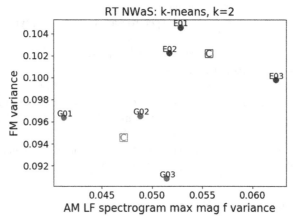

Fig. 7. k-means positioning of readings of The North Wind and the Sun by AM and FM spectral properties.

3.4 Distance Mapping

Other similarity visualisations such as distance maps show links which are compatible with the *k-means* division. In order to make distance relations clearer at a glance, distances above 0.75 (range 0...1) were excluded from the graph.

The first analysis compares utterances on the basis of the LF AM spectrum vectors (Fig. 8), using the Cosine Distance metric. The readings in English are nearer to each other than to the readings in German, and vice-versa, with cluster-internal distances <0.7 in each case. However the first reading in English is nearer to the first reading in German than to the last reading in English, perhaps due to the chronology of the scenario: the last reading in German was produced immediately before the first reading in English. Such ordering effects were intended, as systematic context.

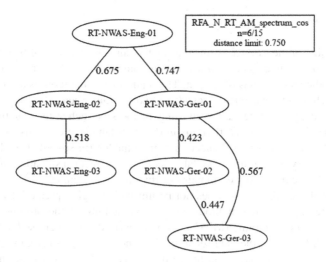

Fig. 8. AM LF spectrum distances (English and German readings by female bilingual).

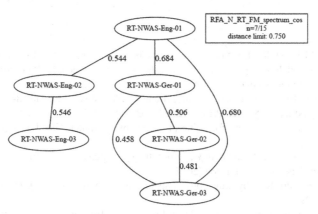

Fig. 9. FM LF spectrogram frequency max peak distances (English and German readings by female bilingual).

The second analysis (Fig. 9) compares utterances on the basis of the LF FM spectrum vectors. The analysis also uses the Cosine Distance metric, and shows the same cluster formation as the analysis based on LF AM spectrum vectors: the readings in English are nearer to each other than to the readings in German, here with cluster-internal distances <0.6 in each case.. In this case, the first reading in English is nearer to both the first and third readings in German than to the third reading in English.

Further analyses based on spectrogram properties rather than spectrum properties, also conducted with the Cosine Distance metric, show the same clear partitions and also similar anomalies. The results are also confirmed by further analysis with hierarchical clustering (cf. Gibbon 2021).

4 Conclusion

The functionality of the *CRAFT* online tutorial tool and the extended offline RFA toolkit for acoustic prosody analysis is demonstrated in some detail, with attention to the rhythmic and melodic modulations of speech. The main uses of the toolkits are in advanced phonetics teaching and in acoustic prosody research. Many open research questions (cf. the discussion in [12]) can be addressed using the toolkits, such as the quantitative analysis of the variability of speech rhythms in different language domains, from varying rhythms of the consonant vowel succession in syllables (in so-called 'syllable-timed languages'), to varying rhythms of syllable sequences in feet (in so-called 'pitch accent languages') and the much longer domains of rhythms in discourse.

Evidently, the small data set does not permit wide-ranging predictions. For larger data sets more sophisticated methods and complementing of the present strategy of unsupervised machine learning (ML) by semi-supervised and supervised ML methods. will be needed. However, the data set fulfils its function of demonstrating the validity of the RFA method itself and the utility of the extended RFA toolkit, and the heuristic value of the method in raising further pertinent questions is clear.

References

1. Arjmandi, M.K., Dilley, L.C., Lehet, M.: A comprehensive framework for F0 estimation and sampling in modeling prosodic variation in infant-directed speech. In: Proceedings of the 6th International Symposium on Tonal Aspects of Language, Berlin, Germany (2018)
2. Barbosa, P.A.: Explaining cross-linguistic rhythmic variability via a coupled-oscillator model of rhythm production. Speech Prosody **2002**, 163–166 (2002)
3. Boersma, P.: Praat, a system for doing phonetics by computer. Glot Int. **5**(9/10), 341–345 (2001)
4. Camacho, A.: SWIPE: A Sawtooth Waveform Inspired Pitch Estimator for Speech and Music. Ph.D. thesis, University of Florida (2007)
5. Cummins, F., Port, R.: Rhythmic constraints on stress timing in English. J. Phon. **1998**(26), 145–171 (1998)
6. De Cheveigné, A., Kawahara, H.: YIN, a fundamental frequency estimator for speech and music. J. Acoust. Soc. Am. **111**(4), 1917–1930 (2002)
7. Garg, D.: SWIPE pitch estimator (2018). https://github.com/dishagarg/SWIPE. [*PySWIPE*]
8. Gaspari, D.: Mandarin Tone Trainer. Master's thesis, Harvard Extension School (2016). https://github.com/dgaspari/pyrapt
9. Gibbon, D.: The future of prosody: It's about time. Proc. Speech Prosody **9** (2018). https://www.isca-speech.org/archive/SpeechProsody_2018/pdfs/_Inv-1.pdf
10. Gibbon, D.: Rhythm zone theory: speech rhythms are physical after all. In: Wrembel, M., Kiełkiewicz-Janowiak, A., Gąsiorowski, P. (eds.) Approaches to the Study of Sound Structure and Speech. Interdisciplinary Work in Honour of Katarzyna Dziubalska-Kołaczyk. Routledge, London (2019). https://arxiv.org/abs/1902.01267
11. Gibbon, D.: CRAFT: A Multifunction Online Platform for Speech Prosody Visualisation (2019). https://arxiv.org/pdf/1903.08718.pdf
12. Gibbon, D.: The rhythms of rhythm. J. Int. Phonetic Assoc. First View 1–33 (2021). https://doi.org/10.1017/S0025100321000086
13. Guyot, P.: Fast Python implementation of the Yin algorithm (2018). https://github.com/patriceguyot/Yin/

14. Hermansky, H.: History of modulation spectrum in ASR. In: Proceedings of the ICASSP 2010 (2010)
15. Hess, W.: Pitch Determination of Speech Signals: Algorithms and Devices. Springer, Berlin (1983). https://doi.org/10.1007/978-3-642-81926-1
16. Inden, B., Malisz, Z., Wagner, P., Wachsmuth, I.: Rapid entrainment to spontaneous speech: a comparison of oscillator models. In: Miyake, N., Peebles, D., Cooper, R.P. (eds.) Proceedings of the 34th Annual Conference of the Cognitive Science Society. Cognitive Science Society, Austin (2012)
17. Jouvet, D., Laprie, Y.: Performance analysis of several pitch detection algorithms on simulated and real noisy speech data. In: 25th European Signal Processing Conference (2017)
18. Klessa, K.: Annotation Pro. Enhancing Analyses of Linguistic and Paralinguistic Features in Speech. Wydział Neofilologii UAM, Poznań (2016)
19. Martin, P.: WinPitch: un logiciel d'analyse temps réel de la fréquence fondamentale fonctionnant sous Windows. Actes des XXIV Journées d'Étude sur la Parole, Avignon 224–227 (1996)
20. Rabiner, L.R., Cheng, M.J., Rosenberg, A.E., McGonegal, C.A.: A comparative performance study of several pitch detection algorithms. IEEE Trans. Acoust. Speech Sig. Process. **ASSP-24**(5), 399–418 (1976)
21. Schmitt, B.J.B.: AMFM_decompy (2014). [PyYAAPT]. https://github.com/bjbschmitt/AMFM_decompy
22. Sjölander, K., Beskow, J.: Wavesurfer – an open source speech tool. Proc. Interspeech 464–467 (2000). http://www.speech.kth.se/wavesurfer/
23. Talkin, D.: A robust algorithm for pitch tracking (RAPT). In: Kleijn, W.B., Palatal, K.K. (eds.) Speech Coding and Synthesis, pp. 497–518. Elsevier Science B.V. (1995)
24. Talkin, D.: Reaper: Robust Epoch And Pitch EstimatoR (2014). https://github.com/google/REAPER
25. Tilsen, S., Johnson, K.: Low-frequency Fourier analysis of speech rhythm. J. Acoust. Soc. Am. **124**(2), EL34–EL39 (2008). [PubMed: 18681499]
26. Tilsen, S., Arvaniti, A.: Speech rhythm analysis with decomposition of the amplitude envelope: characterizing rhythmic patterns within and across languages. J. Acoust. Soc. Am. **134**, 628 (2013)
27. Todd, N.P.M., Brown, G.J.: A computational model of prosody perception. ICSLP **94**, 127–130 (1994)
28. Traunmüller, H.: Conventional, biological, and environmental factors in speech communication: a modulation theory. In Dufberg, M., Engstrand, O. (eds.) PERILUS XVIII: Experiments in Speech Process, pp. 1–19. Department of Linguistics, Stockholm University, Stockholm (1994). [Also in Phonetica **51**, 170–183 (1994)]
29. Xu, Y.: ProsodyPro – a tool for large-scale systematic prosody analysis. In: Tools and Resources for the Analysis of Speech Prosody (TRASP 2013), Aix-en-Provence, France, pp. 7–10 (2013)
30. Zahorian, S.A., Hu, H.: A spectral/temporal method for robust fundamental frequency tracking. J. Acoust. Soc. Am. **123**(6), 4559–4571 (2008). [YAAPT]

Hybridised Deep Ensemble Learning for Tone Pattern Recognition

Udoinyang G. Inyang📷 and Moses E. Ekpenyong$^{(\boxtimes)}$ 📷

University of Uyo, P.M.B. 1017, Uyo 520003, Nigeria
{udoinyanginyang,mosesekpenyong}@uniuyo.edu.ng

Abstract. In this contribution, a multi-classification framework comprising of three heterogeneous classifiers–self-organising map (SOM), ensemble of deep neural networks (DNNs), and adaptive neuro-fuzzy inference system (ANFIS) is proposed for tone pattern discovery in Ibibio (New Benue Congo, Nigeria). The proposed system is a set of localised DNN classifiers that derives input data from cluster classes generated by the SOM. The ANFIS classifier adopts information from a meta-algorithm to create the classification rules and model instances, sufficient to eliminate prediction uncertainties in the DNN output vectors. To demonstrate the feasibility of the framework, six tone features extracted from Ibibio text utterances namely: beginning of sound boundary (B); end of sound boundary (E); vowel count (V); consonant count (C); syllable count (S); and phoneme count (P), were considered. Results of SOM visualisation revealed similar patterns for features: V and S, and C and P; but features B and E, however, showed weak correlation. Next, four cluster vectors discovered by SOM were fed into a 4-DNN ensemble for training and validation. The neuron weights and target vectors obtained from the DNN then provided inputs to the ANFIS for building of fuzzy rules–using the sugeno-inference mechanism and partitioned into 70% training samples, and 30% testing samples. Predicted tone patterns yielded satisfactory mean absolute error of 0.027; and R-value of 0.8315. ANFIS however, improved on the results of the ensemble of DNNs, as the mean squared errors for train and test datasets were respectively lower than those obtained from DNN ensemble.

Keywords: ANFIS · DNN · Ensemble learning · Heterogeneous classifier · SOM · Tone language · Tone recognition

1 Introduction

Overcoming algorithmic shortcomings and achieving syngenetic effects during features classification have prompted the hybridisation of diverse classification methods – as there exists no (single) optimal algorithm for every problem [1]. Hence, exploiting the strengths of two or more specialised algorithms is certain to produce improved and optimal classification results – through combined prediction and solution refinements. Creating stronger learning systems by leveraging unlabelled data and classifier combination in a lightly supervised manner was proposed in [2]. Lightly supervised learning and ensemble learning constitute two state-of-art machine learning paradigms. While the

© Springer Nature Switzerland AG 2022
Z. Vetulani et al. (Eds.): LTC 2019, LNAI 13212, pp. 46–64, 2022.
https://doi.org/10.1007/978-3-031-05328-3_4

former strives to achieve a strong generalisation by exploiting unlabelled data, the later attempts to realise strong generalisation using multiple independent classification learners. This paper therefore presents a confluence of lightly-supervised classification and ensemble learning – sufficient to trigger significant improvements to an intractable inter-disciplinary problem – tone pattern classification. It combines the strengths of integrated multiple classification models into a meta-algorithm that controls state-of-the-art learning algorithms and exploits the potentials of congregating unsupervised and supervised learning (SL) approaches in the discovery of tone features pattern in speech signals. One of the most active research in SL has been to study approaches of constructing good ensembles of classifiers [3]. Unsupervised learning (UL) classification is adopted to visualise speech input datasets and correlate the tone pattern indicators as well as providing labels for the unlabelled input datasets. An ensemble of DNNs is then used to classify the discovered cluster vectors (samples) into different classes required for training the different DNNs. Adaptive neuro-fuzzy inference system (ANFIS), another supervised approach is finally employed to handle imprecision as the tone patterns are predicted. Indeed, combining multiple classifiers has been known to increase predictive performance. Hence, the novelty of this research is in the strength of our framework to reduce cost of computation through the concurrent execution for clusters discovery and ensemble learning.

1.1 Lightly Supervised Learning

In many pattern classification problems, the acquisition of labelled training data is costly and/or labour intensive, while unlabelled data samples are inexpensive to obtain. Hence, the goal of lightly supervised classification is to exploit both labelled and unlabelled data to build better classifiers, as opposed to the use of only labelled data. Lightly supervised algorithms that learn from both labelled and unlabelled examples have undergone intensive research in the last few years [4], and have been found to be very effective in speech technology applications. This paper utilises the functionalities of heterogeneous classifiers – different learning algorithms operating on the same data – to improve performance – through a serial combination of different learning paradigms.

1.2 Ensemble Learning

Ensemble learning is a method for constructing accurate classifiers or predictors from a committee of weak classifiers. Suppose $\tilde{E}$ denotes the expected classification error of ensemble classifiers, then the ensemble theory [5], suggests that,

$$\tilde{E} = \overline{E} - \overline{D} \tag{1}$$

where, $\overline{E}$ and $\overline{D}$, are the average classification error and individual diversity, respectively. Equation (1) implies that accuracy and diversity scale with performance of the ensemble classifier. A combination is an ensemble if the several classifiers are essentially performing the same task [6]. Hence, an ensemble of classifiers is a set of classifiers which individual decisions are combined to classify new exemplars. The necessary and sufficient condition for an ensemble classifiers to outspace individual classifiers are accuracy and diversity [7]. There is no definitive taxonomy of ensemble learning [9], as the success or failure of an ensemble model depends on the member models and the nature of data. Numerous approaches have also proposed accurate and diverse classifiers. Prominent amongst these approaches include: Exploitation of existing knowledge of the problem, Randomisation, Randomisation, Variation, Training data manipulation, Target features manipulation and Input features manipulation [8] and [9]. Three notable strategies separate the training of ensembles; bagging, boosting and averaging [9]. Bagging is widely used for statistical classification and regression while boosting is a apt for constructing multiple classification systems. Notable ensemble implementations include; Ensembles of NNs, random forests, adaboost.

1.3 Problem Statement

Tone creates meaning and enhances the comprehensibility of utterances in a tonal language. A tonal language uses pitch (the auditory correlate of tone) to discriminate lexical or grammatical meaning and is grouped into two broad categories according to this traditional divisions: contour- and register-tone languages [10]. While contour-tone languages are those from the Asian region (e.g., Mandarin, Cantonese, and Thai – from Chinese), register-tone languages are those from the West African region (e.g., Ibibio and Yoruba – from Nigeria). Two main systems are commonly used to transcribe tone. The first applies to register-tone languages: where some symbols representing the tone levels are superimposed on the phonemes (usually the vowel), e.g., á, à, and ¬ā, for high, low, and mid, respectively. Others are realisations or combination of the basic tones: e.g., ǎ, and â, are rising (low+high) and falling (high+low) tones, respectively. The second is iconic, and usually applies to contour-tone languages – to represent pitch movements on a 5-point scale (1=lowest, 5=highest). But it is not always possible to replicate all such sequences typographically as tone letters. Nearly all tone languages have several rules that modify tones when spoken in a sequence, i.e., when spoken in normal phrases rather than in isolation. One of the most well-known cases is in Mandarin Chinese: when two Tone-3 syllables occur in sequence, and the first one is changed to Tone 2. Same goes for Ibibio (new Benue Congo, Nigeria) when two words are combined, and the tone of the first phoneme of the second word changes from high to low tone: e.g., úfọ̀k 'house' and àbàsì 'God', are combined to yield úfọ̀kábàsì 'house of God'. Three mutually exclusive tasks are important for studying a tone system: (i) surface tonal contrasts determination; (ii) possible tonal alterations ('morphotonemics') discovery; (iii) tone analysis: interpreting what has been discovered in (i) and (ii). At this point, theoretical constructs, and formal devices such as auto-segmental notation are necessary, to gain useful insights on how the tone system operates.

Most research works on tone are concerned with stage (iii), which presupposes the first two stages: that the analysis of tone is impossible without an understanding of the surface contrasts and any occurring tonal alterations in the tone system. Again, in tone languages, different tone patterns of the same syllable may convey different meanings, thus making tone modelling intractable. The recognition of tone patterns therefore requires efficient approach to combining speech features into a single measure capable of discriminating among the various input classes [11], and can be regarded as a pattern recognition problem [12]. Pattern recognition is an area of ML that discovers common features among various data exemplars using either supervised, unsupervised or hybrid learning (a combination of both supervised and UL) [13]. The standard formulation of a machine learning problem assumes that the available data is independent and identically distributed (i.i.d) [15]. Although pattern discovery has greatly advanced the field of computational intelligence, and data mining – the discovery of interesting pieces of knowledge in data – and has become the de-facto approach to robust knowledge-driven system design, it is challenging to discover accurate knowledge (or patterns/trends) from collected data. The emergence of numerous key data mining techniques including association rule mining, classification, clustering, prediction, sequential patterning, decision tree, and combinations of these, have been explored in the literature to describe the data mining type, and for mining useful patterns in data; but the effective use and update of the discovered patterns remain an open problem. To robustly drive these techniques, deep learning [15, 16] has evolved to dramatically improve the accuracy of pattern discovery – as several deep architectures and learning methodologies have been developed with distinct strengths and weaknesses in recent years.

2   The Ibibio Language

Many languages spoken in Nigeria are classified under the Benue Congo language group, with phonological features including labial flaps, labial velars, and implosives [17]. Although tone systems vary dramatically across languages, pitch patterns in tone languages appear limited, and phonological features can be exploited to manically distinguish tone [18]. Hence providing phonological evidence is important in the design and modelling of tone patterns. In the following subsections, we succinctly describe the syllable structure and tone system of Ibibio (New Benue Congo, Nigeria), to provide insights into their composition, interaction, and modelling.

2.1   Syllable Structure

An Ibibio syllable may consist of any of the following distinct tones: High (H), Low (L), Downstepped High (!H), Rising (LH) and Falling (HL). The syllable structure of Ibibio is as analysed as follows:

(i)   a vowel (*V syllable*), example, /á-wó/ 'human'
(ii)   a syllabic nasal (*N syllable*), assimilates, or is homorganic with the nasal consonant it occurs with, example, /ń̄ -káñ/ 'charcoal';
(iii) a consonant and a vowel (*CV syllable*): example, in /sé/ 'look';
(iv)   a consonant, vowel and consonant (*CVC syllable*): example, /dèp/ 'to rain';
(v)   a consonant, a vowel and another vowel (*CVV syllable*): example, in the first syllable of verbal items as in: /dèè-mé/ 'share';
(vi)   a consonant, a long vowel and a consonant (*CVVC*): example, in /dɔ́ɔ́k/ 'climb';
(vii) a consonant, a glide and a vowel (*CGV syllable*): example, /frɛ̌/ 'forget';
(viii) a consonant, a vowel and a glide (*CVG syllable*): example, /dáj/ 'lick'. The CVG syllable structure is also found in nominal items like /á-kàj/ 'forest';

(ix)   a consonant, a glide, a vowel and consonant (*CGVC syllable*): example, /fjɔ́k/ 'to block the way'. The CGVCsyllable pattern is also found in nominal items like /ḿ.fjɔ́k/ 'traditional cooking stand';

(x)   a consonant, a glide, a vowel and a glide *(CGVG syllable)*: example, in frequentative verbal items like /twàj/ 'hit (freq)';

(xi)   a consonant, a glide, a long vowel and a consonant (*CGVVC syllable*): example, in the second syllable of nominal items like /á-bjɔ́ɔ́ŋ/ 'hunger'.

The structure has glides modelled as vowels, and the dotted lines indicate syllable boundaries. It is generic for languages with similar structure and has also been used for speech synthesis and recognition designs of African tone languages (c.f., [19–21]; tone modelling and classification systems [11, 14, 22].

2.2 Tone System

Tone is a supra-segmental that spreads over a group of voiced segments, and is commonly associated with vowels. In Ibibio, tone dominates all phonemes of the vowel system and consists of the following phonemes: a, e, i, ḭ, ə, o, o̞, u, ṵ, and ʌ; as well as nasalised consonants: m, n, and ñ. The consonant system of Ibibio, however contains the following phonemes: b, d, f, ʁ (gh), h, j (y), k, kp, m, n, ñ, ny, p, ɾ, s, t, and w. Four tonal constraints necessary to predict the next tone occurrence in an Ibibio include:

(i)   a H tone must occur before a !H tone can occur in any word;

(ii)   a !H tone, however, contrasts with a L tone and terminates in a well-formed utterance. They cannot occur on their own, but are distinguished from the mid tone of a word, and hence, occur sparingly in any utterance;

(iii) H and L pitches are affected by downdrift and could spread across various word positions (initial, medial and final)

(iv)   LH (rising contour-R) and HL (falling contour-F) sequences may occur at any word position, from its first tone level, due to the down-step and downdrift scenarios.

In [32], a context-dependent HMM algorithm for tone and prosody features labelling was proposed. The generic HMM showing the tone and prosodic states, as well as transitions that generate the respective tone patterns is given in Fig. 1.

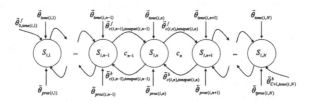

Fig. 1. A generic context-dependent HMM labelling of tone and prosody.

3 Related Works

Recently, deep neural networks (DNNs) have shown high effectiveness in acoustic modelling as regards speech technology [23]. In [24], a DNN for frame-level 5-tone classification and a single-layer neural network at segment (syllable) level were explored. The segment-level models were trained to classify syllables from a Mandarin Broadcast News Speech corpus, using co-articulation features. When provided with only raw MFCCs as input, the method obtained low error rate of 16.86%, but the same DNN architecture scored substantially worse when trained and tested with sub-band autocorrelation change detection (SACD) features [25]. Hence, raising concerns over the theories of Chinese tone. In [26], a DNN model was trained to classify each frame of speech into one of six classes: five tones and one non-tone. Each frame was represented by a 40-D MFCC vector. The tone-bearing units (TBUs) were classified based on tonal features, segment duration and contextual features. Experiments without context segments resulted in a segment error rate (SER) of 17.73%. However, contextual frames (frames before and after the center frames) were still used to classify each frame within the segment. In [27], the efficiency of an ANN in recognising Mandarin tone patterns was examined. Speech data were recorded from 12 children and 15 adults – all native Mandarin Chinese speakers. Feature extracted for the experiment was F0 of each monosyllabic word, which contours served as inputs to a feed-forward back-propagation ANN. After a Levenberg-Marquardt optimisation, the NN was able to successfully classify the tone patterns with an accuracy of about 90% for both children and adults. Their results showed that ANN may provide an objective and effective assessment of tone production. In [14], an unsupervised mining of speech corpora was proposed for the efficient classification of tone features. Input vectors to the experiment were pattern alignments of Ibibio corpus. Their design integrated two unsupervised tools: the k-means clustering and SOM model to evaluate the optimal number of clusters. Their results validated existing claims and demonstrated the importance of vowel-only features in tone pattern recognition.

Within the past three decades, consistent enquiries into ensemble learning approaches have been made, resulting in three different classification systems namely, independent, sequential, and simultaneous [9, 28]. Ensemble constructions have also witnessed progressive refinements from simple methods that rely on training NNs with different initial conditions or starting points [29]; varying the topology algorithm or number of hidden neurons [30]; to varying the data in some manner – using sampling, diverse data sources, diverse pre-processing methods, distortion and adaptive re-sampling. In Liu [31], a balanced ensemble learning approach for defining adaptive learning error function for individual NNs in an ensemble is proposed. Sharkey, Sharkey, Gerecke and Chandroth [6] reviewed available methods for creating ensembles and defined an approach for testing potential ensemble combinations on a validation set, and selecting the best performing ensemble, which was finally tested on the resultant dataset. They demonstrated this approach on two case studies that rely on ensembles of networks trained from three sources, and on the output of trained SOMs. Improved results were obtained by combining the NNs to form ensembles. Jin and Liu [15] proposed an approach that learned the relationship between data and models using a set of switches for routing test instances to appropriate classification models in an ensemble. Study on real-world and benchmark data showed that their approach could achieve significant performance improvement using heterogeneous data. Romsdofer [32] combined an ensemble learning technique using NNs as base learners with feature relevance determination. The weighted NN ensemble model was applied for both phone duration and fundamental frequency (F0) modelling. The NN ensemble model showed improvement when compared with state-of-the-art prosody models such as the classification regression tree (CART), multivariate adaptive regression spline (MARS), or artificial neural network (ANN). Further, their model outperformed similar ensemble model based on gradient tree boosting. Chen, Bunescu, Xu and Liu [33] proposed a method that fully automates tone classification of syllables in Mandarin Chinese. Their model considered raw tone data as input and used convolutional NNs to classify syllables into one of the four tones of Mandarin and does not require manual checking of F0. Their classification system could have significant clinical applications in the speech evaluation of hearing-impaired population.

4   Hybridised Classification Framework

The proposed classification framework is driven by three state-of-the-art learning systems (SOM, DNN and ANFIS) working in synergy. The workflow describing the cooperative system is shown in Fig. 2. The SOM network accepts a speech corpus as input and abstracts significant tone attributes $(x_1, x_2, \ldots, x_n)$, sufficient to generate k distinct clusters $(c_1, c_2, \ldots, c_k; 2 \leq k \leq n)$ as targets. The k-clusters are generated such that data points with similar tone attribute patterns are distinctively grouped. Each SOM target (cluster vectors) then feeds a corresponding DNN within the DNN module. Hence, the number of DNNs generated depends on the cluster classes discovered by the SOM. Each DNN cluster data are then trained for rules extraction by ANFIS.

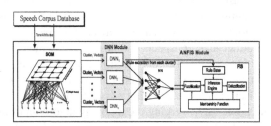

Fig. 2. Proposed system architecture

4.1   Speech Corpus Database

In this research, a total of 16,905-word patterns were populated, to enable the extraction of tone features for this experiment (see Table 1). Six input feature vectors were extracted according to the following categories: (i) beginning of sound (vowel/consonant) boundary; (ii) end of sound (vowel/consonant) boundary; (iii) number of vowels; (iv) number of consonants; (v) number of syllables; and (vi) number of phonemes. For options (i) and (ii), the following coding system was used to code the various tone patterns:

1-Consonants, 2-H, 3-L, 4-DS, 5-LH, 6-HL; where, H represents High tone (´), L represents Low tone (`), DS represents Down-Step tone (!), LH represents Low-High tone (ˇ), HL represents High-Low tone (ˆ). A statistical breakdown of the TBUs (vowels and syllabic nasals) include H (21,980), L (15,739), DS (2,399), LH (1,492), HL (378) – giving a total of 41,988 (56.73%) tone bearing units; while 43.27% (32,028) were non TBUs (consonants). Statistics of the consonants and vowels at the (beginning of; end of) word boundaries are (2439; 6176) and (8893; 5590), respectively. Table 1 shows input classification of the first 15 words. The syllabification of words and input tone patterns were generated using the FST and tone model.

Table 1. Input vectors classification

Word (Syllabified)	Input (tone pattern)										Begin sound boundary	End sound boundary	No. of vowel	No. of consonant	No. of syllable	No. of phoneme
	1	2	3	4	5	6	7	8	9	10						
bọ̀-ŋà-kàm	1	3	1	3	1	3	1				1	1	3	4	3	7
kúù-kpá-m̀-bà	1	2	3	1	1	2	3	1	3		1	3	5	4	4	9
á-ké-fèé-fé-ʁè	2	1	2	1	3	2	1	2	1	3	2	3	6	4	5	10
á-yàk	2	1	3	1							2	1	2	2	2	4
í-kòt	2	1	2	1							2	1	2	2	2	4
á-bà-sì	2	1	3	1	3						2	3	3	2	3	5
è-ɲé	3	1	2								3	2	2	1	2	3
á-mà-á-n!ám	2	1	3	2	1	4	1				2	1	4	3	4	7
á-ŋwá-ŋà	2	1	1	2	1	3					2	3	3	3	3	6
ké	1	2									1	2	1	1	1	2
m̀-m̀e	3	1	3								3	3	2	1	2	3
ó-wó	2	1	2								2	2	2	1	2	3
é-n!í!é	2	1	4	4							2	4	3	1	2	4
n-trèù-bọ̀k	3	1	1	3	3	1	2	1			3	1	4	4	3	8
ké	1	2									1	2	1	1	1	2
ú-sáŋ	2	1	2	1							2	1	2	2	2	4
ọ̀m-mọ̀	6	1	1	5							6	5	2	2	2	4
kèèd	1	3	3	1							1	1	2	2	2	4
kèèd	1	3	3	1							1	1	2	2	1	4

4.2 DNN Modelling

Optimising the DNN training was achieved in the following steps: (i) pre-training each layer, exclusively, using a greedy algorithm; (ii) applying UL at every layer in a way that preserves information from the input and disentangling any factor of variation; (iii) fine-tuning the entire network, subject to the ultimate criterion of interest.

Figure 3 shows the proposed DNN architecture, with L layers (L > 3). Each layer has connection weights, a bias vector ($\mathbf{b_L}$), and an output vector ($\mathbf{O_L}$). The number of neurons in each layer is denoted by m, where, m^1, m^2,..., m^L, are the number of neurons in Hidden Layers 1, 2, ..., L, respectively. Inputs to the system, x_i, are the various tone output vectors discovered by the SOM classifier. In each of the layers, the input vector elements enter the DNN via weights, w_{i,m_n^L}, which represents the weight of the link between the *ith* input neuron and the *nth* neuron of the *Lth* hidden layer.

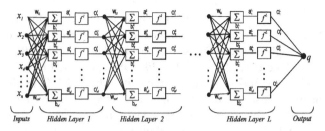

Fig. 3. System architecture of the proposed DNN

The inputs to each layer of the DNN are denoted as, a, with a^1, a^2,..., a^L, representing inputs to layers 1, 2, ..., L, respectively. Therefore, modelled are as follows:

$$a^2 = \sum_{i=1}^{n} f^l \left(w_{L,m^L} a^{L-1} + b_{m^L}^L \right) \qquad (2)$$

where, $i = 1,2...$, n, are the number of input variables (length of the input vector), m, is the number of neurons in *Lth* layer (L > 3), while f^L, is the transfer function of the *Lth* Layer. The output of a proceeding layer is the input of the immediate succeeding layer.

The steps used to implement the DNN-pattern recognition include: (i) initialise the hidden layer size; (ii) create the DNNs; (iii) set up the division of data for training, validation, and testing; (iv) train the network; (v) test the network. As described in [34], a feed-forward multi-layered NN with more than two hidden layers between the input and output layers was created using *tansig* and *purelin* as the transfer functions for the hidden layers and output layer, respectively. At the configuration stage, the number of hidden layers was set to 3, to achieve prediction accuracy. During implementation phase, the input layer ascended to 6 nodes (x_1, x_2,..,x_6), to satisfy the criteria stipulated in [48]. The size of the hidden layers was initialised to 4 and iterated to a maximum of 10. The best performing number of hidden layer neurons was then used as a standard for all the hidden layers. Each of the clustered speech dataset was randomly partitioned into training (70%) validation (15%) and testing (15%).

4.3 DNN Modelling

ANFIS is multi-layered architecture that integrates the strenghts of NN and fuzzy logic for quailitative cognitive reasoning [35, 38]. It implements a five-layered structure of a first order Sugeno model, and is adopted in this paper to process the DNN target vectors by accepting the DNN classification results as input, as well as map the entire input space to their respective membership functions (MFs). Rules are then constructed to drive the mapping of the input MFs, which are subsequently mapped to the output [36]. In our design, the first layer (input layer) consists of six tone features {beginning of sound boundary: B, end of sound boundary: E, vowel count: V, consonant count: C, syllable count: S, and phoneme count: P} as nodes. Each node in this layer are adaptive and executes the fuzzification operation – generates a membership grade for each tone feature. The Guassian MF [37], (Eq. 3) is adopted in predicting the tone quality:

$$\mu_{A^i}(x) = exp \left[-\frac{1}{2} \frac{(1 - \beta_i)^2}{\sigma_i} \right] \qquad (3)$$

where, β_i and σ_i are the center and width of the *ith* linguistic label, respectively, and are the premise parameters governing the Gaussian MF, x, is the input to node $i : 1 \leq i \leq 6$. Layer 2 has fixed nodes and computes the firing strength of rules using fuzzy operators (OR, AND) by fuzzifying the antecedent part of each rule. Normalisation of each rule's firing strength –the ratio of the *ith* rule's firing strength to the sum of all rule's firing strengths, is performed by the nodes of layer 3. The fourth layer consists of adaptive nodes and determines the product of the normalized firing strength of rules and the first order polynomial of Sugeno consequent parameters (r_0, r_1, r_2, ..., r_6). ANFIS uses either back-propagation algorithm or hybrid approach (a combination of least-squares estimation (LSE) and back propagation gradient descent), to identify and tune the MF parameters of the output. The hybrid-learning algorithm involves two phases: forward and backward pass. In the forward pass, each node's output approaches the next layer up to the fourth layer, where LSE is used to identify and tune the consequent parameters [38]. Since accuracy of prediction is our interest and considering the nature of the data, the mean absolute error (MAE) is used as a performance measure [54]. The error signals are configured to propagate backwards while tuning the premise parameters with back propagation gradient descent algorithm during the backward pass phase. Whereas, only the consequent parameters changed during the forward pass, they remained unchanged during the backward pass. The iterations of the forward and backward passes cause the premise and consequent parameters to be identified and generated for the fuzzy inference system (FIS). The single node (summation neuron) in layer 5 determines the final output of the ANFIS model by summing incoming signals of layer four and performing the defuzzification operation using the identified FIS parameters. The output function, O_i^l, for each layer is given in Eqs. 4–8:

$$O_i^1 = \mu_{A^i}(x_i); i = 1, 2, \ldots, 6 \tag{4}$$

$$O_i^2 = \alpha_i = \mu_{A_n}(\beta_i)\mu_{\beta_n} \times (E_i)\mu_{E_n} \times (V_i)\mu_{V_n} \times (C_i)\mu_{C_n} \times (S_i)\mu_{S_n} \times (P_i)\mu_{P_n} \tag{5}$$

$$O_i^3 = \bar{\alpha}_i = O_i^2 = \frac{\alpha_i}{\sum \alpha_i} \tag{6}$$

$$O_i^4 = \alpha_i f_i = O_i^2 = \alpha_i \times \left(r_0^i + r_1^i B_i + r_2^i E_i + r_3^i V_i + r_4^i C_i + r_5^i S_i + r_6^i P_i \right) \tag{7}$$

$$O_i^5 = \sum_i \bar{\alpha}_i f_i = \frac{\sum_i \alpha_i f_i}{\sum_i \alpha_i} \tag{8}$$

5   Results

5.1   SOM Training

The SOM design was implemented using MATLAB 7.7.0 (R2015a). The batch unsupervised weight/bias algorithm (*trainbu*) – where weights and biases are only updated after all the inputs and targets were fed into the network [39], was adopted. The *trainbu* algorithm trains a network with weights and bias learning rules using batch updates, in two stages – rough and fine training phases [39]. The rough training phase spanned 1,000 iterations with initial and final neighbourhood radius of 5 and 2 respectively, in addition to a learning rate in the range (0.5 and 0.1). The fine training phase had a maximum of 1000 epochs, and a fixed learning rate of 0.2. The output layer had k neurons; the weight vector, $v_1^1, v_2^1, v_3^1, \ldots, v_k^1$, of the connections constitutes the prototype of each neuron and has the same dimension as the input vector. Selection of best centroids within each cluster was based on the Euclidean distance criterion. SOM was applied to identify and classify speech features into segments based on tone feature similarity using six input attributes. The tone cluster centers and map topology are given in Tables 2 and 3, respectively. In Table 2, cluster (1,1) – cluster 1, has 5614 (33.21%) data points; while cluster (1,2) – cluster 2, and cluster (2,1) – cluster 3; has 6407 and 1814 members, respectively; cluster (2,2) – cluster 4, has 3070 (18.16%) data points. Other results are a visual representation and overview of the homogeneous segments as well as the approximate number of clusters using the hexagonal grid topology.

Table 2. SOM cluster map topology

	1	2
1	5614 (33.21%)	6407 (37.90%)
2	1814 (10.73%)	3070 (18.16%)

Table 3. Tone attribute cluster centers

Attribute	Cluster centers			
	1	2	3	4
BSB	2.18	2.10	2.36	2.63
ESB	1.09	2.75	1.78	2.48
NV	2.13	1.85	4.25	3.28
NC	2.09	1.03	3.66	2.30
NS	1.93	1.74	3.86	2.95
NP	4.22	2.88	8.12	5.58

A map quality (total sum of squares) of 71.35% was attained from four (4) clusters. In Fig. 4, an 8×13 hexagonal grid visualising the neurons according to their respective clusters is presented. The neurons are distributed descending from top to bottom as follows: cluster 2 – 39 (37.9%), cluster 1 – 35 (33.2%), cluster 3 – 19 (18.2%), and cluster 4 – 11 (10.7%). Figure 7 is a visualisation of Table 3 as a Unified Distance Matrix (U-Matrix) and input component plane maps. The U-Matrix provides a visualisation of the distance between neurons – where computed distance of adjacent neurons and represented with appropriate colours. Dark colours between the neurons correspond to a large distance (a gap) between the codebook values while light colours between the neurons indicate that the codebook vectors are close to each other in the input space. The dark and light colours are regarded as cluster separators and clusters respectively.

Fig. 4. Feature clusters visualisation

We observe in Fig. 5 that all the input features have sharp boundaries with well separated clusters, but the following component (planes) pairs: V and S; and, C and P, exhibit similar patterns and are moderately correlated. In practice, vowels are prominent attributes when modelling syllables; while consonants by its nature are critical to speech phoneme classification, as this explains its importance in modelling and evaluating speech technology systems. However, a closer look at components B and E reveals a weak correlation between the component features with noticeable pattern not situated at the same positions. Perhaps, the variation may not be unconnected with the effect of co-articulation which could interface with the various sound boundaries and proves that sound boundaries carry important information to cue pitch ranges.

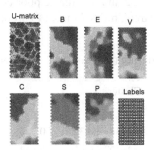

Fig. 5. U-Matrix and input component planes for selected tone features

5.2 DNN Model Training

The DNN training process focused on tuning the weights and biases of each model to optimise its performance. Although the models took longer to train, better generalisations for large or noisy tone feature datasets were obtained. The gradient was computed, and weights update commenced after all the tone features were fed into the DNN. The stopping criterion was achieved through adaptive weight minimisation (regularisation) using the *trainbu* function. The mean square error (MSE) – the average squared error between the DNN outputs and the target; was used to evaluate the performance of the DNNs. Figure 6 is the objective function showing the performance of the network in selecting members and parameters of each cluster depicting decreasing errors as the number of epoch increased, with no evidence of over-fitting in the network. In DNN cluster 1, the validation error reached a minimum at the 14th iteration without any further change, while the best performance of DNN for cluster 2 was at epoch 32. In training the DNN datasets for cluster 3 and 4, validation of the DNNs was achieved at epoch 7 (with a validation error of 0.33308), and epoch 6 (with a validation error of 0.49742), respectively.

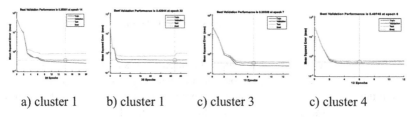

a) cluster 1   b) cluster 1   c) cluster 3   c) cluster 4

Fig. 6. Objective functions of DNN during cluster training, validation and testing

Table 4 presents a summary of the RMSEs for each DNN cluster dataset. Although RMSE values of each DNN model cluster indicated satisfactory models' classifiers for tone pattern prediction, the training, validation and testing errors, on the average, showed that DNN cluster 3 performed better than others, followed by DNN cluster 1. But the high testing errors of DNN clusters 1, 2 and 4, indicate poor classification of new samples, and raise some doubts on the strength DNN model to properly classify new samples.

Table 4. Performance (MSE) of DNN models in each Cluster

	Cluster			
	1	2	3	4
Train	0.2738	0.3070	0.2458	0.4104
Validation	0.3530	0.4294	0.3331	0.4974
Test	0.7641	0.7068	0.3932	0.5818
Average	0.4636	0.4811	0.3240	0.4965

5.3 DNN Validation

Regression analysis was used to validate the performance of DNN – through the assessment of the relationship between the DNN outputs and target of each cluster [39]. In the prediction of speech pattern, the response of DNN models to training, validation and testing datasets for the inputs in each of the cluster was determined. Four (4) regression plots resulting from training, testing, validation datasets and overall (combined) dataset were generated (see Fig. 7a–d). The dotted line in each plot represents the perfect result (i.e., outputs = targets). The solid line represent the best (fit) linear regression line between output and target. The R value gives an indication of the (strength) correlation between the outputs and targets. The validation and test results gave R values of 0.8050 and 0.6262, respectively. The overall performance of the model in cluster 1 dataset yielded an R value of 0.79078 ($R^2 = 0.63$). This implies that 79.08% of the DNN output is explained by the target variables. A summary of DNN models for all clusters is presented in Table 5.

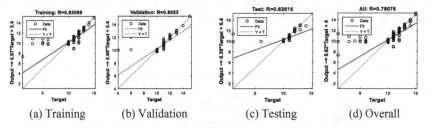

(a) Training   (b) Validation   (c) Testing   (d) Overall

Fig. 7. Regression plots for DNN output relative to targets of DNN Cluster 1

Table 5. Best linear Fit models for tone clusters

| | Cluster | | | | | | | |
| | 1 | | 2 | | 3 | | 4 | |
	Best Linear Fit	R	Best Linear Fit	R	Best Linear Fit	R	Best Linear Fit	R
Train	Y = 0.67T + 3.4	0.831	Y = 0.46T + 4.4	0.761	Y = 0.49T + 4.7	0.797	Y = 0.46T + 4.2	0.800
Validation	Y = 0.68T + 3.4	0.805	Y = 0.53T + 3.8	0.812	Y = 0.42T + 5.3	0.757	Y = 0.67T + 3.4	0.752
Test	Y = 0.39T + 6.4	0.626	Y = 0.34T + 5.4	0.651	Y = 0.37T + 5.8	0.697	Y = 0.67T + 3.4	0.706
Overall	Y = 0.62T + 4.0	0.791	Y = 0.44T + 4.5	0.749	Y = 0.45T + 5.0	0.775	Y = 0.67T + 3.4	0.771

To compensate for the weakness discovered in the DNN ensemble model, ANFIS was made to accept crisp output values obtained DNN ensemble learning, from which fuzzy rules were built using the sugeno-inference mechanism. The neuron weights and outputs of the DNNs were input and target vectors to the ANFIS system and partitioned into: 70% (11834) training samples, and 30% (5071) exemplars for testing the ANFIS model. The Fuzzy c-means (FCM) algorithm generated the FIS with Gaussian membership function *(gaussmf)* using *genfis3*. Out of a maximum of 100 iterations of FCM, the algorithm converged at the 35th iteration with an objective function value of 1.82×10^4 (topmost plot of Fig. 9). The cost optimisation graph (bottom plot of Fig. 9) reached the best value of 0.524, at the 1000th iteration (Fig. 8).

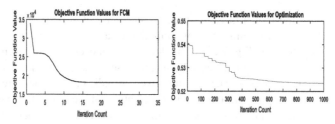

Fig. 8. Objective function plots of FCM

The parameters of the *gaussmf* indicating each input parameter in the 4th clusters are presented in Table 6, while the MF plots is given in Fig. 9. Least-squares and back propagation gradient descent methods were combined, to identify and update the output MF parameters during the FIS training.

Table 6. Tone features cluster centers

Tone Features	Cluster Index			
	1	2	3	4
BSB	[-1.663, 2.164]	[-7.89, -41.227]	[9.079,18.1501]	[3.788 , 3.764]
ESB	[1.293, 3.345]	[-8.654,18.785]	[10.636, -53.546]	[0.639, 1.1928]
NV	[0.7398,1.880]	[-1.648,-10.999]	[2.2701, 15.398]	[0.373, 1.043]
NC	[0.4164, 1.0549]	[0.7038, -61.939]	[0.4030, 1.6031]	[1.2969, 1.952]
NS	[2.7570,13.8367]	[16.0414,84.758]	[-3.064 , -8.401]	[5.750 , -5.5110]
NS	[19.6073,2.543]	[-9.325,2.719]	[0.94842,133.889]	[7.1709,-36.797]

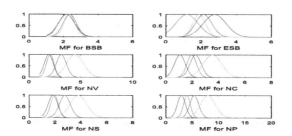

Fig. 9. MF plots of Tone features

The consequent parameters $P\{r_0, r_1, r_2, r_3, r_4, r_5, r_6\}$ of the sugeno type FIS were obtained as given in Eq. (9).

$$P = \begin{bmatrix} -0.022 & 0.044 & 0.667 & 0 & -0.178 & 0.264 & 6.029 \\ -0.020 & 0.121 & 16.682 & 0 & -1.235 & -1.755 & -91.086 \\ 0.193 & -1.106 & -0.260 & 0 & -4.401 & -0.096 & 150.731 \\ -0.060 & 0.394 & 0.667 & 0 & -0.259 & 0.264 & 7.045 \end{bmatrix} \quad (9)$$

Each row is a vector of input parameters – terms of the first order polynomial utilised by layer 4 of ANFIS in deriving its output (see Eq. 11), column 1 is the constant value vector, and tone feature parameters are represented in columns 2–4. The training (see Fig. 10a) and testing (see Fig. 10b) results showed significant positive drift from the results of the DNN in each cluster.

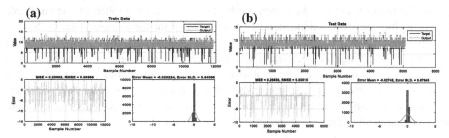

Fig. 10. Error plots and relationship between a) training target and predicted output b) test target and predicted output.

The MAE of the train data is 0.025, while the *MAE* of the test data error is 0.027. This improvement is due to the combine capabilities of the DNN ensembles and ANFIS. The relationship between the target vector and the ANFIS result was also significant at $R = 0.8314$ (see Fig. 11). A hybridised predictive framework composed of three hetero-geneous classifiers operating on same data. Each component generates refined results in the form of target outputs to feed the next classifier. SOM was used to learn the dataset patterns and cluster the dataset to eliminate redundancy at the next processing stage. ANFIS then served as a final collector and optimiser to aggregate the ensem-ble DNNs target vectors and compensate for any weakness exhibited by the ensemble DNNs – which only offered independent description of the tone cluster patterns. Signif-icant progression in performance was exhibited when migrating from each phase of the experiment, justifying the combination of different Ml paradigms. A future direction of this research is to explore the use of mixed features in the experiment and also assess the causal relationship of tone features and tone quality. After enumerating the speech and text features, a dimension reduction technique is also necessary to abstract significant features for analysis and classification.

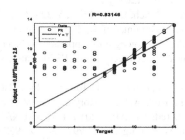

Fig. 11. Relationship between target and predicted output for all datasets.

6 Conclusion and Future Research Direction

A hybridised ensemble comprising three heterogeneous classifiers operating on a single data was proposed. Each component generates refined results in the form of target outputs for the next classification. SOM was used to learn the dataset patterns and cluster the dataset while ANFIS was the final collector and optimiser; aggregating the ensemble DNNs target vectors and compensate for any weakness exhibited by the ensemble DNNs – which only offered independent description of the tone cluster patterns. Significant progression in performance was exhibited when migrating from each phase of the experiment, justifying the combination of different ML paradigms. A future direction of this research is to explore the use of mixed features (combination of text and speech features) and assess the causal relationship of tone features and quality. After enumerating the speech and text features, a dimension reduction technique is also necessary to abstract significant features for analysis and classification.

References

1. Wolpert, D.H., Macready, W.G.: No free lunch theorems for optimization. IEEE Trans. Evol. Comput. **1**(1), 67–82 (1997)
2. Zhou, Z.H.: When semi-supervised learning meets ensemble learning. Front. Electr. Electron. Eng. China **6**(1), 6–16 (2011)
3. Liao, Z., Zhang, Z.: A generic classifier-ensemble approach for biomedical named entity recognition. In: Tan, P.-N., Chawla, S., Ho, C.K., Bailey, J. (eds.) PAKDD 2012. LNCS (LNAI), vol. 7301, pp. 86–97. Springer, Heidelberg (2012). https://doi.org/10.1007/978-3-642-30217-6_8
4. Krishnapuram, B., Williams, D., Xue, Y., Carin, L., Figueiredo, M., Hartemink, A.J.: On semi-supervised classification. In: Advances in Neural Information Processing Systems, pp. 721–728 (2005)
5. Mitchell, H.: Ensemble learning. In: Mitchell, H. (ed.) Multi-Sensor Data Fusion, pp. 221–240. Springer, Heidelberg (2007). https://doi.org/10.1007/978-3-540-71559-7_13
6. Sharkey, A.J., Sharkey, N.E., Gerecke, U., Chandroth, G.O.: The "test and select" approach to ensemble combination. Mult. Classif. Syst. **30–44**, 2000 (1857)
7. Hansen, L.K., Salamon, P.: Neural network ensembles. IEEE Trans. Pattern Anal. Mach. Intell. **12**(10), 993–1001 (1990)
8. Reid, A., Ramos, F., Sukkarieh, S.: Proceedings of IEEE International Conference on Robotics and Automation, Shanghai, China (2011)
9. Breiman, L.: Random forests. Mach. Learn. **45**(1), 5–32 (2001)
10. Ayuninjam, F.F.: A Reference Grammar of Mbili. University Press of America Inc. (1998)
11. Atal, B.S., Rabiner, L.R.: A pattern recognition approach to voiced—Unvoiced—Silence classification with applications to speech recognition. IEEE Trans. Acoust. Speech Sig. Process. **24**(3), 201–212 (1976)
12. Ekpenyong, M.E., Inyang, U.G., Umoren, I.J.: Towards a hybrid learning approach to efficient tone pattern recognition. In: Rutkowski, L., Korytkowski, M., Scherer, R., Tadeusiewicz, R., Zadeh, L.A., Zurada, J.M. (eds.) ICAISC 2016. LNCS (LNAI), vol. 9692, pp. 571–583. Springer, Cham (2016). https://doi.org/10.1007/978-3-319-39378-0_49
13. Deng, L., Li, X.: Machine learning paradigms for speech recognition: an overview. IEEE Trans. Audio Speech Lang. Process. **21**(5), 1–30 (2013)

14. Ekpenyong, M.E., Inyang, U.G.: Unsupervised mining of under-resourced speech corpora for tone features classification. In: Proceedings of IEEE International Joint Conference of Neural Networks, Vancouver Canada. IEEE Publishers, USA (2016)
15. Jin, R., Liu, H.: SWITCH: a novel approach to ensemble learning for heterogeneous data. In: Boulicaut, J.-F., Esposito, F., Giannotti, F., Pedreschi, D. (eds.) ECML 2004. LNCS (LNAI), vol. 3201, pp. 560–562. Springer, Heidelberg (2004). https://doi.org/10.1007/978-3-540-30115-8_51
16. Hinton, G.E., Salakhutdinov, R.R.: Reducing the dimensionality of data with neural networks. Science 313(5786), 504–507 (2009)
17. Clements, N., Rialland, A.: Africa as a phonological area. In: Heine, B., Nurse, D. (eds.) Africa as a Linguistic Area. CUP, Cambridge (2006)
18. Wang, W.: Phonological features of tones. Int. J. Am. Linguist. 33, 93–105 (1967)
19. Ekpenyong, M.E.: Speech synthesis for tone language systems. Ph.D. Thesis, University of Uyo in Supervision Collaboration with The Centre for Speech Technology Research (CSTR), University of Edinburgh (2013)
20. Ekpenyong, M., Urua, E.-A., Watts, O., King, S., Yamagishi, J.: Statistical parametric speech synthesis for Ibibio. Speech Commun. J. 56, 243–251 (2014)
21. Ekpenyong, M.E., Inyang, U.G., Ekong, V.E.: A DNN framework for robust speech synthesis systems evaluation. In: Zygmunt, V., Mariani, H. (eds.) Proceedings of 7th Language and Technology Conference (LTC), Poznan, Poland, pp. 256–261. Fundacja Uniwersytetu im. A. Mickiewicza (2015)
22. Ekpenyong, M.E., Udoh, E-O.: Tone modelling in Ibibio speech synthesis. Int. J. Speech Technol. 17(2), 145–159 (2014). http://www.link.springer.com/content/pdf/10.1007%2Fs10772-013-9216-2.pdf
23. Hinton, G., et al.: Deep neural networks for acoustic modeling in speech recognition: the shared views of four research groups. IEEE Sig. Process. Mag. 29(6), 82–97 (2012)
24. Ryant, N., Slaney, M., Liberman, E., Shriberg, E., Yuan, J.: Highly accurate Mandarin tone classification in the absence of pitch information. In: Proceedings of 7th Speech Prosody Conference, Dublin, pp. 673–677 (2014)
25. Slaney, M., Shriberg, E., Huang, J-T.: Pitch-gesture modeling using subband autocorrelation change detection. In: Proceedings of INTERSPEECH, Lyon, France, pp. 1911–1915 (2013)
26. Ryant, N., Yuan, J., Liberman, M.: Mandarin tone classification without pitch tracking. In: Proceedings of IEEE International Conference on Acoustics, Speech and Signal Processing (ICASSP), pp. 4868–4872 (2014)
27. Xu, L., et al.: Mandarin Chinese tone recognition with and artificial neural network. J. Otol. 1(1), 30–34 (2006)
28. Sarkar, D.: Randomness in generalization ability: a source to improve it. IEEE Trans. Neural Netw. 7(3), 676–685 (1996)
29. Perrone, M.P., Cooper, L.N.: When networks disagree: ensemble methods for hybrid neural networks (chap. 10). In: Mammone, R.J. (ed.) Neural Networks for Speech and Image Processing. Chapman-Hall (1993)
30. Partridge, P., Yates, W.B.: Engineering multiversion neural-net systems. Neural Comput. 8(4), 869–893 (1996)
31. Liu, Y.: A Balanced ensemble learning with adaptive error functions. In: Kang, L., Cai, Z., Yan, X., Liu, Y. (eds.) ISICA 2008. LNCS, vol. 5370, pp. 1–8. Springer, Heidelberg (2008). https://doi.org/10.1007/978-3-540-92137-0_1
32. Romsdofer, H.: Weighted neural network ensemble models for speech prosody control. In: Proceedings of Interspeech 2009, Brighton, UK, pp. 492–495 (2009)
33. Chen, C., Bunnescu, R., Xu, L., Liu, C.: Tone classification in Mandarin Chinese using convolutional neural networks. In: Proceedings of INTERSPEECH, San Francisco, USA, pp. 2150–2154 (2016)

34. Karsoliya, S.: Approximating number of hidden layer neurons in multiple hidden layer BPNN architecture. Int. J. Eng. Trends Technol. **3**(2012), 714–717 (2012)
35. Silarbi, S., Abderrahmane, B., Benyettou, A.: Adaptive network based fuzzy inference system for speech recognition through subtractive clustering. Int. J. Artif. Intell. Appl. (IJAIA) **5**(6), 43–52 (2014)
36. Vaidhehi, V.: The role of dataset in training ANFIS system for course advisor. Int. J. Innov. Res. Adv. Eng. (IJIRAE) **1**(6), 249–253 (2014)
37. Yadav, R.S., Soni, A.K., Pal, S.: Academic performance evaluation using soft computing techniques. Curr. Sci. **106**(11), 1504–1517 (2011)
38. Inyang, U.G., Akinyokun, O.C.: A hybrid knowledge discovery system for oil spillage risks pattern classification. Artif. Intell. Res. **3**(4), 77–86 (2014)
39. Inyang, U.G., Akpan, E.E., Akinyokun, O.C.: A hybrid machine learning approach for flood risk assessment and classification. Int. J. Comput. Intell. Appl. **19**(02), 2050012 (2020)

ANNPRO: A Desktop Module for Automatic Segmentation and Transcription

Katarzyna Klessa[1]([✉]) [iD], Danijel Koržinek[2] [iD], Brygida Sawicka-Stępińska[1] [iD], and Hanna Kasperek[1] [iD]

[1] The Faculty of Modern Languages and Literatures,
Adam Mickiewicz University in Poznań, Al. Niepodległości 4, Poznań, Poland
`{klessa, brygida.sawicka-stepinska,hanna.kasperek}@amu.edu.pl`
[2] The Chair of Multimedia, Polish-Japanese Academy of Information Technology,
ul. Koszykowa 86, 02-008 Warszawa, Poland
`danijel@pjwstk.edu.pl`

Abstract. This paper describes an automatic segmentation and transcription module for Polish and its integration with the Annotation Pro software tool. The module is an extended desktop version of the CLARIN-PL online tool and has been named ANNPRO. Thanks to developing the module, it becomes possible to combine the functionality of Annotation Pro desktop program and the web-based automatic aligner. The results can be immediately used as the input for further acoustic-phonetic analyses with Annotation Pro native functions or annotation mining plugins. Annotation Pro enables using any number of external alignment modules, provided that certain basic format requirements are kept. We discuss these requirements and exemplify them with the ANNPRO module functionality and the integration steps. As an illustration, we present a brief report on experiences gained in the process of annotation of a multimodal corpus with the use of ANNPRO. Both Annotation Pro and the ANNPRO module are publicly available for download and can be freely used for research.

Keywords: Automatic segmentation · Grapheme-to-phoneme conversion · CLARIN-PL Mowa · Annotation Pro · Corpus annotation · ANNPRO

1 Introduction

The investigation of the phenomena occurring in spoken utterances is inevitably dependent on information derived from many different levels of analysis. That

The development of the ANNPRO module was funded by the Clarin-PL project. The use-case for MumoStance corpus was supported by the grant from National Science Centre, MuMoStance: Multimodal Stancetaking: Expressive Movement and Affective Stance. Political Debates in German Bundestag and Polish Sejm, Project ID: 2018/31/G/HS2/03633.

The original version of this chapter was revised: errors in the affiliations of three authors have been corrected. The correction to this chapter is available at https://doi.org/10.1007/978-3-031-05328-3_25

may include such high-level information as the register of the spoken text, the information about the context of the communication, the prosodic as well as multimodal characteristics of utterances. However, before many of the high-level descriptions can be completed, it may be needed to involve with low-level descriptions. For example, segment-level information is usually a starting point for examining the variability of spoken utterances, the properties of phrases, sentences or discourse units realized within monologues or in conversational contexts.

Manual procedures aimed at creating such descriptions are time-consuming, costly, and prone to human error. On the other hand, human expert verification of the automatic result is often performed [5] and recommended to ensure a high level of confidence in the data [4,20]. The main purpose of automation is thus not to eliminate manual work entirely but to greatly reduce the workload. A further goal is to introduce an aspect of objectivity to the process. Among other requirements formulated for the contemporary speech resources are their re-usability and interoperability [6]. The multilayered annotations and rich metadata are aimed to reflect the multi-layered structure of speech communication [15] as precisely as possible. Therefore, the ideal annotations would include the description of both phonetic-acoustic features of speech and higher level including e.g. the paralinguistic one [16].

A number of tools have been created to automatize corpus preparation and corpus-related research e.g. automatic grapheme-to-phoneme (G2P) conversion of speech transcripts as well as speech signal labelling and segmentation into phones, syllables or words. Many of the tools are freely and openly available for research, and what is more, they enable data processing for more than just one language, cf. SPPAS desktop application [1] or WEBMaus on-line services [11]. The tools are constantly improved, e.g. to make them more effective for spontaneous speech data [10]. Tools dedicated specifically to Polish that can be openly accessed and used on-line are e.g. CLARIN-PL services [14] or ORTFON [22]. The respective G2P conversion options were developed using different approaches, i.e. statistical learning (CLARIN-PL), and rule-based method for Polish (ORTFON); cf. also [26,28].

In this paper, two integrated openly available desktop tools are presented that can be used for both automatic and manual speech segmentation and labelling. In addition, we report on a use-case for the tools' usage. In Sect. 2 we discuss the tools and their options that have been integrated in order to combine their functionality within one framework. In Sect. 3, the tools' integration and its outcome, named ANNPRO, is presented. Section 4 provides a brief summary of observations gained within the process of annotation of MumoStance, a multimodal corpus of parliamentary speeches (see Acknowledgements and https://mmstance.home.amu.edu.pl/) using ANNPRO. We conclude the work and schedule our future steps in Sect. 5.

2   The Tools

2.1   Annotation Pro

Annotation Pro [13] is a freely available software tool for annotation of linguistic and paralinguistic features in speech. Currently, the program is available under Windows OS only, porting to other platforms is planned for the future (see also: http://annotationpro.org/).

An innovative feature, compared with other existing tools is that apart from a multilayer annotation interface, configurable spectrogram and waveform display of the speech signal, the program offers a graphical control window that supports annotation based on continuous, categorial or mixed rating scales. Additionally, the same interface can be used in a perception test mode. Including the graphical representation as a built-in component of the annotation tool was motivated by the need to annotate emotions and other paralinguistic or non-linguistic features in spoken utterances for which categorial labelling may not always be sufficient. The actual graphics used as the representations can be adjusted by the user depending on the specific use case and are not constrained to any specific project. This reflects one of the initial assumptions of the tool design, namely: the universal character of the built-in features and flexibility with respect to different use cases. Another functionality, also complying with the above assumption, is the possibility of extending the tool's built-in functionality with plugins and external modules. In this study, we make use of the possibility in order to develop an automatic segmentation and G2P module for Annotation Pro.

2.2   CLARIN-PL Align

The CLARIN-PL speech package includes several different tools [14], including ones supporting automatic phonetization and alignment of speech signal but also automatic speech recognition and text/audio normalization (see also: https://mowa.clarin-pl.eu/).

Feature Extraction. The initial step of each speech analysis tool is the extraction of distinctive information from the raw audio signal. Most often this is based on some form of Fourier spectral analysis in order to produce Short-Term Fourier Transform (STFT) mel-filterbank features or Mel-Frequency Cepstral Coefficient (MFCC) feature sets. In the present solution we analyze files at the sampling frequency of 16 kHz and use a set of 13 MFCC features, spliced with a context of 3 frames before and 3 frames after, thus giving 91 features that are subsequently processed through a Latent Dirichlet Allocation (LDA) transform to give 40 features that are presented for acoustic modeling [19].

Acoustic Modeling. The purpose of the acoustic model (AM) is to estimate the likelihood of phonetic events occurring at specific times in the signal, which are usually defined in some form of sub-phoneme units (e.g. tri-phones). Regardless of the actual choice of phonetic units, this step requires the definition of the set of phonemes we intend to search within the data. Altering this set requires retraining of the whole acoustic model to the new definition. The presented work uses a set of phonemes based on the Polish SAMPA phonetic alphabet [27].

The AM can be constructed in many ways, but it is always a result of machine learning process. The presented system uses a Gaussian Mixture Model (GMM) [21]. Even though neural network based solutions exist [18], the difference in the segmentation performance is not significant enough for our purpose, but GMMs are considerably faster and easier to develop.

An important factor to the performance of the AM is the data it is trained on. The presented system was trained on a medium-sized studio corpus, which means that the model will respond best to the data of equal or similar quality. It is impossible to predict all possible uses of the system, so it is important to make some initial assumptions about the analyzed data. Given a few examples of the real data, it is possible to adapt the model, thus increasing its performance [19].

Grapheme-to-Phoneme Conversion. A key component to the system is the mapping from the orthographic transcription provided by the users into its phonetic form as modeled by the AM. The simplest form of such a mapping is a lexicon: a document which provides a phonetic transcription (as a sequence of phonemes) for each word in the transcript. We also generally allow multiple possible pronunciations for each word. The acoustic model decides which transcription most likely matches the audio.

Even though it is possible to prepare a lexicon with many words in a language, it is unlikely to predict every possible word a user may want to process. We therefore need a grapheme-to-phoneme (G2P) tool to generate the lexicon for any word. In general, there are two approaches to automatic lexicon generation in Polish: rule-based and statistical systems. While rule-based systems are very robust and quite feasible for Polish, in practice there are many exceptions to the standard pronunciation and foreign words present in most real-life corpora. For convenience, the presented solution uses a statistical system trained by the Phonetisaurus G2P toolkit [17].

Similarly to the AM, this step requires a definition of the set of phonemes, as well as the rules of their pronunciation. Fortunately, training a statistical G2P model is not too time consuming and simply requires a properly constructed lexicon to learn from. The creation of such a lexicon is the most time consuming part of the process. The trained model can then be used to re-process the acoustic data used to re-train the AM.

Alignment Process. The actual alignment of the transcription is contingent of the presence of the aforementioned information: the phonetic likelihoods computed by the AM from the audio and the phoneme sequence computed by the G2P system from the transcription. The purpose of the aligner is to match (i.e. align) one sequence to the other. This is achieved by constructing a state graph

akceleracji	a k ts e l e r a ts j i
akceleratora	a k ts e l e r a t o r a
akcentach	a k ts e n t a x
akcentem	a k ts e n t e m
akcentują	a k ts e n t u j o m
akcentują	a k ts e n t u j o~
akcentująca	a k ts e n t u j o n ts a
akcentująca	a k ts e n t u j o~ ts a

Fig. 1. Examples of grapheme-to-phoneme conversion.

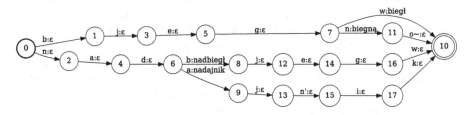

Fig. 2. Example of a finite state transducer for converting phonemes into words. The purpose of the alignment process is to find an optimal path through a graph of similar design.

representing the transcription and applying some form of Viterbi search of the most likely sequence of states given the acoustic likelihoods [21]. This approach can also account for the alternative pronunciations of individual words, by constructing a graph with multiple pathways (see Fig. 2).

It is impractical to make an exhaustive search of all the possible outcomes of this alignment process, so a heuristic is applied by pruning the least likely hypotheses at each time step. This method is known as beam-search [7] and it relies on a parameter known as beam width that denotes the maximum difference between the likelihoods of the best and worse hypotheses being analyzed. Using a large beam will slow down the process, but ensure a better quality result, while using a small beam can significantly speed up the process, but can also accidentally prune the optimal solution during alignment. In some cases this can also cause the algorithm to fail altogether. In fact, there are situations where the alignment is impossible (e.g. if we provide transcription that doesn't match the audio). In such cases, the algorithm will explore all the likely hypotheses, but if it doesn't find a reasonable match, it will fail with an error message.

This process is also known as forced alignment. That means, that the algorithm will attempt to insert tokens provided by the user even if they don't exist in the audio. This can cause unpredictable behavior even with small mismatches between the audio and the transcription. To solve the problems, there exist other algorithms that perform a more "lenient" form of alignment [9] and are especially useful with very long recordings which have a greater chance of failing the forced alignment process. These algorithms rely on performing speech recognition using limited vocabulary in order to find the key points in the audio which match the transcription. The present module supports only forced alignment, at the moment.

3 CLARIN-PL Align Module for Annotation Pro: ANNPRO

3.1 Module Integration

The CLARIN-PL alignment module has been integrated with Annotation Pro in the form of an external segmentation module. Other such modules had been implemented earlier, but were not available for distribution. An alternative was to import the results of automatic alignment obtained with programs such as SPPAS [1]. Apart from the phone-level and word-level segmentation, the output in Annotation Pro can also include syllable-level segmentation made available by the option that inserts syllable boundaries into the Annotation Pro layer using the grapheme-to-phoneme level segmentation as input. The syllabification process is based on a set of fixed rules. The starting point for the rule list was the work of Daniel Sledzinski, e.g. [23,24] with further modifications, cf. also [2].

It is possible to integrate any number of segmentation modules and to select the desired one via the Annotation Pro interface, i.e. the user can use different segmentation engines for various pronunciation variants or languages. The modules work independently from Annotation Pro and none of their functions

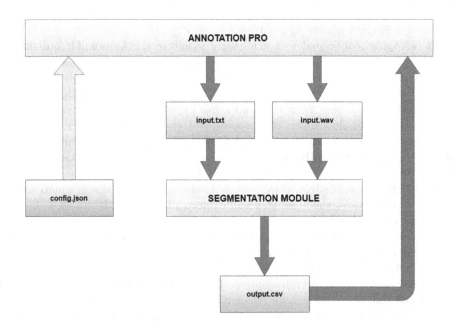

Fig. 3. Auto Segmentation module integration components for Annotation Pro

are integrated natively. Therefore, the modules themselves can be implemented using different technologies and approaches to the segmentation (and transcription) tasks (Fig. 3).

The required format for the module is either a .bat or .exe format. Before launching the segmentation process for a recording, Annotation Pro places two files in the module folder: input.wav (the recording) and input.txt. The input.txt file is a temporary plain text file containing the orthographic transcript corresponding to the speech included in the input.wav file. When the module is launched in Annotation Pro, the transcript comes from the previously selected annotation layer or individual segment(s), depending on the user's choice. In the next step, the segmentate.bat or (if not found) segmentate.exe is launched. If the process ends successfully, an output.csv file is created in the same folder. The file contains the list of TAB-separated timestamps and transcription labels in the format:

SegmentStart [**TAB**] SegmentDuration [**TAB**] Label

Finally, the output is saved as the Annotation Pro .ant file in XML-based [12, pp. 131–133]. In case if the output.csv file is missing, an error log is reported and also the respective segment(s) are highlighted with red color in the Annotation Pro GUI. The crucial settings for each module are defined in a config.json file saved in the same folder as the rest of the module files.

3.2 Using the ANNPRO Module

In order to use the combined version of the tool, it is needed to download both Annotation Pro (http://annotationpro.org/downloads) and the ANNPRO module (https://clarin-pl.github.io/speech-annotationpro-plugin/).

ANNPRO should be saved to the Documents/Annotation Pro/Segmentation Modules folder. Once that is done, it becomes possible to choose the aligner from a drop-down list in the Auto Segmentation window of Annotation Pro (Fig. 4). The input for the processing will be the orthographic transcriptions (segment labels) in Annotation Pro. Two modes of processing are available: for selected segment(s) only or for the whole annotation layer. The latter implements the same algorithm as the former, but simply iterates all the segments in a track sequentially.

4   Example Use Case

This section reports on an example usage of ANNPRO in the process of multilayer transcription of speech corpus data. For this particular example, the output of ANNPRO is a subset of a larger number of annotation layers because the corpus in question is a multimodal one and its description involves multiple annotation layers referring not only to speech signal (primarily phonetic and prosodic features) but also to the visual component of communication (e.g.

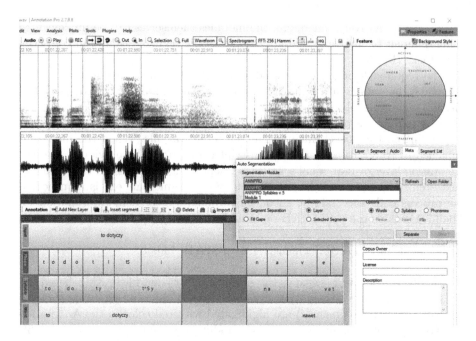

Fig. 4. Annotation Pro interface. Auto Segmentation options (segmentation module selection) and a sample segmentation result: phone, syllable and word level.

annotation of gesture, head movement, face expression). All annotation layers are time-aligned.

Noteworthy, here we focus exclusively on those properties of speech and respective annotations that are relevant for the functioning of the ANNPRO module (G2P and automatic segmentation of speech signal).

The MumoStance Corpus. Transcripts Preparation. The Polish subset of the MumoStance corpus (cf. Acknowledgements and https://mmstance.home. amu.edu.pl/) was constructed on the basis of parliamentary speeches delivered by the members of the Polish lower house of parliament (Sejm) during the first reading of the government's draft budget act for 2020 (January 8, 2020 and February 12, 2020).

Altogether, 69 Polish parliamentary speeches were used for the corpus construction, the average speech duration was 4:45 min. However, the length of speeches varied significantly, lasting from 00:51 to 26:07. Most of the sessions were shorter than 2 min; the median is 01:39. All the material was automatically transcribed and segmented using ANNPRO.

Before the automatic segmentation and transcription, however, it was necessary to prepare the orthographic input so that it fulfilled the requirements of the segmentation module, i.e. had a form of a normalized text input within the Annotation Pro layer. The initial step was to download and inspect the original stenographic reports of the speeches publicly available as PDFs from the official website of the Sejm. Then, the PDFs were converted to text format and normalized. The normalization process included manual replacement of all the number representations, abbreviations, and special characters with the respective transliterations. Such normalized texts were subsequently imported to an annotation layer in Annotation Pro (using the tool's *Import text* option).

Since the original stenographic reports had been stylistically edited, they were not matching exactly the speakers' utterances. In order to provide the exactly matching transcripts, human annotators listened to the recordings and compared them with texts, applying adjustments wherever they were necessary. At that stage, the material was also manually segmented into time-aligned phrase-level fragments. The phrase boundaries were inserted based on grammatical and prosodic rules following the paradigms used e.g. by [8].

Multilayer Annotation Procedures and the Use of ANNPRO. The automatic segmentation was carried out for three different layers: Phone, Syllable and Word (see also Fig. 4). The segmentation process was performed in three cycles for each recording, by launching ANNPRO separately for each layer. The input was always the same, i.e. the orthographic transcription of the speech segmented into phrases. Since ANNPRO does not automatically identify pauses as separate segments, it was crucial to eliminate them manually (by indicating pause boundaries) in advance, to avoid excessively long segments in the Phone, Syllable and Word layers. The same procedure applies to non-speech events such as fillers or grunting.

Three expert phoneticians manually evaluate the outcomes of the segmentation obtained with ANNPRO. The evaluation is performed individually by each expert and is based on visual inspection of multilayer annotation and spectrograms. So far, the following problems were detected:

- inaccurate boundary segmentation,
- reducing lengthened vowels in the final position of interpausal units,
- phone segmentation of some units in the Syllable layer.

More cases of the inaccurate boundary segmentation are observed when background noise appears (especially periodic noise, such as the alarm announcing the end of the speaking time for a parliament member). Unclear pronunciation or articulatory simplifications also appear to be a factor (e.g., "pierwszy"/pjerSI/).

Secondly, the Module tends to reduce pre-pausal vowels that in regular pronunciation are usually lengthened in the final position of the speech sequences, especially those of fall-rise intonation. Since the corpus is designed for prosodic research, it is fundamental that the segmentation reflects correctly the phone duration.

Last but not least, a number of issues have been observed in the process of segmentation on the syllable level. The errors result from the fact that the G2P outcome is based on a statistical process where certain atypical pronunciation variants are allowed, while, on the other hand, the syllabification is based on a fixed set of rules where only canonical phonotactic sequences have been fore-seen. The evaluation procedure described above is mostly qualitative in char-acter; however, each segment for which an error has been detected is marked and labeled using a time-aligned comment layer. A follow-up step will be the detailed documentation of the problems that will further be used as the basis for developing a systemic solution to reduce the number of errors. That task falls within the scope of the future work.

Preliminary Evaluation of the Scope of Manual Adjustments. In order to infer the scope of the manual changes for the current version of the tool, a con-trol experiment was performed using a sample subset of 20 min of recordings from the MumoStance corpus. Namely, we automatically generated transcription and segmentation data with ANNPRO and compared them with the annotations of exactly the same material manually inspected and corrected by human experts. The results of the comparison are shown in Tables 1 and 2.

Table 1. Summary of errors produced by the automated system. These are the errors that would need to be manually corrected by the user. The first set of rows denote errors that require the user to add, remove or fix the transcription of a segment, and the last two rows show places were a user only has to move a segment boundary. Depending on the specific needs, different levels of accuracy may be required.

Measure	Value
Number of phoneme segments in the sample	12489
G2P substitution errors	490
G2P insertion errors	43
G2P deletion errors	81
G2P error rate	4.92%
Percentage of boundaries shifted > 10 ms	47.40%
Percentage of boundaries shifted > 30 ms	16.11%

Table 2. Summary of distances between automatic and manually created boundaries. These are the average amounts of corrections that would need to be done by the user.

Measure	Value
Mean Absolute Error (MAE)	19.66 ms
Root Mean Square Error (RMSE)	42.84 ms
MAE StdDev	38.07 ms
RMSE StdDev	111.38 ms
Maximum Error	627.81 ms

5   Discussion and Conclusions

We present a freely available desktop workbench enabling both manual and automatized transcription and segmentation of Polish speech recordings. The automation of the tasks is an unquestionable advantage itself. The advantage is even greater when we consider the fact that the outcomes of the automatic process are saved within a multilayer workspace composed of one or more annotation files. The workspace may include information about both linguistic and paralinguistic features of speech, and more than speech, e.g. gesture or face expression annotation or extralinguistic context. Having the variety of labels within a common workspace facilitates speech research and enhances the scope of data analyses and applications. Thanks to the off-line mode of operation, the tool may be used for various kinds of data, including the recordings of restricted access (e.g., due to security reasons, sensitive data etc.).

The ANNPRO module allows to automate certain steps in the course of creating corpus annotations for various purposes. However, because it relies on a machine learning process, it is susceptible to introducing errors. For example, the G2P process can introduce errors by erroneously phonetizing certain words, especially those having foreign or otherwise atypical pronunciation. A user should have the option to modify the G2P lexicon before it is used by the alignment process. The aligner itself can recognize speech in non-speech segments (e.g. due to noise) or miss recognizing speech (due to low signal-to-noise levels), thus place the word boundaries in incorrect places. Finally, the phoneme boundaries can also be placed incorrectly, due to low quality of speech or atypical (from the acoustic model point of view) speaker voice or pronunciation. The results obtained with ANNPRO can be verified by the user at each processing stage. For example, the lexicon file can be accessed and checked by the user in the Segmentation Modules folder. The verification of segment boundary placement and G2P results is made easier thanks to using the graphical user interface of Annotation Pro.

The usage of the segmentation and transcription results is not limited to ANNPRO environment. Annotation Pro is highly compatible with other popular annotation and speech analysis tools regarding data exchange. A wide range of import/export options have been built in into Annotation Pro, e.g.: Praat [3], ELAN [25] or SPPAS [1] as well as plain text or CSV database formats. At any stage, the results of the automatic segmentation and transcription may be exported/imported back from/to Annotation Pro.

The so-far example uses of ANNPRO include the process of creating multilayer annotations of speech data for MumoStance, a multimodal corpus of parliamentary speeches. The preliminary qualitative evaluation based on expert judgements confirms the usefulness of the tool and the reduction of the workload (as compared to fully manual procedures), even though human verification is still needed, as reported by expert evaluators, and also illustrated by the result of the control experiment for a sample data subset.

Further tests are currently being conducted using the recordings of Polish spontaneous and controlled speech in an experimental setting, as well as a teaching material for students of experimental phonetics at Adam Mickiewicz University in Poznań. The authors have received very positive feedback in terms of the usefulness and the ease of installation. For the word-level segmentation, a need to accelerate the processing was reported.

Among other future works, we consider including automatic speech recognition component to allow automating the initial transcription step of the whole process. Likewise, automatic voice activity detection can be used to automatically segment the audio into short speech segments. Combined with automatic speaker diarization, a reasonable dialog speech segmentation can be achieved. Since these components are already available as online tools [14], the work would mainly involve preparing the desktop variants and designing the integration procedure for Annotation Pro. The G2P conversion can be further optimized, allowing the user to both inspect and modify the generated output at any time. The alignment process itself can benefit greatly from parallelization in a multiprocessing environment. Finally, adaptation of the individual models to the user-provided data could greatly improve the performance of the system.

References

1. Bigi, B.: SPPAS-multi-lingual approaches to the automatic annotation of speech. Phonetician **111**, 54–69 (2015). ISSN:0741–6164
2. Bigi, B., Klessa, K.: Automatic syllabification of Polish. In: Proceedings of the 7th Language Technology Conference, pp. 262–266. Poznań (2015)
3. Boersma, P., Weenink, D.: Praat: doing phonetics by computer (ver. 4.3.14) [computer program] (2014). http://www.praat.org. Accessed Dec 2017
4. Castro, A.D., Ramos, D., Gonzalez-Rodriguez, J.: Forensic speaker recognition using traditional features comparing automatic and human-in-the-loop formant tracking. In: Proceedings of the 10th INTERSPEECH Conference (2009)
5. Gibbon, D., Moore, R., Winski, R.: Handbook of standards and resources for spoken language systems. Walter de Gruyter (1997)
6. Ide, N., Pustejovsky, J. (eds.): Handbook of Linguistic Annotation. Springer, Dordrecht (2017). https://doi.org/10.1007/978-94-024-0881-2
7. Jelinek, F.: Statistical Methods for Speech Recognition. MIT Press (1997)
8. Karpiński, M., Kleśta, J., Baranowska, E., Francuzik, K.: Interphrase pause realization rules for the purpose of high quality Polish speech synthesis. In: Speech Analysis, Synthesis and Recognition (SASR) in Technology, Linguistics and Medicine, Szczyrk 2003, pp. 85–89. AGH Kraków (2005)
9. Katsamanis, A., Black, M., Georgiou, P.G., Goldstein, L., Narayanan, S.: Sailalign: robust long speech-text alignment. In: Proceedings of the Workshop on New Tools and Methods for Very-Large Scale Phonetics Research (2011)
10. Keller, E., Terken, J., Huckvale, M., Gailly, G., Monaghan, A.: Improvements in Speech Synthesis. Wiley (2001)
11. Kisler, T., Reichel, U.D., Schiel, F., Draxler, C., Jackl, B.: BAS Speech Science Web Services - An Update of Current Developments. In: LREC 2016. Portorož (2016)
12. Klessa, K.: Annotation Pro. Enhancing Analyses of Linguistic and Paralinguistic Features in Speech, Wydział Neofilologii UAM (2016)

13. Klessa, K., Karpiński, M., Wagner, A.: Annotation Pro - a new software tool for annotation of linguistic and paralinguistic features. In: Proceedings of the TRASP Workshop, pp. 51–54. Aix en Provence (2013)

14. Koržinek, D., Marasek, K., Brocki, L., Wołk, K.: Polish read speech corpus for speech tools and services. In: Selected Papers from the CLARIN Annual Conference 2016, Aix-en-Provence, 26–28 Oct 2016, CLARIN Common Language Resources and Technology Infrastructure, vol. 136, pp. 54–62. Linköping University Electronic Press (2017)

15. Laver, J.: Principles of Phonetics. Cambridge University Press (1994)

16. Marasek, K., Gubrynowicz, R.: Multi-level annotation in SpeeCon Polish speech database. In: Bolc, L., Michalewicz, Z., Nishida, T. (eds.) IMTCI 2004. LNCS (LNAI), vol. 3490, pp. 58–67. Springer, Heidelberg (2005). https://doi.org/10.1007/11558637_7

17. Novak, J.R., Minematsu, N., Hirose, K.: WFST-based grapheme-to-phoneme conversion: open source tools for alignment, model-building and decoding. In: Proceedings of the 10th International Workshop on Finite State Methods and Natural Language Processing, pp. 45–49 (2012)

18. Peddinti, V., Povey, D., Khudanpur, S.: A time delay neural network architecture for efficient modeling of long temporal contexts. In: Proceedings of the 16th INTERSPEECH Conference (2015)

19. Povey, D., Saon, G.: Feature and model space speaker adaptation with full covariance Gaussians. In: Proceedings of the 9th ICSLP Conference (2006)

20. Poznyakovskiy, A.A., Mainka, A., Platzek, I., Mürbe, D.: A fast semiautomatic algorithm for centerline-based vocal tract segmentation. BioMed Res. Int. (2015)

21. Rabiner, L.R.: A tutorial on hidden Markov models and selected applications in speech recognition. Proc. IEEE **77**(2), 257–286 (1989)

22. Skurzok, D., Ziółko, B., Ziółko, M.: Ortfon2-tool for orthographic to phonetic transcription. In: Proceedings of the 7th LTC Conference. Poznań (2015)

23. Sledzinski, D.: Fonetyczno-akustyczna analiza struktury sylaby w języku polskim na potrzeby technologii mowy. Unpublished Ph.D. Thesis. Adam Mickiewicz University, Poznan (2007)

24. Sledzinski, D.: Podział korpusu tekstów na sylaby-analiza polskich grup spółgłoskowych. Kwartalnik Jezykoznawczy **3**(15), 48–100 (2013)

25. Sloetjes, H., Wittenburg, P.: Annotation by category-elan and ISO DCR. In: Proceedings of the 6th LREC (2008)

26. Steffen-Batogowa, M.: Automatyzacja Transkrypcji Fonematycznej Tekstów Polskich. PWN, Warszawa (1975)

27. Wells, J.C.: SAMPA computer readable phonetic alphabet. In: Gibbon, D., Moore, R., Winski, R. (eds.) Handbook of Standards and Resources for Spoken Language Systems, Part IV, Section B. Mouton de Gruyter, Berlin, New York (1997)

28. Wypych, M., Baranowska, E., Demenko, G.: A grapheme-to-phoneme transcription algorithm based on the SAMPA alphabet extension for the Polish language. In: Proceedings of the 15th ICPhS, pp. 2601–2604. Barcelona (2003)

Language Resources and Tools

Analysis and Processing of the Uzbek Language on the Multi-language Modelled Computer Translator Technology

Mersaid Aripov ⓘ, Muftakh Khakimov ⓘ, Sanatbek Matlatipov$^{(\boxtimes)}$ ⓘ,
and Ziyoviddin Sirojiddinov ⓘ

National University of Uzbekistan of name Mirzo Ulugbek, University Street 4, 100104
Tashkent, Republic of Uzbekistan
s.matlatipov@nuu.uz

Abstract. It is known, that a process of any machine translation system is, decoding of sense of the entrance text in a natural language and re-encoding this meaning in the target language to inform the user in semantic conformity with the entrance/source text.

One of variants of achievement of this purpose is the formalization of grammatical structures of natural languages participating in translation systems. More specifically, grammatical structures mean linguistic rules of word structures and its offers on word types. The analysis of word structures will give the chance to construct the logically correct linguistic models.

In the given work the new developed logical-linguistic models of words and offers on types of the Uzbek language are offered for created system of the multilanguage modelled computer translator (MMCT). The system which is developed as multilingual on the basis of the new technology MMCT consists seven stages. The logical-linguistic models of the Uzbek language is described as one of the seven stages. The models are described using an extensible input language. Validation over the nouns of the Uzbek language was carried out and the truth of one variant of the logical-linguistic model was proved.

Keywords: Uzbek language · Extensible input language · Logical-linguistic model · Mathematical model · Validation · Processing · Word types · Multilingual modelled computer translator technology · Algorithm · System

1 Introduction

Each natural language (NL) is the difficult system, consisting mathematical not structured and not as not formalized components. However, researches on NLP show what it is possible to lead NL not structured, and to the formalized form using linear methodology – revealing of structure of a word, construction of logical-linguistic models on types of words and sentences, and further construction of mathematical models by means of special meta language. The given methodology can be defined as degree of language formalization. Formalization degree in turn defines semantics formalization degree of

NL and accuracy of the algorithm. The superficial understanding of NL formalization degree, that the formalized language – abstract, the design completely torn off from the maintenance with simple logic structure leads to low technology of machine translation [3]. Formalization allows to allocate its various parts and to investigate dynamics of their communications, and mainly the description of its semantic structure.

Mathematicians and linguists had been developed various models. In contrast, in this work, the model implies a clear transformation of language possibilities by describing them with the help of a new tool – an expandable input language. Thus, it is possible to generate and simulate different classes of languages. The essence transformation of language possibilities – from formalization before modelling, consists in the following. At first the analysis of words of a certain natural language on the types is made, in what ways it is possible to "construct" a noun, an adjective, a verb, a pronoun, an adverb and the numeral. For each type of words, some variants of certain chains (A_1, $A_{2,...}$, A_n), consisting of prefixes, roots of words and the various suffixes. Further analysis, only logically and semantically correct chains are selected from them. On the basis of these chains written in the expanded source language, logical-linguistic models of words on types are developed by its type. The expanded source language promotes correct reflection of semantic - designs on words formation.

Further, the analysis is carried out to obtain an answer to the question: How can one construct narrative, interrogative and exclamatory sentences using words by types? For each type the sentences it is also possible to receive chains B_1, then B_2 etc. B_n. Then the analysis is carried out, and only logically and semantically correct chains are selected from them. On the basis of these chains written in the expanded source language, logical-linguistic models of developed by type. The expanded source language promotes correct reflection of semantic designs on sentences formation.

The functionality of word in NL is manifested in its ambiguity. In concrete cases the concrete value gets each word in a phrase and-or in the sentence. The recognition of functionality of a word leads to semantic unambiguity, except for some concrete cases followed from NL. It leads two essentially various approaches at construction of NL models, NL – or to develop uniform system of linear processing of words and sentences, or to consider each word and the sentences as individual structure according to which it is processed. In our case the first approach which performance provides transfer from language A to language B, belonging to the class 0 by Chomshy classification [2].

We characterize multilingual situation of computer transfer, as set of languages potentially participating in translation process from one language on another. For example, if three language (A, B, C) are available in the translating system, translation can be carried out in 6 directions (A $\rightarrow$ B, A $\rightarrow$ C, B $\rightarrow$ A, B $\rightarrow$ C, C $\rightarrow$ A, C $\rightarrow$ B). By technology of the multi-language modelled computer translator NL and mathematical models of language A, allow to find if not identical yet almost identical model of language B. In this case if the mathematical model is identifying (source language), it characterizes language A, otherwise as generating (target language) is characterized by language B.

The technology of the multi-language modelled computer translator [5] including above stated concept, consists of several stages, such as research of natural languages on the given technology, creation expanded entrance language, building semantic databases

NL, developing of multi-language bases of terms and phrases on subject domains, modelling NL, algorithmization and working out of program circle of the translator.

Overall objectives reached in the present work:

1. The theoretical analysis of structure of words and sentences of the Uzbek language (UL) is done by the multi-language modelled computer translator technology.
2. Logical-linguistic models on types of words with application of two operations of the expanded source language – nouns, adjectives, a verb, a pronoun, an adverb and numerals are constructed. It has been constructed seven models of the nouns, seven models of adjectives, and seven models of a verb by affixation rules. It has been constructed six models of the nouns, eight models of adjectives and five models of a verb by composition rules. Also six models of the pronoun, six models of an adverb and one model of numerals have been constructed. Total of the constructed logical-linguistic models of words equally to fifty-four.
3. Logical-linguistic models of UL sentences – are constructed and they as following:
• narrative sentences models are eighteen;
• question sentences models are nineteen;
• exclamatory sentences models are sixteen.

In total were 53 variants of logical-linguistic models of UL sentences are constructed.

4. For further understanding of models, we described one example for word structure of logical-linguistic model and one example for sentences structure of logical-linguistic model in *Experiments and Results* section. Additionally, for validation (truth) of the structure of logical-linguistic models, the program[1] has developed that operates on the basis of two algorithms. Experiments were conducted with mathematical models of Uzbek nouns.

The remainder of this paper is organized as follows: after this Introduction, Sect. 2 describes related work that has been done so far. It is followed by a description of the methodology in Sect. 3 and continues with Sect. 4 which focuses on Experiments and Results. The final Sect. 5 concludes the paper and highlights the future work.

2 Related Work

UL unlike Turkic languages is considered low-resource language and consists very deep agglutinativity. One word can form a sentence. To our knowledge, there is no well-established rule based machine translation resources for Uzbek. But, Turkish and Kazakh same family languages made considerable results in the field. For instance, sentiment analysis in work [1] has been carried out.

But during a globalization epoch, despite all difficulties UL adequately should be active the participant in information community. Here it is possible to result some articles [7, 8] where the morphological analysis and stemming of the Uzbek language is carried out.

---

[1] https://github.com/zsirojiddinov/WordValidationProgram.

Extremely seldom, there are articles on formalization of NL. But sufficiently there are articles on the Turkic languages, studying their various aspects. For example, in work [9] the sentiment analysis of feelings Kazakh and Russian is carried out, and in work [11] sentiment analysis of sentences in the Kazakh language, on a basis ontologics is carried out.

In mathematical model of Russian is stated in the book [10]. As mathematical model the functions displaying certain words are considered. This approach demands creation of a large quantity of functions and difficult from the point of view of application for the computer translator.

3 Methodology

3.1 The General Logical-Linguistic Model of Formation of a Word

UL lexical analysis shows, that words share on four types of components – a root, affixes forming words, affixes forming the form and affixes changing words. According to it, it is possible to construct the general logical-linguistic model of formation of word UL using signs on the expanded source language for processing NL [4]:

root $\oplus$ $\Downarrow$ affixes forming words $\oplus$ $\Downarrow$ affixes forming the form $\oplus$ $\Downarrow$ changing words.

Hereinafter, signs mean: - $\oplus$ joining operation, $\Downarrow$- operation of possible "connection" or "not connections" a component following it [4].

3.2 Logical-Linguistic Models of Words on Types

Logical-linguistic models of a conclusion *of nouns* UL by *affixation rules* consist of seven variants:

1) a noun $\oplus$ $\Downarrow$ an affix $\oplus$ $\Downarrow$ a suffix of plural forming a word $\oplus$ $\Downarrow$ $\oplus$ $\Downarrow$ a suffix of subjects $\oplus$ $\Downarrow$ a suffix of a case $\oplus$ $\Downarrow$ a particle
2) $\Downarrow$ an affix forming a word $\oplus$ a noun $\oplus$ $\Downarrow$ an affix forming a word $\oplus$ $\Downarrow$ a suffix of plural $\oplus$ $\Downarrow$ a suffix of subjects $\oplus$ $\Downarrow$ a suffix of a case $\oplus$ $\Downarrow$ a particle
3) a verb $\oplus$ an affix forming a word $\oplus$ $\Downarrow$ a suffix of plural $\oplus$ $\Downarrow$ a suffix of subjects $\oplus$ $\Downarrow$ a suffix of a case $\oplus$ $\Downarrow$ a particle
4) an adjective $\oplus$ an affix forming a word $\oplus$ $\Downarrow$ a suffix of plural $\oplus$ $\Downarrow$ a suffix of subjects $\oplus$ $\Downarrow$ a suffix of a case $\oplus$ $\Downarrow$ a particle
5) the numeral $\oplus$ an affix forming a word $\oplus$ $\Downarrow$ a suffix of plural $\oplus$ $\Downarrow$ a suffix of subjects $\oplus$ $\Downarrow$ a suffix of a case $\oplus$ $\Downarrow$ a particle
6) a pronoun $\oplus$ an affix forming a word $\oplus$ $\Downarrow$ a suffix of plural $\oplus$ $\Downarrow$ a suffix of subjects $\oplus$ $\Downarrow$ a suffix of a case $\oplus$ $\Downarrow$ a particle
7) an adverb $\oplus$ an affix forming a word $\oplus$ $\Downarrow$ a suffix of plural $\oplus$ $\Downarrow\oplus$ $\Downarrow$ a suffix of subjects $\oplus$ $\Downarrow$ a suffix of a case $\oplus$ $\Downarrow$ a particle

Logical-linguistic models of a conclusion *of nouns* UL by *composition rules* consist of six variants:

1) a noun ⊕ a noun ⊕ ⇊ an affix forming a word ⊕ ⇊ a suffix of plural ⊕ ⇊ a suffix of subjects ⊕ ⇊ a suffix of a case ⊕ ⇊ a particle

2) an adjective ⊕ a noun ⊕ ⇊ an affix forming a word ⊕ ⇊ a suffix of plural ⊕ ⇊ a suffix of subjects ⊕ ⇊ a suffix of a case ⊕ ⇊ a particle

3) the numeral ⊕ a noun ⊕ ⇊ an affix forming a word ⊕ ⇊ a suffix of plural ⊕ ⇊ a suffix of subjects ⊕ ⇊ a suffix of a case ⊕ ⇊ a particle

4) a noun ⊕ a verb ⊕ ⇊ an affix forming a word ⊕ ⇊ a suffix of plural ⊕ ⇊ a suffix of subjects ⊕ ⇊ a suffix of a case ⊕ ⇊ a particle

5) a verbal adverb ⊕ a participle ⊕ ⇊ a word designating actions ⊕ ⇊ a suffix of plural ⊕ ⇊ a suffix of subjects ⊕ ⇊ a suffix of a case ⊕ ⇊ a particle

6) a noun ⊕ a participle ⊕ ⇊ a word designating actions ⊕ ⇊ a suffix of plural ⊕ ⇊ a suffix of subjects ⊕ ⇊ a suffix of a case ⊕ ⇊ a particle

At construction *of adjectives* by *rules of affixation* the noun, an adjective, a verb, an adverb can be roots of words. Logical-linguistic models of a conclusion of adjectives UL by affixation rules consist of seven variants:

1) a noun ⊕ an affix forming a word ⊕ ⇊ diminutive degree of an adjective ⊕ ⇊ a suffix of plural ⊕ ⇊ a suffix of subjects ⊕ ⇊ a suffix of a case ⊕ ⇊ a particle

2) a noun ⊕ diminutive degree of an adjective ⊕ ⇊ a suffix of plural ⊕ ⇊ a suffix of subjects ⊕ ⇊ a suffix of a case ⊕ ⇊ a particle

3) an adjective ⊕ ⇊ diminutive degree of an adjective ⊕ ⇊ a suffix of plural ⊕ ⇊ a suffix of subjects ⊕ ⇊ a suffix of a case ⊕ ⇊ a particle

4) magnifying degree of an adjective ⊕ an adjective ⊕ ⇊ a suffix of plural ⊕ ⇊ a suffix of subjects ⊕ ⇊ a suffix of a case ⊕ ⇊ a particle.

5) a verb ⊕ an affix forming a word ⊕ ⇊ a suffix of plural ⊕ ⇊ a suffix of subjects ⊕ ⇊ a suffix of a case ⊕ ⇊ a particle

6) an adverb ⊕ an affix forming a word ⊕ ⇊ a suffix of plural ⊕ ⇊ a suffix of subjects ⊕ ⇊ a suffix of a case ⊕ ⇊ a particle

7) an affix-pretext for a noun ⊕ a noun ⊕ ⇊ diminutive degree of an adjective ⊕ ⇊ a suffix of plural ⊕ ⇊ a suffix of subjects ⊕ ⇊ a suffix of a case ⊕ ⇊ a particle

Logical-linguistic models of a conclusion *of adjectives* UL by *composition rules* consist of eight variants:

1) an adjective ⊕ an adjective ⊕ ⇊ diminutive degree of an adjective ⊕ ⇊ a suffix of plural ⊕ ⇊ a suffix of subjects ⊕ ⇊ a suffix of a case ⊕ ⇊ a particle

2) an adjective ⊕ a noun ⊕ ⇊ diminutive degree of an adjective ⊕ ⇊ a suffix of plural ⊕ ⇊ a suffix of subjects ⊕ ⇊ a suffix of a case ⊕ ⇊ a particle

3) a noun ⊕ a noun ⊕ ⇊ diminutive degree of an adjective ⊕ ⇊ a suffix of plural ⊕ ⇊ a suffix of subjects ⊕ ⇊ a suffix of a case ⊕ ⇊ a particle

4) an adverb ⊕ a noun ⊕ ⇊ diminutive degree of an adjective ⊕ ⇊ a suffix of plural ⊕ ⇊ a suffix of subjects ⊕ ⇊ a suffix of a case ⊕ ⇊ a particle

5) an adverb ⊕ an adjective ⊕ ⇊ diminutive degree of an adjective ⊕ ⇊ a suffix of plural ⊕ ⇊ a suffix of subjects ⊕ ⇊ a suffix of a case ⊕ ⇊ a particle

6) an adjective ⊕ a noun ⊕ ⇓ a suffix of plural ⊕ ⇓ a suffix of subjects ⊕ ⇓ a suffix of a case ⊕ ⇓ a particle
7) a noun ⊕ an affix «apo» ⊕ ⇓ a particle
8) an affix «умум» ⊕ a noun ⊕ ⇓ a particle

Logical-linguistic models of a conclusion *of verb* UL by *affixation rules* consist of seven variants:

1) a verb ⊕ an affix forming a word ⊕ ⇓ a negative particle ⊕ ⇓ a particle specifying in time of a verb ⊕ ⇓ a particle of the person ⊕ ⇓ a particle
2) a noun ⊕ an affix forming a word ⊕ ⇓ a negative particle ⊕ ⇓ a particle specifying in time of a verb ⊕ ⇓ a particle of the person ⊕ ⇓ a particle
3) a noun ⊕ an affix forming a word ⊕ ⇓ a negative particle ⊕ ⇓ an inclination particle
4) an adjective ⊕ an affix forming a word ⊕ ⇓ a negative particle ⊕ ⇓ an inclination particle
5) an adverb ⊕ an affix forming a word ⊕ ⇓ a negative particle ⊕ ⇓ a mortgaging particle ⊕ ⇓ a particle
6) exclamatory words ⊕ an affix forming a word ⊕ ⇓ a negative particle ⊕ ⇓ a mortgaging particle ⊕ ⇓ a particle
7) words of imitation ⊕ an affix forming a word ⊕ ⇓ a negative particle ⊕ ⇓ a mortgaging particle ⊕ ⇓ a particle

Logical-linguistic models of a conclusion *of verb* UL by *composition rules* consist of five variants:

1) a noun ⊕ a verb ⊕ ⇓ a negative particle ⊕ ⇓ a particle of the person ⊕ ⇓ an inclination particle
2) a noun ⊕ a verb ⊕ ⇓ an affix forming a word ⊕ ⇓ a negative particle ⊕ ⇓ a mortgaging particle ⊕ ⇓ a particle
3) a noun ⊕ a verb ⊕ ⇓ an affix forming a word ⊕ ⇓ a negative particle ⊕ ⇓ a particle of the person ⊕ ⇓ a particle
4) a verb ⊕ a verb ⊕ ⇓ an affix forming a word a negative ⊕ ⇓ particle ⊕ ⇓ a particle of the person ⊕ ⇓ an inclination particle ⊕ ⇓ a mortgaging particle ⊕ ⇓ a particle
5) a verb ⊕ a verb ⊕ ⇓ a negative particle ⊕ ⇓ a particle of the person ⊕ ⇓ a mortgaging particle ⊕ ⇓ a particle

Logical-linguistic models of a conclusion *of pronouns* UL consist of ten variants:

1) a personal pronoun ⊕ ⇓ a suffix of plural ⊕ ⇓ a suffix of a case ⊕ ⇓ a particle
2) a demonstrative pronoun ⊕ ⇓ a suffix of plural ⊕ ⇓ a suffix of a case ⊕ ⇓ a particle
3) an interrogative pronoun ⊕ ⇓ a suffix of plural ⊕ ⇓ a case suffix
4) an attributive pronoun ⊕ ⇓ a case suffix
5) a negative pronoun ⊕ ⇓ a case suffix
6) a reflexive pronoun ⊕ ⇓ a suffix of plural ⊕ ⇓ a suffix of a subject ⊕ ⇓ a suffix of a case ⊕ ⇓ a particle

7) an affix «алла» ⊕ ⇓ an interrogative pronoun ⊕ ⇓ a suffix of a subject ⊕ ⇓ a case suffix
8) an interrogative pronoun ⊕ ⇓ affixes of an interrogative pronoun
9) a demonstrative pronoun ⊕ ⇓ affixes of a demonstrative pronoun ⊕ ⇓ a particle
10) an uncertain pronoun ⊕ ⇓ a suffix of plural ⊕ ⇓ a suffix of a subject ⊕ ⇓ a case suffix

Logical-linguistic models of a conclusion *of adverbs* UL consist of six variants:

1) a noun ⊕ affixes forming an adverb
2) an adjective ⊕ affixes forming an adverb
3) a pronoun ⊕ affixes forming an adverb
4) the numeral ⊕ affixes forming an adverb
5) an adverb ⊕ ⇓ affixes forming an adverb
6) a prefix of an adverb ⊕ a noun

The logical-linguistic model of a conclusion *of numerator* UL will be expressed as:
numerator ⊕ ⇓ an affix forming a word ⊕ ⇓ suffixes of plurals a suffix ⊕ ⇓ of a case ⊕ ⇓ a particle.

3.3 Logical-Linguistic Models of Sentences

Logical-linguistic models of a conclusion *of narrative sentences* UL consists of twenty one variants:

1. pronoun ⊕ ⇓ a postposition ⊕ ⇓ a noun ⊕ a verb ⊕ ⇓ the union ⊕ ⇓ an adjective ⊕ ⇓ a noun ⊕ ⇓ a verb
2. noun ⊕ an adverb ⊕ an adjective ⊕ ⇓ a noun ⊕ ⇓ a pronoun ⊕ ⇓ an adjective
3. noun ⊕ the numeral ⊕ ⇓ an adverb the union ⊕ ⇓ ⊕ ⇓ an adjective ⊕ ⇓ a noun an adjective ⊕ ⇓ ⊕ ⇓ the numeral
4. adverb ⊕ a noun ⊕ a pronoun ⊕ ⇓ the union ⊕ ⇓ an adjective ⊕ ⇓ a noun ⊕ ⇓ a pronoun ⊕ ⇓ an adjective
5. modal word ⊕ a noun ⊕ an adjective ⊕ ⇓ a noun ⊕ ⇓ a verb
6. adverb ⊕ a noun ⊕ ⇓ a modal word ⊕ ⇓ an adjective ⊕ ⇓ a noun ⊕ a verb
7. adjective ⊕ a noun ⊕ ⇓ an adverb ⊕ ⇓ a verb ⊕ ⇓ the union ⊕ ⇓ an adjective ⊕ ⇓ a verb
8. noun ⊕ a pronoun ⊕ a verb ⊕ ⇓ the union ⊕ ⇓ a noun ⊕ ⇓ a verb
9. ⇓ particle ⊕ an adjective ⊕ a noun ⊕ a verb
10. ⇓ modal word ⊕ a noun ⊕ ⇓ a postposition ⊕ a verb
11. ⇓ modal word ⊕ a pronoun ⊕ ⇓ a postposition ⊕ a verb
12. pronoun ⊕ ⇓ a postposition ⊕ a noun ⊕ a verb ⊕ ⇓ the union ⊕ ⇓ an adverb ⊕ ⇓ an adjective
13. noun ⊕ the numeral ⊕ a verb ⊕ ⇓ an adverb ⊕ ⇓ a noun ⊕ ⇓ a verb
14. ⇓ modal word ⊕ a noun ⊕ the numeral ⊕ ⇓ a pronoun ⊕ ⇓ a verb
15. ⇓ particle ⊕ a noun ⊕ ⇓ a postposition ⊕ a pronoun ⊕ ⇓ a particle ⊕ ⇓ the numeral ⊕ ⇓ an adjective ⊕ ⇓ a noun ⊕ ⇓ a verb

16. adverb ⊕ ⇓ a pronoun ⊕ ⇓ the union ⊕ a noun ⊕ a verb
17. pronoun ⊕ ⇓ an adverb ⊕ ⇓ an adjective ⊕ ⇓ a noun ⊕ a verb
18. adjective ⊕ a noun ⊕ ⇓ an adverb ⊕ a pronoun ⊕ a verb
19. ⇓ particle ⊕ a noun ⊕ ⇓ a postposition ⊕ a pronoun ⊕ ⇓ a particle ⊕ ⇓ an adjective ⊕ ⇓ a noun ⊕ ⇓ a verb
20. pronoun ⊕ ⇓ a postposition ⊕ a noun ⊕ a verb ⊕ ⇓ the union ⊕ ⇓ an adjective
21. adverb ⊕ a noun ⊕ a pronoun ⊕ ⇓ the union ⊕ ⇓ an adjective ⊕ ⇓ a noun ⊕ ⇓ a pronoun ⊕ ⇓ an adjective

Logical-linguistic models of a conclusion *of questions* UL consists of twenty-two variants:

1. adverb ⊕ an interrogative pronoun ⊕ a verb
2. interrogative pronoun ⊕ ⇓ an adverb ⊕ ⇓ an adjective ⊕ a verb ⊕ a noun
3. pronoun ⊕ ⇓ a noun ⊕ ⇓ an adverb ⊕ ⇓ an adjective ⊕ ⇓ a noun ⊕ ⇓ a pronoun ⊕ ⇓ a noun ⊕ a verb
4. ⇓ particle ⊕ a noun ⊕ an adverb ⊕ a verb
5. ⇓ particle ⊕ a pronoun ⊕ ⇓ an adverb ⊕ a verb ⊕ ⇓ the union ⊕ ⇓ a noun ⊕ ⇓ a verb
6. an adverb ⊕ a pronoun ⊕ ⇓ an adjective ⊕ ⇓ a noun ⊕ a verb
7. the union ⊕ ⇓ a noun ⊕ ⇓ a postposition ⊕ a verb
8. ⇓ particle ⊕ a pronoun ⊕ ⇓ a postposition ⊕ ⇓ a noun ⊕ a verb
9. ⇓ particle ⊕ a pronoun ⊕ ⇓ an adjective ⊕ a noun ⊕ a verb ⊕ ⇓ an adverb
10. ⇓ particle ⊕ an adverb ⊕ a pronoun ⊕ ⇓ a noun ⊕ ⇓ an adjective ⊕ a verb
11. ⇓ particle ⊕ a pronoun ⊕ ⇓ an adjective ⊕ ⇓ a noun ⊕ a verb
12. the union ⊕ ⇓ a pronoun ⊕ ⇓ a postposition ⊕ a verb
13. noun ⊕ ⇓ a pronoun ⊕ ⇓ an adverb ⊕ a verb
14. ⇓ particle ⊕ the numeral ⊕ ⇓ a noun ⊕ a verb
15. ⇓ particle ⊕ the numeral ⊕ ⇓ a noun ⊕ ⇓ an adjective ⊕ ⇓ a noun ⊕ a verb
16. adjective ⊕ ⇓ a noun ⊕ the numeral ⊕ a verb
17. pronoun ⊕ ⇓ the union ⊕ ⇓ a noun ⊕ ⇓ the numeral ⊕ ⇓ an adjective ⊕ ⇓ a noun ⊕ a verb
18. the union ⊕ ⇓ an adverb ⊕ ⇓ a pronoun ⊕ ⇓ the numeral ⊕ ⇓ a noun ⊕ ⇓ an adjective ⊕ ⇓ a noun ⊕ ⇓ a verb
19. the union ⊕ ⇓ an adverb ⊕ a noun ⊕ a verb
20. pronoun ⊕ ⇓ the union ⊕ ⇓ a noun ⊕ ⇓ the numeral ⊕ ⇓ an adjective ⊕ a verb
21. the union ⊕ ⇓ an adverb ⊕ ⇓ a pronoun ⊕ ⇓ the numeral ⊕ ⇓ an adjective ⊕ ⇓ a noun ⊕ ⇓ a verb
22. pronoun ⊕ ⇓ a noun ⊕ ⇓ an adverb ⊕ ⇓ an adjective ⊕ ⇓ a noun ⊕ ⇓ a pronoun ⊕ a verb

Logical-linguistic models of a conclusion *of exclamatory sentences* UL consists of nineteen variants:

1. exclamatory word ⊕ a noun ⊕ ⇓ an adverb ⊕ ⇓ a noun ⊕ ⇓ the numeral ⊕ ⇓ a noun ⊕ a verb

2. ⇓ particle ⊕ a noun ⊕ a verb
3. ⇓ particle ⊕ an adjective ⊕ a verb
4. ⇓ particle ⊕ a pronoun ⊕ a verb
5. pronoun ⊕ ⇓ the numeral ⊕ a verb ⊕ ⇓ the union ⊕ ⇓ a noun ⊕ ⇓ a verb
6. the numeral ⊕ a verb ⊕ ⇓ a noun ⊕ ⇓ an adverb
7. verb ⊕ a noun
8. the numeral ⊕ ⇓ a noun ⊕ a verb
9. pronoun ⊕ ⇓ a noun ⊕ a verb
10. verb ⊕ ⇓ a pronoun ⊕ ⇓ a noun ⊕ a verb
11. exclamatory word ⊕ ⇓ a noun ⊕ a verb
12. pronoun ⊕ ⇓ a noun
13. adjective ⊕ ⇓ a noun ⊕ a verb
14. verb ⊕ ⇓ a verb
15. the numeral ⊕ ⇓ a noun ⊕ ⇓ a pronoun ⊕ a verb
16. pronoun ⊕ ⇓ the numeral ⊕ ⇓ a noun ⊕ a verb
17. exclamatory word ⊕ a noun ⊕ ⇓ an adverb ⊕ ⇓ a noun ⊕ ⇓ the numeral ⊕ a verb
18. pronoun ⊕ ⇓ the numeral ⊕ a verb ⊕ ⇓ the union ⊕ ⇓ a verb
19. exclamatory word ⊕ a noun ⊕ ⇓ a noun ⊕ ⇓ the numeral ⊕ a verb

4 Experiments and Results

We show the following noun to illustrate the logical-linguistic model of words.

$$
\text{Берилган} \Big| \oplus \Downarrow \text{лар} \left|
\begin{array}{ll}
\oplus \Downarrow \text{им} & \oplus \Downarrow \text{нинг} \\
\oplus \Downarrow \text{инг} & \oplus \Downarrow \text{ни} \\
\oplus \Downarrow \text{и} & \oplus \Downarrow \text{га} \\
\oplus \Downarrow \text{имиз} & \oplus \Downarrow \text{да} \\
\oplus \Downarrow \text{ингиз} & \oplus \Downarrow \text{дан}
\end{array}
\right.
$$

It is example, 2, 3 and 4th columns are suffixes. Here "Берилган" is translated as *date*. Noun word is created by combining above suffixes.

We show the following narrative sentences to illustrate the logical-linguistic model.

Берилганларнинг ⊕ ташқи ⊕ тузилиши ⊕ ⇓ дастурчилар ⊕ ⇓ тасаввуридаги ⊕ тақдимотни ⊕ билдиради.

Additionally, for validation (truth) of the structure of logical-linguistic models, a special program was developed that operates on the basis of two algorithms. The first function of the program is to automatically compile algorithms based on mathematical models created by categories of words. The second function of the program is to create software modules on automatically generated algorithms. Analysis of the semantic correct composition of words is performed through these modules in the program.

In order to analyze a mathematical model, it is necessary to first transfer a logical-linguistic model to a mathematical model, and then to a mathematical model that the computer understands. Logical-linguistic model to transfer the to a mathematical model, we again use the extensible input language, which is a natural language processing. Extensible input language consists of special characters, numbers and letters [4], and is also an important tool for transferring the models of vocabulary into a computer-savvy view. Elements extensible input language are not repeated; each element is applied to perform a certain task. Elements extensible input language are formed in the program as the KKT base. Even if a new element is entered into the extensible input language, the program will work. Also included in the program will be word categories, bases of suffixes.

The program performs the first function on the basis of the following enlarged 1-th algorithm:

1. Check the integrity of the mathematical model entered in the word categories.
2. According to the separating sign "V" in the mathematical model, each model should be allocated in Aloha, and they should be temporarily stored in the array called "Algorithm List". Let the length of the array be equal to the number of models (algorithms).
3. To compile the algorithm, let each line in the array with the name "AlgoritmList" be processed.
4. Looking at each of the elements of the array "AlgoritmList" as an object, divide them by the sign "*" and keep them in the array with the name "AlgoritmList[i]".
5. Based on Step 4, determine the number of steps of the algorithm that will be automatically generated.
6. Elements of the array "AlgoritmList[i]" should be in one of the following cases:

   a) $\downarrow\$[t,1/n]$ CZ(AZ[t]) or $\downarrow\$[t,1/n]$ (AZ[t])
   b) $\downarrow$ CZ(AZ)
   c) $\downarrow$XZ

7. Let it be processed according to the circumstances listed in Step 6.
8. Let "AlgortmList" be processed on massive elements starting from Step 4.

The program performs the second function according to the following enlarged 2nd algorithm:

1. Analyzing the given word, it is necessary to determine the root and semantic suffixes according to the semantic bases of the Uzbek language [6].
2. Let the constellation of the word uzak be determined.
3. Formulate a mathematical model of the given word.
4. Let the resulting model be compared with the models in the array "AlgoritmList".
5. If the mathematical models are equal, the process should be completed and information about the uniformity of the models should be released.
6. If the models are not equal, then finish the process. Let the given word "Stem-NotFound" be kept on the base and information about the diversity of models is released.

Of the logical-linguistic model, created according to the rules of the noun category affiksatsya, given above, we will describe the collection of mathematical models in the program for analysis as follows:

$[i,1/h1](C[i])*↓$[j,1/33]C(A1[j])*↓X*↓$[s,1/10](X2[s])*↓$[t,1/5](X3[t])*
↓$[r,1/6]U(A1[r]))V(↓C(A1)*$[i,1/h1]C[i]*↓$[j,1/33]C(A1[j])*↓X*
↓$[s,1/10](X2[s])*↓$[t,1/5](X3[t])*↓$[r,1/6]U(A1[r]))V($[i,1/h3](G[i])*
$[j,1/46]C(A1[j])*↓X*↓$[s,1/10](X2[s])*↓$[t,1/5](X3[t])*↓$[r,1/6]U(A1[r]))
V($[i,1/h2](P[i])*$[j,1/6]C(A1[j])*↓X*↓$[s,1/10](X2[s])*
↓$[t,1/5](X3[t])*↓$[r,1/6]U(A1[r]))V($[i,1/h4](F[i])*$[j,1/6]C(A1[j])*
↓X*↓$[s,1/10](X2[s])*↓$[t,1/5](X3[t])*↓$[r,1/6]U(A1[r]))V($[i,1/h4](M[i])*
$[j,1/6]C(A1[j])*↓X*↓$[s,1/10](X2[s])*↓$[t,1/5](X3[t])*↓$[r,1/6]U(A1[r]))
V($[i,1/h4](N[i])*$[j,1/6]C(A1[j])*↓X*↓$[s,1/10](X2[s])*↓$[t,1/5](X3[t])*
↓$[r,1/6]U(A1[r])

where $h31 = \overline{1, h20}$, h31 is the word order, h1 is the number of words where the base is a noun, h2 is the number of words where the base is an adjective, h3 is the number of words where the base is a verb, h4 is the number of words where the base is a numeral, h5 is the number of words where the base is a pronoun, all from the corresponding databases, h20 is the number of words that can be derived based on this pattern h20 = {min(6 * h6), max(105600 * h1)}, h6 is the number of words where the base is an adverb [6].

According to the 1st step of the 1st algorithm, we will analyze over a set of these mathematical models of noun-type words. According to the 2nd step of the 1st algorithm, we will analyze over a set of mathematical models and obtain the following types of models:

- $[i,1/h1](C[i])*↓$[j,1/33]C(A1[j])*↓X*↓$[s,1/10](X2[s])*
  ↓$[t,1/5](X3[t])*↓$[r,1/6]U(A1[r])
- ↓C(A1)*$[i,1/h1]C[i]*↓$[j,1/33]C(A1[j])*↓X*
  ↓$[s,1/10](X2[s])*↓$[t,1/5](X3[t])*↓$[r,1/6]U(A1[r])
- $[i,1/h3](G[i])*$[j,1/46]C(A1[j])*↓X*↓$[s,1/10](X2[s])*↓$[t,1/5](X3[t])*
  ↓$[r,1/6]U(A1[r])
- $[i,1/h2](P[i])*$[j,1/6]C(A1[j])*↓X*↓$[s,1/10](X2[s])*↓$[t,1/5](X3[t])*
  ↓$[r,1/6]U(A1[r])
- $[i,1/h4](F[i])*$[j,1/6]C(A1[j])*↓X*↓$[s,1/10](X2[s])*↓$[t,1/5](X3[t])*
  ↓$[r,1/6]U(A1[r])
- $[i,1/h4](M[i])*$[j,1/6]C(A1[j])*↓X*↓$[s,1/10](X2[s])*↓$[t,1/5](X3[t])*
  ↓$[r,1/6]U(A1[r])
- $[i,1/h4](N[i])*$[j,1/6]C(A1[j])*↓X*↓$[s,1/10](X2[s])*↓$[t,1/5](X3[t])*
  ↓$[r,1/6]U(A1[r])

At the first stage, a computer analysis of multi-type mathematical models of words is carried out. Where each type of mathematical model will be highlighted separately, and all of them will be saved in the database of word models.

At the second stage, there is a computer analysis of the database of word models. Below is a mathematical model of the first type according to the above logical-linguistic model of nouns of the first type:

$[i,1/h1](C[i])*↓$[j,1/33]C(A1[j])*↓X*↓$[s,1/10](X2[s])*↓$[t,1/5](X3[t])* ↓$[r,1/6]U(A1[r])

The first algorithm is to divide the mathematical model by the Step 4 by the "*" sign:

1) $[i,1/h1](C[i])
2) ↓$[j,1/33]C(A1[j])
3) ↓X
4) ↓$[s,1/10](X2[s])
5) ↓$[t,1/5](X3[t])
6) ↓$[r,1/6]U(A1[r])

Each of the selected six parts of the mathematical model will be compared with each part of the model outlined in the sixth step of the 1st algorithm. Based on the results of the comparisons, we get that variant *a)* of the sixth step of the first algorithm corresponds to 1, 2, 4–6 parts of this model; variant *c)* corresponds to the 3rd part of this model, and in accordance with the introduced mathematical model of the word, the algorithm is automatically developed (Fig. 1); variant *b)* does not correspond to any of the parts of this model.

Algorithm 1
1. The corresponding noun phrase belonging to the entered noun phrase should be selected from the (C, i=1, h1) database.
2. If there are word-forming affixes that make up a noun phrase, in the database of noun-forming affixes, select one (A1, j=1,33).
3. If there is a plural suffix (x) belonging to the entered noun phrase, the corresponding plural should be selected from the suffix base.
4. If the corresponding possessive suffix belonging to the entered word group is selected from the (X2, s=1, 10) database.
5. If the suffix corresponding to the entered word group is selected from the database, (X3, t=1, 5).
6. If there are load-forming, word-forming affixes belonging to the entered noun, select one in the database of word-forming affixes (A1, r=1, 6).

Fig. 1. 1-th algorithm of the noun category by the rules of declension

By the same method, from the remaining six parts of the mathematical models, comparing each separately with each part of the model set out in the sixth step of the 1st algorithm, we obtain automatically developed six algorithms (Figs. 2, 3, 4, 5 and 6).

Algorithm 2
1. Word-forming affixes belonging to the noun phrase.
2. The corresponding noun phrase belonging to the entered noun phrase should be selected from the (C, i=1, h1) database.
3. If there are word-forming affixes that make up a noun phrase, select one in the word-formation affix database (A1, j=1, 33).
4. If there is a plural suffix (X) for the entered noun phrase, select the corresponding plural from the suffix base.
5. If the word belonging to the entered word group is the corresponding possessive suffix, select it from the database (X2, s=1, 10).
6. If the suffix corresponding to the entered word group is selected from the database, (X3, t=1, 5).
7. if there are load-forming, word-forming affixes belonging to the entered noun, select one in the database of word-forming affixes (A1, r=1, 6).

Fig. 2. 2-th algorithm of the noun category by the rules of declension

---

Algorithm 3

1. The word entered belongs to the noun phrase, the corresponding verb, selected from the database (G, i=1, h3).
2. If there are word-forming affixes that make up a noun phrase, select one in the word-formation affix database (A1, j=1, 47).
3. If there is a plural suffix (X) for the entered noun phrase, select the corresponding plural from the suffix base.
4. If the word belonging to the entered word group is the corresponding possessive suffix, select it from the database (X2, s=1, 10).
5. If the suffix corresponding to the entered word group is selected from the database, (X3, t=1, 5).
6. If there are load-forming, word-forming affixes belonging to the entered noun, select one in the database of word-forming affixes (A1, r=1, 6).

Fig. 3. 3-th algorithm of the noun category by the rules of declension

---

Algorithm 4

1. The corresponding adjective phrase belonging to the entered word noun phrase should be selected from the database (P, i=1, h2).
2. If there are word-forming affixes that make up a noun phrase, select one in the word-formation affix database (A1, j=1, 6).
3. If there is a plural suffix (X) for the entered noun phrase, select the corresponding plural from the suffix base.
4. If the word belonging to the entered word group is the corresponding possessive suffix, select it from the database (X2, s=1, 10).
5. If the suffix corresponding to the entered word group is selected from the database, (X3, t=1, 5).
6. If there are load-forming, word-forming affixes belonging to the entered noun, select one in the database of word-forming affixes (A1, r=1, 6).

Fig. 4. 4-th algorithm of the noun category by the rules of declension

---

Algorithm 5

1. The selected word belongs to the word group, the corresponding number phrase, selected from the database (F, i=1, h4).
2. If there are word-forming affixes that make up a noun phrase, select one in the word-formation affix database (A1, j=1, 6).
3. If there is a plural suffix (X) for the entered noun phrase, select the corresponding plural from the suffix base.
4. If the word belonging to the entered word group is the corresponding possessive suffix, select it from the database (X2, s=1, 10).
5. If the suffix corresponding to the entered word group is selected from the database, (X3, t=1, 5).
6. If there are load-forming, word-forming affixes belonging to the entered noun, select one in the database of word-forming affixes (A1, r=1, 6).

Fig. 5. 5-th algorithm of the noun category by the rules of declension

---

Algorithm 6

1. The selected word belongs to the word group, the corresponding pronoun, (M, i=1, h4) is selected from the database.
2. If there are word-forming affixes that make up a noun phrase, select one in the word-formation affix database (A1, j=1, 6).
3. If there is a plural suffix (X) for the entered noun phrase, select the corresponding plural from the suffix base.
4. If the word belonging to the entered word group is the corresponding possessive suffix, select it from the database (X2, s=1, 10).
5. If the suffix corresponding to the entered word group is selected from the database, (X3, t=1, 5).
6. If there are load-forming, word-forming affixes belonging to the entered noun, select one in the database of word-forming affixes (A1, r=1, 6).

Fig. 6. 6-th algorithm of the noun category by the rules of declension

---

Algorithm 7

1. The selected word belongs to the word group, the corresponding adverb, selected from the database (N, i=1, h4).
2. If there are word-forming affixes that make up a noun phrase, select one in the word-formation affix database (A1, j=1, 6).
3. If there is a plural suffix (X) for the entered noun phrase, select the corresponding plural from the suffix base.
4. If the word belonging to the entered word group is the corresponding possessive suffix, select it from the database (X2, s=1, 10).
5. If the suffix corresponding to the entered word group is selected from the database, (X3, t=1, 5).
6. If there are load-forming, word-forming affixes belonging to the entered noun, select one in the database of word-forming affixes (A1, r=1, 6).

Fig. 7. 7-th algorithm of the noun category by the rules of declension

For the purity of the computer experiment on at least one type of output of words of the Uzbek language, the above six types of logical-linguistic models of nouns of the Uzbek language were processed according to the rules of composition. According to the first algorithm, their mathematical model was processed by the same method. Six automatically developed algorithms have been obtained that process Uzbek nouns according to the rules of composition (Fig. 7).

Next, the nouns of the Uzbek language were analyzed in the amount of 7245, using algorithms 1 and 2. The verification results are shown in Table 1.

Table 1. .

The number of words proving the correctness of the models	The number of words for which the models matched the automatically developed algorithms		As a percentage of the total number of words	Relative to the number of correct models	
	According to the rules of affixation	According to the rules of composition		According to the rules of affixation	According to the rules of composition
6715	6271	444	92,68	93,39	6,61

5 Conclusion and Future Work

In the given work we have constructed logical-linguistic models for words and sentences UL and have received primary models with application of two operations of the expanded source language.

The multilanguage system of the modelled computer translator gradually can be expanded by subsystems processing other languages. It is also clear, analyses should be carried out and models according to application of technology of the multilanguage modelled computer translator are constructed.

References

1. Adali, E., Adamov, A.Z.: Sentiment analysis for agglutinative languages. In: 2016 IEEE 10th International Conference on Application of Information and Communication Technologies (AICT). IEEE (2016)
2. Chomshy, N.: Formal properties of grammars. In: Handbook of Mathematical Psychology, vol. 2, pp. 323–418. Wiley, New York (1963)
3. Khakimov, M.H.: Formal machine translation systems in a multi-languages situation. In: Materials of Republican – Scientific Conference «Modern Problems of Mathematics, Mechanics and Information Technologies», NUUz, Institute of Mathematics and IT AS RUz, pp. 297–301, Tashkent (2008)
4. Khakimov, M.H.: The extensible source language of mathematical modelling of a natural language for a multi-languages situation of machine translation. In: News NUUz, vol. 1, pp. 80–85, Tashkent (2009)

5. Khakimov, M.H.: The computer translator on the basis of modeled technology. In: Questions of Computing and Applied Mathematics. The Collection of Proceedings. Release, vol. 129, pp. 87–106, Tashkent (2013)
6. Khakimov, M.H.: Semantic bases of the Uzbek language for machine translation. In: Proceedings of the International Conference "Actual Problems of Applied Mathematics and Information Technologies - Al-Khorezmi-2014", vol. 2, pp. 106–110, Samarkand (2014)
7. Marciniak, M., Mykowiecka, A.: Representation of Uzbek morphology in prolog. In: Marciniak, M., Mykowiecka, A. (eds.) Aspects of Natural Language Processing. LNCS, vol. 5070. Springer, Heidelberg (2009). https://doi.org/10.1007/978-3-642-04735-0_4
8. Matlatipov, S., Tukeyev, U., Aripov, M.: Towards the Uzbek language endings as a language resource. In: Hernes, M., Wojtkiewicz, K., Szczerbicki, E. (eds.) ICCCI 2020. CCIS, vol. 1287, pp. 729–740. Springer, Cham (2020). https://doi.org/10.1007/978-3-030-63119-2_59
9. Sakenovich, N.S., Zharmagambetov, A.S.: On one approach of solving sentiment analysis task for Kazakh and Russian languages using deep learning. In: Nguyen, N.-T., Manolopoulos, Y., Iliadis, L., Trawiński, B. (eds.) ICCCI 2016. LNCS (LNAI), vol. 9876, pp. 537–545. Springer, Cham (2016). https://doi.org/10.1007/978-3-319-45246-3_51
10. Tuzov, V.A.: Matematicheskaja's ases language model, p. 176. LGU (1984)
11. Yergesh, B., Bekmanova, G., Sharipbay, A., Yergesh, M.: Ontology-based sentiment analysis of Kazakh sentences. In: Gervasi, O., et al. (eds.) ICCSA 2017. LNCS, vol. 10406, pp. 669–677. Springer, Cham (2017). https://doi.org/10.1007/978-3-319-62398-6_47

Computer Application of Georgian Words

Irakli Kardava[1,2](✉) (iD), Nana Gulua[2], Jemal Antidze[2], Beka Toklikishvili[2],
and Tamta Kvaratskhelia[1]

[1] Adam Mickiewicz University in Poznań, 61-712 Poznań, Poland
irakar@amu.edu.pl
[2] Sokhumi State University, 0186 Tbilisi, Georgia

Abstract. Morphological synthesis of Georgian words requires to compose the word-forms by indication unchanged parts and morphological categories. Also, it is necessary by using a stem of the given word to get by the computer all grammatically right word-forms. In case of morphological analysis of Georgian words, it is essential to decompose the given word into morphemes and get the definition each of them. For solving these tasks we have developed some specific approaches and created software. Its tools are efficient for a language, which has free order of words and morphological structure is like Georgian. For example, a Georgian verb (in Georgian: " წერა " - ts'era, in English: Writing) has several thousand verb-forms. It is very difficult to express morphological analysis' rules by finite automaton and it will be inefficient as well. Splitting of some Georgian verb-forms into morphemes requires non-deterministic search algorithm, which needs many backtracks. To minimize backtracking, it is necessary to put constraints, which exist among morphemes and verify them as soon as possible to avoid false directions of search. Sometimes the constraints can be as a description type of specific cases of verbs. Thus, proposed software tools have many means to construct efficient parser, test and correct it. We realized morphological and syntactic analysis of Georgian texts by these tools. Besides this, for solving such problems of artificial intelligence, which requires composing of natural language's word-form by using the information defining this word-form, it is convenient to use the software developed by us.

Keywords: Morphemes · Morphological analysis · Morphological synthesis · Rules · Word-forms

1 Introduction

As it is mentioned in the abstract, the root of the given Georgian word can have thousands of correct forms. The existence of every form in the database is necessary for the complete computer processing of Georgian words. However, it is very difficult for a human to define the correct forms of all Georgian words and then record them as data. In addition to the complexity, this approach is not flexible in practical terms. Because morphological synthesis and analysis still requires the existence of morpheme classes, it is more convenient to automatically generate the correct forms of the word by a computer.

Z. Vetulani et al. (Eds.): LTC 2019, LNAI 13212, pp. 96–111, 2022.
https://doi.org/10.1007/978-3-031-05328-3_7

The created software consists of the various modules, by which it is possible to get the appropriate grammatically correct word-form or word-forms if such exist based on the unchanged part of a word and choosing morphological categories. Otherwise we shall get notification, which grammatical categories are not correct or the unchanged part of the word is not in data base. Also with this software it is possible by unchanged part of a word get all possible grammatically right word-forms (The menu language of the software is Georgian). This used approach is based on description of natural language morphology by using formal grammar [1] and characterizing symbols of grammar with feature structures [2, 3]. For description natural language morphology we use special type of context free grammar, which describes all correct word-forms. With given unchanged part and its features existed in database (we use MS SQL Server and mysql server), also with given morphological categories (in the case of first problem) we compose morpheme classes and their representatives, which must be in related word-forms [4]. We demonstrate application of the software for Georgian language. Using of the software is effective for languages, which have developed morphology like Georgian [5]. For composing features of unchanged part of the word we use Georgian language grammar and D. Melikisvili classification for verbs [6, 7]. In order to be able to decompose the Georgian words into morphemes by a program for morphological analysis using by a formal grammars, we took all the existing morphemes from the parts of speech. Then we classified and put them in one of the sets of terminal symbols. From the semantic point of view, there is a need for morphemes to be the sub-sets of the major set. After that, according to the formal rules created in advance, the words will be parsed and the information about each morpheme will be returned. Since the Georgian language is not yet perfectly computerized, these issues are very relevant.

2 Brief Information About Georgian Language and Alphabet

We would like to bring here a brief information about the Georgian language and alphabet.

It is known that the Georgian language and script are unique in the world. It is a South Caucasian or Kartvelian (Kartveli- ქართველი, in Georgian means "Georgian") language. It is spoken mainly in Georgia, where it is the official language. Georgian is related to Mingrelian, Laz, and Svan, all of which are spoken mainly in Georgia and are written with the Georgian (Mkhedruli) alphabet (Omniglot - the online encyclopedia of writing systems and languages).

Georgian is thought to share a common ancestral language with the other South Caucasian languages. It started to develop as a separate language during the 1st millennium BC in an area that became the Kingdom of Iberia (c. 302 BC - 580 AD). It was first referred to in writing in the 2nd century AD by the Roman grammarian, Marcus Cornelius Fronto, in a letter to the emperor Marcus Aurelius (Omniglot - the online encyclopedia of writing systems and languages). However, it is considered an even more ancient language. The Georgian alphabet had three stages of evolution. See Fig. 1 and Fig. 2.

Fig. 1. Old Georgian alphabets (Asomtavruli & Nuskha-khucuri). Created by Simon Ager, Omniglot.com – the guide to writing systems and languages.

Fig. 2. Modern Georgian alphabet (Mkhedruli). Created by Simon Ager, Omniglot.com – the guide to writing systems and languages.

3   Software Architecture

Software uses the database created by us. In it the information is represented according to each type of a word. By the hierarchical viewpoint, the tables in the database are divided into two parts: 1. The main members of the hierarchy are the tables of the basis of the verbs and nouns; 2. The second row members are tables of type, or morpheme tables. In main tables there are written down unchanged parts of words and their features. For each part of speech there are used different tables. For example, in main tables of verbs there is written down root, type, proverbs and vowel prefixes, see Table 1 (rt = root, tp = type, pv = pre verb, vp = vowel prefix).

Table 1. A fragment of the table for verbs.

rt	tp	pv1	pv2	Pv3	pv4	pv5	pv6	vp	...
ნთ (Eng.: "nTh")	1	ა (Eng.: "a")	და (Eng.: "da")	ჩა(Eng.: "cha")	-	-	-	ვ (Eng.: "v")	...

To this table is connected the table, which consists of persons, numbers and so on - all of those elements, which are necessary to get grammatically correct word-form (each morpheme table represents a set of terminal symbols and instructions for their positioning in a word.), see Table 2.

Table 2. A fragment of the table for morphemes

Morph1	Morph2	Morph3	Morph4	Morph5
ვ (Eng.: "v")	ები (Eng.: "ebi")	-	-	თ (Eng.: "T^{h}")
-	ები (Eng.: "ebi")	-	-	თ (Eng.: "T^{h}")
-	ებ (Eng.: "eb")	ა (Eng.: "a")	-	ოან (Eng.: "ebi")
ვ (Eng.: "v")	ებოდი (Eng.: "ebodi")	-	-	თ(Eng.: "T^{h}")

The same principle is used for other parts of speech. In order to, that software will give us the result we must give to it unchanged part of searched word. If it is found in the database, it applies to appropriate table, process it and gives all grammatically right forms. If the entered unchanged part can't be found in database or it is indicated incorrectly, there will appear appropriate notification, which will request to change the sent unchanged part. By using the software, it is possible get all grammatically right forms of any Georgian word and one of its advantage is, that the information is compactly written down in computer memory. In addition, the program is linked to grammar file, which contains a special formal rule. Also semantics files containing descriptions of morphemes and functions of relevant semantics. It should be noted that the software was not initially built on a web platform. But now it is possible to insert the data from the network.

4 Software Sections

The main software is created as a Windows Forms Application. Its design is simple to use. User can write down unchanged part of interesting word in the special field for this purpose. After clicking on the confirm button software should give result or all results. See Fig. 3.

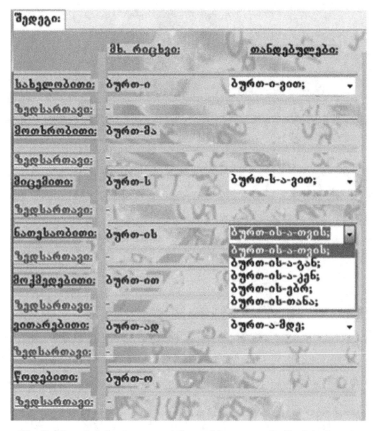

Fig. 3. The result for the nouns (all possible grammatically right forms).

As you can see, the picture shows all possible grammatically correct words-forms (both singular and plural). In addition, words are generated with specific suffixes as well (both singular and plural). There are cases when a noun can be in the context of an adjective. In this case, also adjective forms will be generated by the appropriate algorithm. This approach is very convenient because we do not have to insert manually each form of each word in the database (as it is mentioned at the beginning of the paper, the Georgian word can have thousands of different correct forms. As the words are generated automatically, using this technique we save a lot of time, computer memory and reduce the amount of work to be done. Also, we increase the speed of the program and minimize the probability of morphological mistakes).

If the unchanged part of the word will be simultaneously the verb's root, there will opened a new window, where this given verb should be conjugated and arranged in accordance of diathesis, class, series, rows, person and number. Each new window consists of generated verb's type name. Herewith it is possible change of prefixes and vowel prefixes, if the given verb gives the opportunity. See Fig. 4.

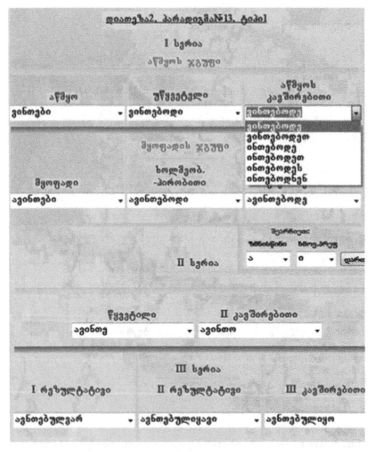

Fig. 4. The result for the verbs.

If the verb has several types, then software will open the window for each type. There are cases, when the verb is used as the noun. For example let's take the verb shown in the picture below – "write/ წერა", if we take it as noun "write", then it is considered exactly as noun. Our software at this time declines the unchanged part "write - წერ" as a verb, with foreseeing all its types and in parallel it declines it as noun. In this particular case a given word has almost 5000 grammatically correct forms. To this can be added the old Georgian plural style. It is obvious that this can also be added to the whole number, although in today's modern Georgian language these forms are almost not used (but it should also be noted that these forms is known to all Georgians. Its name is: "nar-Taniani - ნარ-თანიანი"). See Fig. 5.

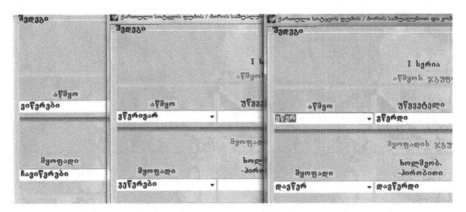

Fig. 5. The result for the verbs.

Analogically the same should we get for other parts of speech. Software can solve the second problem; such is to compose word-form with indication unchanged part of the word and morphological categories. See Fig. 6.

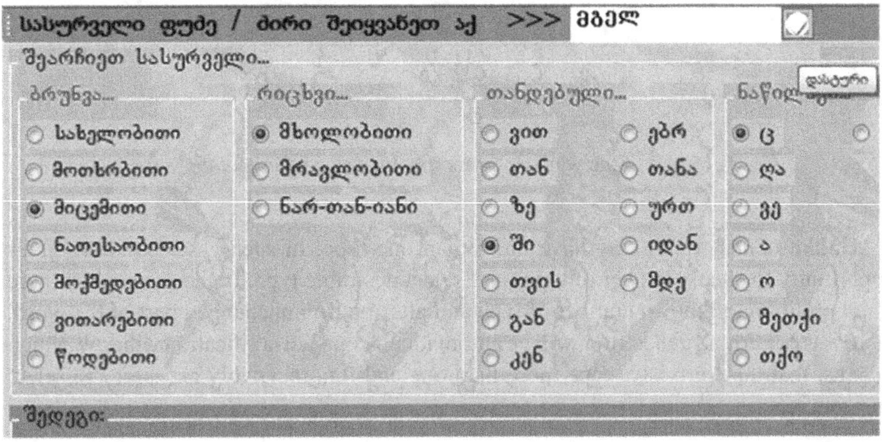

Fig. 6. A part of the window, for composition of the desired word form and the window for morphological synthesis of the nouns.

As it is shown from the table, there is given two opportunities, first is to get concrete forms of noun, and the second to get concrete forms of the verb. Let's discuss each of them successively. The first line gives opportunity to get the word-forms of the noun which is interested for us..

If right preposition or particle is not added to any of case (which is marked by the user from the list), software will give notification, that the given information is not correct. Analogically, we can recall interface which gives concrete forms of verbs, by clicking button "verbs" in the list. See Fig. 7.

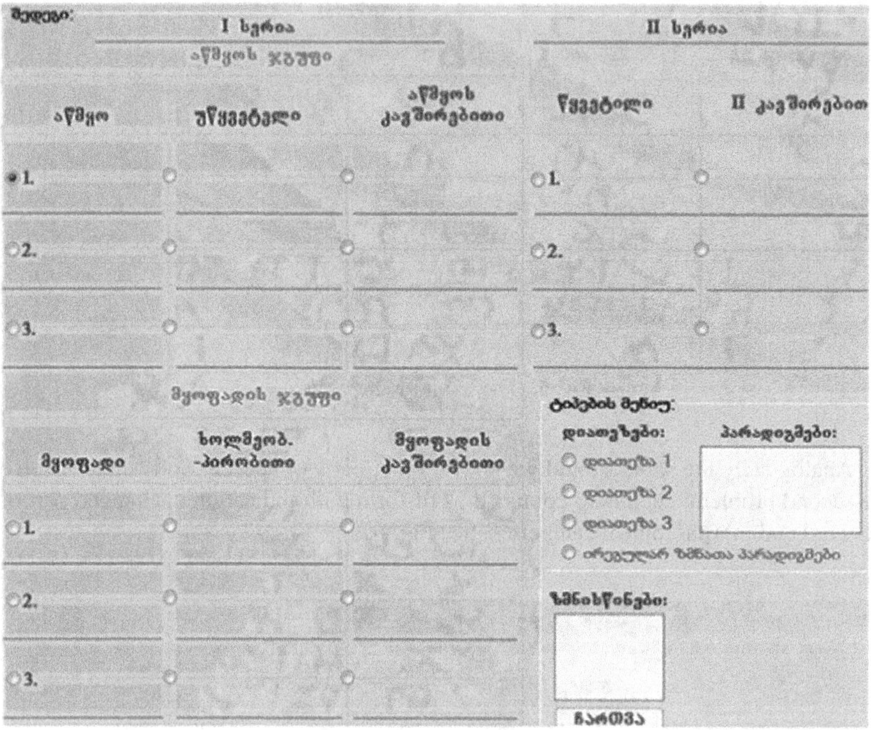

Fig. 7. A part of the window for morphological synthesis of the verbs.

Unlike nouns, verbs may have more than one type. In such cases software gives opportunity to check number of types and select desirable type. In the case, when there is not preliminary known number of type and after putting unchanged part of the word, there was confirmed generation, will be automatically opened notification window, which says us, that this verb have type more than one and it is inevitably necessary to select one concrete.

Also, we can perform the inverse manipulation. We call it morphological analysis. In this case the terminal symbols in the formal grammar we have created will be the morphemes we have previously inserted in the database. It should be noted that the corresponding terminal symbols are automatically changed as the type changes.

The main window of software designated for conducting morphological analysis of Georgian words looks as follows. See Fig. 8.

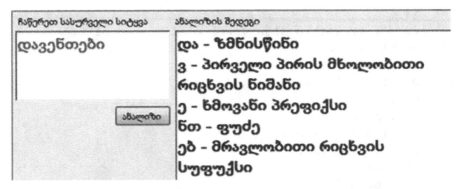

Fig. 8. The result of morphological analysis of Georgian word.

In the upper left hand side of the image is given a field where should be written word/s and you need to press the "Analysis" button to take the analysis. The result is on the bottom of the picture. If the word is incorrectly written, the program will give you a text message about the error and will show the part of the word that is incorrect. It should also be noted that the analysis is carried out only for the words written by Georgian Unicode. Any other Unicode use will be considered as a mistake. Here is shown the part of the formal grammar for one particular type:

```
%token ა %token ჰჰ %token და
%token ე %token ემ %token ემა
%token ემაინ %token კ
%token თბ %token ი %token ით
%token ნთ %token ს %token ჩა
%start word

%%
word : pr_verb p_sign vowel_prefix root pl_sign
compl_sign ;
pr_verb : და | ჩა | ა ;
p_sign : p1 | p2 ;
p1 : კ | ა | ;
p2 : ჰჰ ;
vowel_prefix : ი | ე | ა | ;
root : f1 | f2 ;
r1 : ნთ ;
r2 : თბ ;
pl_sign : r1 | r2 ;
r1 : ემ ;
r2 : ემა | ემაინ ;
compl_sign : s1 | s2 ;
c1 : ი | ს | ;
c2 : ით ;
```

Purpose of morphological analyzer is to split an input word into the morphemes and figure out morphological categories of the word. We may invoke morphological analyzer manually or automatically by the syntactic analyzer.

We used special formalism to describe morphology of natural language and pass it to the morphological analyzer. There are two main constructions in the grammar file of morphological analyzer: morphemes' class definition, and morphological rules [11, 12]. Class definition is used to list all possible morphemes for a given morphemes' class. For example (pseudocode):

@M1 = {"morpheme_1" [... features ...]"morpheme_2" [... features ...]
. . . "morpheme_N" [... features ...]}

It is possible to declare empty morpheme, which means that we may omit the morphemes' class in morphological rules [3]. Below is formal syntax for morphemes' class definition:

```
<morphem_definition> ::= "@" <identifier> "="
"{"<list_of_morphemes> "}"
<list_of_morphemes> ::= <morpheme> { "," <morpheme> }
<morpheme> ::= <string> <feature_structure>
```

We define morphological rules following way:

$$word \rightarrow M1C1\ M2C2\ \ldots Mn\{Cn\} \tag{1}$$

where M_i are morpheme classes, and

$$Ci(i = 1, \ldots, n) \tag{2}$$

are constraints (optional).

5 Web Version

In addition to the above given issues, we have created a web service that can be used by anyone. This will help us to improve our system faster and to correct possible inaccuracies. The web service is a system designed for the processing of natural language texts. We used the system to find syntactic and morphological structure of Georgian language texts. Using specific formalism, which we created for this purpose, allow us to write down syntactic and morphological rules defined by particular natural language grammar. This formalism represents the new, complex approach, which solves problems of morphological and syntactic analyses for some natural language. We implemented a software system according to this formalism [13]. One can realize syntactic analysis of sentences and morphological analysis of word-forms with this software system. We designed several special algorithms for this system. Using of the formalism, which is described in [14, 15], is very difficult to use for Georgian language, as far as expressing of some morphological rules is very complicated and understanding of such writing is difficult. The system consists of two parts: syntactic analyzer and morphological analyzer. Purpose of the syntactic analyzer is to parse an input sentence, to build a parsing tree, which describes relations between the individual words within the sentence, and to collect information about the input sentence, which the system figured out during the analysis process. It is necessary to provide a grammar file for the syntactic and morphological analyzers. There must be recorded syntactic or morphological rules of particular natural language grammar. Syntactic analyzer also needs information about the grammar categories of the word-forms. It uses the information during analysis process. Basic methods and algorithms, which we used to develop the system, are: operations defined on features' structures; trace back algorithm (for morphological analyzer); general syntactic parsing algorithm for context free grammar with constraints. Features' structures are widely used on all levels of analysis. We use them to hold various information about dictionary entries and information obtained during analysis. Each symbol defined in a morphological or syntactic rule has an associated features' structure, which we initially fill from the dictionary, or the system fill them by the previous levels of analysis. Features' structures and operations defined on them we use to build up features' constraints.

With general parsing algorithm, it is possible to get a syntactic analysis of any sentence defined by a context free grammar and simultaneously check features' constraints, which may be associated with grammatical rules. Features' constraints are logical expressions composed by the operations, which we defined on the features' structures. We attach features' constraints to rules, which we defined within a grammar file. If the constraint is not satisfied during the analysis, then the system will reject current rule and the search process will go on. We can attach features' constraints also to morphological rules. However, un-like the syntactic rules, we can attach constraints at any place within a morphological rule, only not at the end. This speeds up morphological analysis, because the system checks constraints early and it rejects incorrect word-form's division into morphemes in a timely manner. Formalism, which we developed for the syntactic and morphological analysis is highly comfortable for human. It has many constructions that make it easier to write grammar file. Morphological analyzer has a built-in preprocessor [3].

The main idea of syntactic analyzer is to analyze sentences of natural language and produce parsing tree and information about the sentence. In order to accomplish this task, syntactic analyzer needs a grammar's file and a dictionary. We write grammar rules for syntactic analyzer like CFG rules. However, they may have constraints and symbol position regulators. We can write a rule according to these conventions:

write a rule according to these conventions:

$$S \rightarrow A1C1\ A2C2\ \ldots An\{Cn\}; \tag{3}$$

$$S \rightarrow A1\ A2 \ldots An : RC; \tag{4}$$

where S is an LHS non-terminal symbol, $A_i(i = 1, \ldots, n)$ are RHS terminal or non-terminal symbols, C and $C_i(i = 1, \ldots, n)$ are constraints, and R is a set of symbol position regulators. Position regulators declare order of RHS symbols in the rule, consequently making non-fixed word ordering. There are two types of position regulators:

$$Ai < Aj; \tag{5}$$

Which means, that symbol A_i must be placed somewhere before the symbol A_j;

$$Ai - Aj; \tag{6}$$

means that symbol A_i must be placed exactly before the symbol A_j. It is not formal difference be-tween syntactic and semantic rules. Therefore, we can provide syntactic and semantic analyses in parallel or separately with the same analyzer [3].

We have implemented the above mentioned approaches basically using GOLD Parser Builder and the C# programming language. First, we translated the formalism we created and built a special class. Then we modified this class to make it compatible with a pre-made web service [16]. Here is given the part of the main function of the web service:

```
[WebService(Namespace = "system")]
[WebServiceBinding(ConformsTo = WsiPro-
files.BasicProfile1_1)]
    [System.ComponentModel.ToolboxItem(false)]
    [System.Web.Script.Services.ScriptService]
    public class WebService : Sys-
tem.Web.Services.WebService
    {
        [WebMethod]
        public string My_Online_Parser(string
Start_Symbol)
        {
            own_parser_online.MyParser own_parser = new
own_parser_online.MyParser(Server.MapPath("/") +
"\\multy_sent.cgt");
            own_parser.Parse(Start_Symbol);
            return own_parser.return_result_to_main();

        }
    }
```

After these stages, everyone can use this system and also, can participate in its development (As you know, nowadays the perfect model of Georgian language still does not exist).

Based on the presented approaches, we have created a demo version of an additional module which checks the morphological correctness of the pronounced words and then

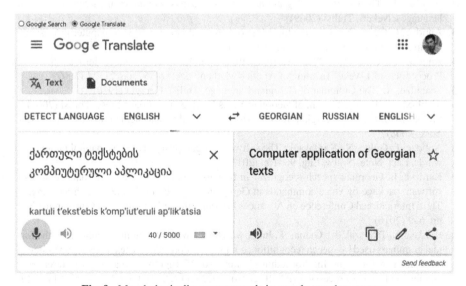

Fig. 9. Morphologically correct words in google translate system.

writes them in the appropriate fields of google search and google translate system. See Fig. 9.

This is very important, because often the words pronounced in Georgian are written in Latin letters in google translate system and in some cases even incorrectly.

6 Conclusion

Thus, created software is used to get all word-forms from unchanged part of the word, also to get the word-form in accordance unchanged part of the word and morphological categories (in the data base of roots at this moment are not all roots of Georgian words). If the presented approach works correctly for particular type of a root, then it will give a result without mistakes for all roots of the given type. It is very important for full parsing of all type of the words. Using our approach simplifies the lemmatization and stemming process. If the initial information required by the program algorithm and the structure is correctly entered into the database, we will get any kind of result without error. In our opinion, the presented paper will help to create a complete computer model of the Georgian language and will give inspiration to new research. The next steps will be linking the presented software with a pattern recognition abilities [8, 9, 10]. Software parts are written mainly in C# language and its design is realized by using visual studio. Also we used GOLD Parser Builder. Is should be noted that web version of our project is system independent and device independent.

References

1. Acho, A., Ulman, J.: The Theory of Parsing, Translation and Compiling, vol. 1. Englewood Cliffs, USA (1972)
2. Antidze, J.: Theory of formal grammars and languages, computer modeling of natural languages. Nekeri, Tbilisi (2009)
3. Antidze, J., Gulua, N.: Software tools for some natural language texts computer processing. Comput. Technol. Appl. 3(3), 219–225 (2012)
4. Antidze, J., Gulua, N.: Machine translation from georgian language into another. In: Proceedings of I. Vekua Institute of Applied Mathematics, vol. 57, pp. 21–34 (2007)
5. Shanidze, A.: The Grammar of Georgian Language, Tbilisi, Georgia (1930)
6. MelikiShvili, D.: The System of Georgian Verbs Conjugation. Logos Press, Tbilisi (2001)
7. MelikiShvili, D.: The Georgian Verb: A Morphosyntactic Analysis. Dunwoody Press, New York (2008)
8. Antidze, J., Gulua, N., Kardava, I.: The software for composition of some natural languages' words. Lect. Notes Softw. Eng. 96–100 (2013)
9. Kardava, I.: Georgian speech recognizer in famous searching systems and management of software package by voice commands in Georgian language. In: Conference Proceedings – Third International Conference on Advances in Computing, Electronics and Communication, pp. 6–9 (2016)
10. Kardava, I., Tadyszak, K., Gulua, N., Jurga, S.: The software for automatic creation of the formal grammars used by speech recognition, computer vision, editable text conversion systems, and some new functions. In: Proceedings Volume 10225, Eighth International Conference on Graphic and Image Processing (ICGIP 2016); 102251Q (2017). https://doi.org/10.1117/12.2267687

11. Melikishvili, D.: System of Georgian Verbs Conjugation. Logos Press, Tbilisi (2001)
12. Melikishvili, D.: On Georgian verb-forms classification and qualification principles. Probl. Linguist. **1**, 30–35 (2008)
13. Antidze, J., Gulua, N., Mishelashvili, D., Nukradze, L.: On complete computer morphological and syntactic analysis of Georgian texts. In: Reports of International Symposium – Natural Language Processing, Georgian Language and Computer Technologies, Institute of Linguistics of Georgian Academy of Sciences, Tbilisi, pp. 17–19 (2009)
14. http://www.sil.org/pcpatr/manual/pcpatr.html
15. http://www.sil.org/pckimmo/v2/pckimmo_v2.html
16. Kardava, I., Gulia, N., Toklikishvili, B., Meshveliani, N., Kvaratskhelia, T., Vetulani, Z.: Individual management of MySQL server data protection and time intervals between characters during the authentication process. Lecture Notes in Engineering and Computer Science, Newswood Limited – International Association of Engineers, pp. 213-217 (2021)

Community-Led Documentation
of Nafsan (Erakor, Vanuatu)

Ana Krajinović[1,2(✉)] (iD), Rosey Billington[2,3] (iD), Lionel Emil[2,4],
Gray Kaltaρ̃au[2,4], and Nick Thieberger[2,5] (iD)

[1] Heinrich-Heine-Universität Düsseldorf,
Universitätsstraße 1, 40225 Düsseldorf, Germany
ana.krajinovic@hhu.de
[2] ARC Center of Excellence for the Dynamics of Language, Canberra, Australia
[3] The Australian National University, Canberra, ACT 0200, Australia
rosey.billington@anu.edu.au
[4] Nafsan Language Team, Erakor Village, Vanuatu
[5] The University of Melbourne, Parkville, VIC 3010, Australia
thien@unimelb.edu.au

Abstract. We focus on a collaboration between community members
and visiting linguists in Erakor, Vanuatu, aiming to build the capacity
of community-based researchers to undertake and sustain documentation
of Nafsan, the local indigenous language. We focus on the technical and
procedural skills required to collect, manage, and work with audio and
video data, and give an overview of the outcomes of a community-led
documentation after initial training. We discuss the benefits and challenges of this type of project from the perspective of the community
researchers and the external linguists. We show that community-led documentation such as this project in Erakor, in which data management
and archiving are incorporated into the documentation process, has crucial benefits for both the community and the linguists. The two most
salient benefits are: a) long-term documentation of linguistic and cultural practices calibrated towards community's needs, and b) collection
of larger quantities of data by community members, and often of better quality and scope than those collected by visiting linguists, which,
besides being readily available for research, have a great potential for
training and testing emerging language technologies for less-resourced
languages, such as Automatic Speech Recognition (ASR).

We wish to thank all the speakers of Nafsan who participated in this documentation
project and we are also grateful for the feedback we received at the Vanuatu Languages
Workshop, 25–27 July 2018 in Port Vila, Vanuatu. We also benefited from discussions
at the 9th Language & Technology Conference, May 17–19, 2019 in Poznań, Poland,
and Language Technologies for All, 4–6 Dec 2019, UNESCO, Paris. This work has been
funded by the ARC Centre of Excellence for the Dynamics of Language (Australia)
(project ID: CE140100041) and the German Research Foundation DFG (MelaTAMP
project with number 273640553).

Z. Vetulani et al. (Eds.): LTC 2019, LNAI 13212, pp. 112–128, 2022.
https://doi.org/10.1007/978-3-031-05328-3_8

Keywords: Community-led language documentation · Technical training · Less-resourced languages · Technology for indigenous languages · Nafsan · Vanuatu · Automatic speech recognition

1 Introduction

There has been increasing recognition that greater collaboration between external linguists and language communities can be mutually beneficial, and aid language maintenance efforts (e.g. [8,12,33]). Approaches incorporating technical training and empowering people to undertake community-led projects are noted to be vital for inclusive collaboration (e.g. [48,49]). In Vanuatu, there are many examples of productive collaborations on language and cultural documentation projects (e.g. [1,18,32,36,43]). In this paper we describe one process of building community capacity to engage in language maintenance and corpus building through linguist-community collaboration. We focus on the community of Erakor, on the island of Efate, Vanuatu (Figs. 1 and 2),[1] near the capital, Port Vila.

Fig. 1. Location of Vanuatu and the island of Efate.

The language of the community in Erakor, as well as nearby Eratap and Pango, is Nafsan (also known as South Efate), a Southern Oceanic language with an estimated 5,000–6,000 speakers [26]. Nafsan is one of 130+ indigenous languages in Vanuatu, and is spoken alongside Bislama, an English-based creole, which is one of three official languages and a lingua franca across the archipelago. Education is mainly carried out in English and French. Vanuatu is undergoing an information and communications technology revolution [10,14], and around 86% of households now have home access to mobile networks [46]. Access to technologies other than mobile phones is still limited, though increasing, and both mobile and internet use is claimed to be linked to changing patterns of language

---

[1] All the maps in this article were produced by using the Generic Mapping Tools [47].

use, such as greater use of Bislama [45]. However, new modes of communication also offer new environments in which indigenous languages of Vanuatu can be used. For example, social media platforms like Facebook include a number of pages in which Nafsan is the main language. This is a positive development, as speakers are actively using Nafsan in its written form on social media, which is not done in other areas of life, where English, French, and Bislama predominate as written languages. In our collaboration, Facebook is also the main means of communication between the researchers and the community members.

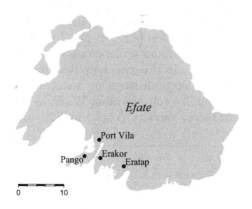

Fig. 2. The island of Efate indicating the places where Nafsan is spoken

Records of Nafsan extend back to the mid-1800s, in materials produced by missionaries (see [41]). Modern linguistic research began with a focus on the phonology and genetic classification of Nafsan (e.g. [11,25,44]). A comprehensive reference grammar of Nafsan has been produced by Thieberger [38], accompanied by corpus data, a book of stories [40], and a dictionary [39], which is regularly updated at community workshops and was published in 2021 [42]. All of this previous research laid the groundwork for the more recent activity, first by creating a corpus that new researchers could use to begin work on the language, and, second, by demonstrating a quid pro quo of returning materials to the village in forms that could be used there.

The main aim of this paper is to demonstrate ways in which linguists can support community efforts in language documentation and maintenance through building capacity, and how these collaborations can result in larger quantities of quality data. We describe the process of training community members in using technology for recording, transcribing and building a corpus (§2), and discuss the outcomes (§3) and the benefits and challenges (§4) of a documentation project undertaken by the third and fourth authors. We also identify ways that both the language community and the wider linguistics community can benefit from community-led documentation, especially if there is greater consideration of how the data will be conserved, archived, and made accessible in different formats (e.g. WAV files, time-aligned transcribed text etc.), which are used in linguistic research and in the development of language technologies. In §5 we argue that the community-led documentation leads to the collection of high-quality data in larger quantities than those collected by visiting researchers, and is thus of crucial importance for emerging language technologies for indigenous and less-resourced languages, such as Automatic Speech Recognition (ASR). We note some promising results with new speech recognition and forced-alignment technologies applied to Nafsan data. We conclude in §6.

2   Technical and Procedural Training

To build on the previous documentation and description of Nafsan, the first two authors (AK & RB) began fieldwork in 2017 in Erakor, aiming to collect new Nafsan data for targeted semantic and phonetic analyses (e.g. [3,4,21,22,30]). In the beginning of their field trip, they participated in a dictionary workshop in Erakor led by the fifth author (NT), focused on checking, correcting and adding to entries for the Nafsan dictionary through group discussions with community members. During the workshop sessions, it became clear that besides the work on the dictionary, there was community interest in collecting more narratives in Nafsan. NT gave a Zoom H1N recorder to a community member, GK, the fourth author, who partnered with the third author, LE, to develop ideas for a recording project. Given that there was an intention of data collection in the absence of linguists, AK & RB realized that there was a need for training in data collection and management. During their semantic and phonetic experiments, they started familiarizing GK and LE with the process of making a recording, transcribing it, and managing the data. GK & LE assisted AK & RB in different types of fieldwork tasks, such as transcription and video recording, and a computer was made available for them to use for independent transcription, using ELAN [37]. As GK & LE became more comfortable with transcribing pre-segmented audio files in ELAN, AK & RB organized more formal training of linguistic tools.

The training focused on four indispensable activities in a language documentation workflow: planning and discussing a recording with participants (including archival access conditions), making a recording, data management, and transcription. For the recording process, GK & LE practiced using the Zoom H1N and including basic spoken metadata at the beginning of each audio recording, and we discussed some basic principles of video recording. The data management component was slightly more challenging as it involved familiarizing the community members with the use of spreadsheets and file-naming practices. We practiced the workflow as a routine of making a recording, transferring it to a computer, entering metadata in a spreadsheet, and backing up the data. This process was easily followed as each activity was understood as an essential part of the workflow. The last step was learning how to use ELAN (see Fig. 3). Until this point, GK & LE were already familiar with transcribing spoken Nafsan in a single pre-segmented tier. These skills were extended to creating a new file and importing audio files together with a template [17] that facilitates exporting into FieldWorks [35], in which it can be semi-automatically glossed. The use of a more

Fig. 3. Training in ELAN transcription

complex template required some explanation of the hierarchical organization of tiers, e.g. that the translation tier depends on the tier of the original text. The focus of transcription efforts was filling in the first tier with orthographic Nafsan, as in Fig. 4. In this training we focused on highlighting the structure of the workflow, and making sure that the community members understood the importance of data management that follows the creation of each recording. Understanding the technical aspects of using different types of software proved to be relatively easy. However, documenting instructions in a simple text was also helpful.

Our training also included a discussion on the importance of explicitly explaining to the speakers who will be recorded and how the recordings will be stored and used, and making sure it is understood that they have a choice to select their own preferences regarding the access rights to the recording. Through their previous collaborations with NT, many community members, including GK & LE, were aware of the concept of an archive and the benefits of archiving the collected language data for posterity, as NT has been providing the community with their own copy of previously collected data, both locally and through online open access to the PARADISEC archive.[2] All the recordings made by AK [24] and RB [6] are also archived in PARADISEC. All the physical language materials created as a result of our research and collaboration, such as the storyboard booklet in Nafsan [23], have also been deposited in the *Vanuatu Kaljoral Senta* (Vanuatu Cultural Centre).[3]

3   Outcomes of the Community-Led Documentation

Between July 2017 and June 2018, GK and LE, as community researchers, collected audio and video data in 21 recording sessions. Some sessions were recorded

---

[2] Available at https://www.paradisec.org.au.
[3] https://vanuatuculturalcentre.gov.vu.

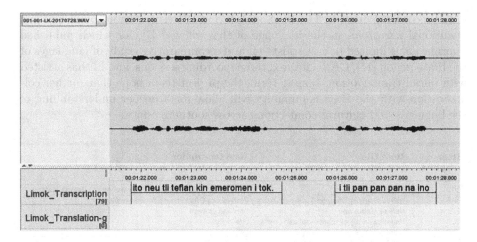

Fig. 4. Orthographic transcription of *Nap̃re nig Taler* (a story about a demon) told by Limok Kaltap̃au (GKLE-001)

only using either video or audio, and others were recorded with simultaneous video and audio, for later synchronization. In total, the collected data comprised 17 audio files totaling five and a half hours, and 25 video files totalling four and a half hours. Recording sessions took place primarily in Erakor, but some took place in Eton, a village further to the north on the coast of Efate, with strong ties to Erakor. The recordings were all of natural speech and related to diverse topics, driven by the interests of the community researchers and the community members they engaged with for their project. Among the recordings which were primarily audio, two were 'kastom' (traditional) stories, four were personal life histories, and three were stories about people and events in Erakor and Eton (see Fig. 5). Among the recordings which were primarily video, there was a story about the first permanent house in Erakor, and many videos demonstrating techniques for weaving baskets, fans and mats using coconut and pandanus leaves. The community researchers chose weaving as a focal topic because of concern that traditional weaving skills are not being passed on to younger generations, and a desire to document these skills and develop educational resources. All of the recordings have been archived in PARADISEC[4] with accompanying metadata, and apart from one, all are open-access [20].

Good progress has also been made on transcribing these recordings in ELAN: seven recordings have been fully transcribed, one partially transcribed, and one long recording has been fully segmented and made ready for transcription. The project is ongoing, and future plans include engaging more community members as participants, recording material for a documentary about Nafsan, and identifying ways to use collected videos for educational purposes. The skills gained by GK and LE have also since been extended to new projects; building on the

---

[4] https://catalog.paradisec.org.au/collections/GKLE.

118   A. Krajinović et al.

initial recordings made by GK in Eton, GK and RB visited Eton to record some traditional narratives in the language of that village[5] [7], for which published information is limited to a wordlist [44] and a comparative study of languages of the Efate region [11]. GK facilitated these recording sessions, and LE has assisted with initial transcription. Longer term, the project team hope that further collaboration with the Eton community will allow for a deeper understanding of the linguistic and cultural connections across southern Efate.

Item Identifier (e.g.	Item Title (e.g. Introductory Materials)	Item Description (e.g. Four text stories for interviews)
001	Napre nig Taler	Story about Taler's encounter with the erakor demon (natopu)
002	Natrauswen nig Oftau go Tiawi nig tutufur	Story about Oftau and his friend from tutufur
003	Nafsan nfapuswen	Farewell speech
004	Natrauswen nig ati touraan teflaan I patlas	Life story
005	Natrauswen ni limok go kaltpau	Life story
006	Natrauswen ni tesa nmatu ralim iskei go	Custom story
007	Natrauswen namolien ni apu abel naar	History of Abel Naar , an evangelist to Eton village
008	Natrausen ni Linmas kalsilik	Life story
009	Linmas i traus natrauswen nig apu samuel	Story of late chief samuel
010	Nfauwen ni likat	Weaving instruction
011	Natiltaewen ni nafkaworwen	Explanation of the nafsan word "pkawor"
012	Nfauwen ni niif	Fan weaving
013	Nfawen ni naal pool	Ball basket weaving
014	Nfawen ni toofrak	Plate weaving
015	Nafeifeien nig naal	Display of baskets
016	Nfawen ni naal	Basket weaving
017	Nfawen ni likat	Likat basket weaving
018	Napnotien ni likat	Finishing likat
019	Natfagien ni nasum pei ni natkoon Erakor	Building of the first permanent house in Erakor
020	Nfawen ni tefkau	Tefkau weaving
021	Nfawen ni tefkau	Tefkau weaving

Fig. 5. A part of the metadata of the recordings made by GK and LE

4 Benefits and Challenges

4.1 Benefits: The Engagement of Community Researchers Improves the Results of Language Documentation

From the perspective of the community researchers, there are a number of advantages to language and cultural documentation projects led by community members. One clear advantage is first-hand knowledge of the language. In most documentation projects, linguists are visitors, and while they may acquire the language of study to varying extents, in most cases they are unlikely to acquire competence approaching that of native speakers. Native knowledge of Nafsan facilitates more accurate and efficient transcription, and also facilitates the process of undertaking recording sessions with different community members.

Community researchers also have a significant advantage in that they have better knowledge of the linguistic and cultural practices which may feature in documentation recordings. They are well-placed to decide which activities are

---

[5] The language is also known as Eton.

better documented with video rather than audio, based on the type of activity and also what participants are most comfortable with. They are also able to use their knowledge of particular activities to more effectively plan and capture these using video. For example, if the goal of a recording is to document the process for weaving a particular type of basket (e.g. Fig. 6), and the community researchers are familiar with what this entails, they can choose the most appropriate framing and zoom level at different stages, so that viewers can identify exactly what the participant is doing. In comparison, an external researcher may focus on capturing the whole scene in every frame, perhaps to include gestures or background interlocutors, but this will be less useful to someone wanting to watch the recording to study the weaving technique. Community researchers are also better able to identify which activities are most important to document, and of the greatest interest to the community, particularly in contexts where a project aims to support language and cultural maintenance.

Another benefit of community-led documentation is the quantity of collected data. In a relatively short time frame, GK and LE were able to collect large quantities of spoken Nafsan data, without any interference of another language, and with a minimized observer's bias. On the other hand, visiting researchers are outsiders, who maximize the observer's bias, and often communicate in another language, adding to the complexity of linguistic influences in the recorded data. Moreover, linguists often collect language data catered towards their specific research questions, either through experiments or elicitation, without necessarily prioritizing community interests. Usually only certain types of data commonly collected for research, such as telling of traditional stories, can be made immediately useful for the language community. Linguists typically undertake additional activities outside of their research topics in order to produce materials for community-wide use, such as the Nafsan dictionary [42]. In contrast, the Nafsan data collected by the community researchers can serve multiple purposes from the start: among others, education in language or cultural practices, entertainment, promotion of the local culture and products, linguistic research, and use in developing language technologies for indigenous and less-resourced languages (see §4).

4.2   Challenges: Sustainability

Challenges noted by GK & LE relate to both the practicalities of using equipment and technology as well as the logistics of managing a project. While the actual transcription process in ELAN was relatively manageable, making a new .eaf file could be difficult. The template provided by AK & RB was helpful, and consistently used, but the main issue was remembering how to navigate the ELAN interface and access the template when starting a new transcription. Sharing one laptop also limited the ability of the community researchers to undertake transcription and data management tasks at the times most convenient to them. Similarly, it was often difficult to find time to spend on recording and transcription among other family and community commitments. It was also not always easy to find people who were willing and available to participate.

Fig. 6. Marian Kalmary weaving *naal pool* (GKLE-013)

In some cases people were interested but had limited time, and in other cases people were intimidated by the prospect of being in an audio or video recording. A particular challenge when recording video was shakiness caused by camera movement. Activities such as weaving required GK & LE to be able to move around in order to best capture different parts of the process, and this proved to be difficult to do without excessive movement caused by using a handheld video camera. Some of the challenges noted here have since been addressed, for example by acquiring a tripod to reduce camera shakiness even if carrying by hand, and an additional laptop, allowing an easier division of tasks between the two community researchers.

The internet in Vanuatu is most readily accessible via mobile data. While this means that the internet is theoretically available even in many remote areas, in our experience, this has been both a benefit and a challenge. While GK & LE can stay in contact with researchers and even transfer ELAN .eaf files, the expense of mobile data and limited connection speed means sharing actual audio and video recordings is hard. Thus, the sharing of the recordings still needs to happen in person. This problem has been especially challenging during the COVID-19 pandemic, which made it impossible for researchers to travel to Vanuatu.

The COVID-19 pandemic has shown that now more than ever we need to improve the sustainability of these types of collaborations. As researchers might be unable to travel to distant fieldwork sites, where communities are especially vulnerable to potential outbreaks, these communities need to be able to independently carry out language documentation and associated activities in order to continue making progress towards particular community goals. However, so far it has been hard to ensure the sustainability of such collaborations, primarily because of the lack of a fast and affordable internet connection that would allow for long-distance communication, and the lack of resources to maintain the hardware necessary for language and cultural documentation. Nevertheless, we have been able to transfer some of the discussion regarding our ongoing research on Nafsan to Facebook chats. The information collected in these chats typically concerns words that can be included in the dictionary [42] and grammaticality judgments for semantic research [22]. Unfortunately, phonetic data cannot be collected this way.

4.3   Benefits Outweigh the Challenges

From the perspective of the visiting researchers, there is no doubt that building local capacity to undertake language and cultural documentation offers benefits in terms of both the scale and quality of documentation. The community-led project contributes to a more comprehensive record of Nafsan, and allows for new research questions to be explored and existing research questions to be addressed more thoroughly. Importantly, the resulting materials are more representative of community priorities and interests, and more useful for developing materials supporting language and cultural maintenance. These and many other ways that collaborative and community-led projects benefit both the specific goals of a community, and the scientific endeavor of linguistic research, have been discussed in detail elsewhere (e.g. [8,12,33]). An additional benefit of the particular approach taken in the current project is that data management and metadata collection was built into the initial training, as were strategies for discussing archiving and access conditions with community participants. While data and metadata management has required some ongoing support, and can be difficult when internet access is limited, the result is that not only is there a rich set of materials collected by the community researchers, but that these materials have been easily archived along with details of their content, and are accessible and therefore usable by others, including community members who have some previous experience accessing Nafsan materials collected by NT via PARADISEC. Other researchers discussing collaborative language documentation acknowledge that there can be logistical, institutional, and interpersonal challenges to the sustainability of community-led projects (e.g. [49]), but we find, as they do, that the benefits of community-led documentation far outweigh the challenges.

5   Increased Potential for Applications of Language Technology to Less-Resourced Languages

One problem arising, which may not seem like a problem at first, is too much data. Scaling up documentation in the way described here leads to more audio and video recordings than would otherwise have been collected thus far, but not all have been transcribed. While engaging community members in transcription is often seen as a way to speed up the process, and to transcribe a higher percentage of recordings than a solo linguist (with less fluency in the language) could manage, community members are generally not able to work on these tasks to the exclusion of other responsibilities, or other interests within a project. Manual transcription, whether it is done by a native speaker or otherwise, is also extremely time consuming, and depending on the familiarity with the language and the nature and level of detail of the transcription, can take anywhere from 2.5 h [16] to 200 h [9] per hour of recorded speech. As long as transcription is fully reliant on human effort, there remains an issue of the 'transcription bottleneck' [9], whereby more data is recorded than can feasibly be transcribed and added to a corpus within time and resource limitations.

At the same time, corpora are most useful for linguistic research, and many community goals, when they are transcribed. For instance, the Nafsan data that has been transcribed has been used in several cross-linguistic projects focusing on different linguistic domains, e.g. phonetics and morphology [28], inferring grammar from texts [19], and semantics [30, 31]. Finding ways to improve transcription workflows is therefore vital to being able to extend the scale and usability of language corpora. Speech and language technologies, such as tools for automatic speech recognition and automatic transcription, can significantly aid the process of transcribing spoken language. However, there is limited availability of usable tools of this sort for many languages (e.g. [2]). In part, this is due to the limited quality and scope of language material available for less-resourced languages.

Regarding quality, one challenge in developing ASR for less-resourced languages is the sometimes variable audio quality of language documentation corpora, where recordings are often made in noisy fieldwork conditions. Recent discussions argue that field linguists should modify their practice to assist the task of machine learning, for example by making high-quality recordings using head-mounted microphones [34]. To add to this discussion, we note that community researchers may be better placed than visiting linguists to collect high-quality audio recordings, given appropriate training opportunities. External researchers typically visit for a set time frame, and generally have specific goals, for example related to collecting a certain number of hours of particular data types, with a range of participants. This means that recordings are often undertaken opportunistically, where and when community members are available, and it is not always possible to have a great deal of control over factors such as environmental noise. Figure 7 shows a sample waveform and spectrogram of a recording made by the second author in one such opportunistic setting. The recording was made with a hypercardioid head-mounted microphone in a location with as much sound attenuation as possible within the available options, but unfortunately took place exactly at dusk, which meant substantial noise from a flock of birds settling in to roost in a tree nearby. As can be seen, the signal-to-noise ratio is not ideal; there is a lot of additional noise in the higher frequency range. While this recording would still be fairly usable for phonetic analyses of fundamental frequency or duration, it would be less useful for analyses of fricative energy or formant transitions, and would also present more of a challenge to ASR.

In comparison, community researchers are able to be more flexible in their project schedules, and can choose to make audio recordings in a quiet environment at a preferred time of day, and to make video recordings under optimal weather and lighting conditions. They may also be better able to negotiate a recording situation which prioritizes both the comfort of the participant and the quality of the recording (in ways that visitors are not always equipped to do appropriately). Figure 8 shows a sample waveform and spectrogram[6] of a recording collected by GK. He chose to record this late at night, after the noise of people, birds and vehicles and generators had stopped, in a small room with closed windows. He also sat close to the speaker in order to hold the recorder at a constant and appropriate distance from her mouth. This recording was made with the inbuilt stereo microphone of the Zoom H1N, which, being less directional, would pick up more background noise than the microphone used for Fig. 7, but as can be seen this is clearly the cleaner recording. Recordings like this are much better suited to training of ASR models.

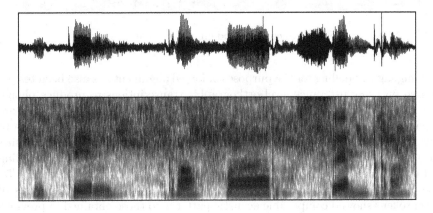

Fig. 7. Recording made in noisy conditions

The amount of data available for developing new speech and language tools is another challenge. Training a language model for adequate speech recognition generally requires very large speech corpora, but these are not typically available for languages which are relatively under-described. In recent years, there has been increasing interest in finding ways to adapt automatic speech recognition and transcription methods to work more effectively with small corpora of the sort typically collected during language documentation. Preliminary tests of developing a speech recognition model and semi-automated transcription for Nafsan have been undertaken using the Kaldi speech recognition engine [29], via the in-development Elpis pipeline, and show promising results [15]. A model based on just 3 h of audio as training data was applied to untranscribed data and returned a word error rate of 42.7%; a 'reasonably decent' result for a first pass using sample data with limited coverage and limited tuning of parameters in the pronunciation model.

---

[6] Figures 7 and 8 correspond to samples of 200 ms; spectrograms show frequencies up to 5000 Hz with a 60 dB dynamic range.

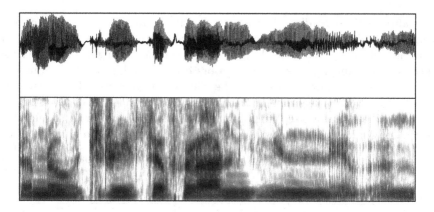

Fig. 8. Recording made in quiet conditions

Language modelling for the purposes of forced alignment has also been tested, focused on using utterance-level orthographic transcriptions to produce phone-level and word-level alignments. These more granular annotations of speech are of particular interest in phonetic research, where it is necessary to be able to take acoustic measurements with reference to individual speech sounds. Using the Montreal Forced Aligner (MFA) [27], which also uses Kaldi, speech modelling and forced alignment was undertaken based on just over 2 h of Nafsan data [5]. The output phone alignment was very accurate, and preliminary analyses of vowel tokens showed comparable acoustic patterns to those obtained in previous experimental datasets. It is important to note that the quality of the output of automatic transcription and forced alignment processes depends on the quality of the data used in the modelling, not just in terms of audio but also the accompanying transcriptions. In cases where the amount of data is limited, language corpora which have been carefully developed in collaboration with community members, including contributions to transcriptions and analyses, will lead to better results, and in turn aid the expansion and enrichment of the corpus.

The potential offered by these kinds of technologies, as they continue to be refined for use in documentation contexts, is clear. There is also scope to draw on language models for a given language to develop models for related or phonetically similar languages which may have even more limited speech material available. This is currently being explored for several languages of the Efate region [5]. In addition, there are various natural extensions of these speech and language technology toolkits which would not only further aid data processing and analysis, but also better support the use of less-resourced languages in digital domains [13].

6   Conclusion

In this paper we described the process and outcomes of building capacity for community-led documentation in Erakor, Vanuatu. We showed that archiving research materials provides a base for reciprocity with the speakers of the language, and then permits further research to be built on existing work in ways that were not previously possible. We highlighted the benefits of direct community involvement in language documentation and maintenance efforts for both the community and the external linguists. We showed that the community researchers are able to contribute to overall larger quantities of linguistic data than that collected only by visiting linguists during fieldwork. Moreover, in some cases the data gathered by community researchers is better than that collected by external linguists, in terms of either content or audio quality. This happens mainly for two reasons: a) the community members are best placed to decide what linguistic and cultural practices to document, and how, thus making the resulting materials more useful for the community, and b) they may have greater choice in and control over recording conditions, resulting in better acoustic quality of audio recordings (and image quality in video recordings). The former aspect is crucial for supporting language maintenance efforts and the latter aspect allows for favorable results from applications of ASR technologies to less-resourced languages. The potential scope for language technology applications is expanded when data of good technical quality is combined with well-maintained corpus materials. More generally, both linguists and the community benefit greatly from an archival collection of the materials, which become available for linguistic research and to the community now and in the future.

References

1. Barbour, J.: Neverver: A study of language vitality and community initiatives. In: Florey, M. (ed.) Endangered languages of Austronesia, pp. 225–244. Oxford University Press, Oxford (2010)
2. Besacier, L., Barnard, E., Karpov, A., Schultz, T.: Automatic speech recognition for under-resourced languages: a survey. Speech Commun. **56**, 85–100 (2014)
3. Billington, R., Fletcher, J., Thieberger, N., Volchok, B.: Acoustic evidence for right-edge prominence in Nafsan. J. Acoust. Soc. Am. **147**(4), 2829–2844 (2020). https://doi.org/10.1121/10.0000995
4. Billington, R., Thieberger, N., Fletcher, J.: Nafsan. J. Int. Phonetic Assoc. 1–21 (2021). https://doi.org/10.1017/S0025100321000177. Published online by Cambridge University Press, 10 August 2021
5. Billington, R., Stoakes, H., Thieberger, N.: The Pacific Expansion: Optimizing phonetic transcription of archival corpora. In: Proceedings of INTERSPEECH 2021, pp. 2021–2167. International Speech Communication Association, Brno (2021). https://doi.org/10.21437/Interspeech
6. Billington, R.: Rosey Billington Nafsan materials. Collection BR1 at catalog.paradisec.org.au [Open Access] (2017). https://dx.doi.org/10.26278/GXDM-J159
7. Billington, R.: Recordings of the language of Eton. Collection BR2 at catalog.paradisec.org.au [Open Access] (2019). https://dx.doi.org/10.26278/TRS1-XP03
8. Bowern, C., Warner, N.: 'Lone Wolves' and collaboration: A reply to Crippen & Robinson (2013). Lang. Documentation Conserv. **9**, 59–85 (2015)

9. Brinckmann, C.: Transcription bottleneck of speech corpus exploitation. In: Lyding, V. (ed.) Proceedings of the Second Colloquium on Lesser Used Languages and Computer Linguistics (LULCL II) Combining Efforts to Foster Computational Support Of Minority Languages, pp. 165–179. Europäische Akademie (EURAC book, 54) (2009)

10. Cave, D.: Digital islands: How the Pacific's ICT revolution is transforming the region. Tech. rep. Lowy Institute for International Policy (2012)

11. Clark, R.: The Efate dialects. Te Reo **28**, 3–35 (1985)

12. Czaykowska-Higgins, E.: Research models, community engagement, and linguistic fieldwork: Reflections on working within Canadian Indigenous communities. Lang. Documentation Conserv. **3**, 15–50 (2009)

13. van Esch, D., Foley, B., San, N.: Future directions in technological support for language documentation. In: Proceedings 3rd Workshop on Computational Methods for Endangered Languages, vol. 1, pp. 14–22 (2019). https://doi.org/10.33011/computel.v1i.341

14. Finau, G., et al.: Social media and e-democracy in Fiji, Solomon Islands and Vanuatu. In: Twentieth Americas Conference on Information Systems. Association for Information Systems, Savannah (2014). http://hdl.handle.net/1885/75381

15. Foley, B., et al.: Building speech recognition systems for language documentation: the CoEDL endangered language pipeline and inference system (ELPIS). In: 6th International Workshop on Spoken Language Technologies for Under-Resourced Languages, pp. 200–204 (2018)

16. Foley, B., Durantin, G., Rakhi, A., Wiles, J.: Transcription survey. In: Paper Presented at the Australian Linguistic Society Annual Conference (2019). Retrieved on 15 October 2021. http://bit.ly/ALS-survey

17. Gaved, T., Salffner, S.: Working with ELAN and FLEx together: An ELAN-FLEx-ELAN teaching set (2014). Retrieved on 15 October 2021. https://www.scribd.com/document/357359102/Working-with-ELAN-and-FLEx-together-pdf

18. Guérin, V., Lacrampe, S.: Trust me, I am a linguist! Building partnership in the field. Lang. Documentation Conserv. **4**, 22–33 (2010)

19. Howell, K.: Inferring grammars from interlinear glossed text: Extracting typological and lexical properties for the automatic generation of HPSG grammars. Ph.D. thesis, University of Washington (2020)

20. Kaltap̄au, G., Emil, L.: Nafsan recordings (GKLE), Digital collection managed by PARADISEC (2017). http://catalog.paradisec.org.au/collections/GKLE

21. Krajinović, A.: Comparative study of conditional clauses in Nafsan. In: Boerger, B.H., Unger, P. (eds.) SIL Language and Culture Documentation and Description 41, Proceedings of COOL 10, pp. 39–61. SIL International (2018). https://www.sil.org/resources/publications/entry/82335

22. Krajinović, A.: Tense, mood, and aspect expressions in Nafsan (South Efate) from a typological perspective: The perfect aspect and the realis/irrealis mood. Ph.D. thesis, Humboldt-Universität zu Berlin and The University of Melbourne (2019). https://minerva-access.unimelb.edu.au/handle/11343/237469

23. Krajinović, A., et al.: Natrauswen ni tesa nen rumtri ki nafsan ni Erakor. ISBN 978-1721654246 (2018)

24. Krajinović, A., (collector).: Nafsan recordings (AK1). Digital collection managed by PARADISEC. [Open Access] http://catalog.paradisec.org.au/collections/AK1 (2017). https://doi.org/10.4225/72/5b2d1d0a315a2

25. Lynch, J.: South Efate phonological history. Oceanic Linguistics **39**(2), 320–338 (2000)

26. Lynch, J., Ross, M., Crowley, T.: The Oceanic Languages. Routledge, London (2002)
27. McAuliffe, M., Socolof, M., Mihuc, S., Wagner, M., Sonderegger, M.: Montreal Forced Aligner: Trainable text-speech alignment using Kaldi. In: Lacerda, F., House, D., Heldner, M., Gustafson, J., Strombergsson, S., Wlodarczak, M. (eds.) Proceedings of Interspeech 2017, pp. 498–502. ISCA, Stockholm (2017)
28. Paschen, L., Delafontaine, F., Draxler, C., Fuchs, S., Stave, M., Seifart, F.: Building a time-aligned cross-linguistic reference corpus from language documentation data (DoReCo). In: Proceedings of the 12th Language Resources and Evaluation Conference, pp. 2657–2666 (2020)
29. Povey, D.: The Kaldi speech recognition toolkit. In: Proceedings of IEEE 2011 Workshop on Automatic Speech Recognition and Understanding. IEEE Signal Processing Society (2011)
30. von Prince, K., Krajinović, A., Krifka, M.: Irrealis is real. Language (in press)
31. von Prince, K., Krajinović, A., Krifka, M., Guérin, V., Franjieh, M.: Mapping Irreality: storyboards for eliciting TAM contexts. In: Gattnar, A., Hörnig, R., Störzer, M., Featherston, S. (eds.) Proceedings of Linguistic Evidence 2018: Experimental Data Drives Linguistic Theory. University of Tübingen, Tübingen (2019). https://publikationen.uni-tuebingen.de/xmlui/handle/10900/87132
32. Regenvanu, R.: Afterword: Vanuatu perspectives on research. Oceania 70(1), 98–100 (1999)
33. Rice, K.: Documentary linguistics and community relations. Lang. Document. Conserv. 5, 187–207 (2011)
34. Seifart, F., Evans, N., Hammarström, H., Levinson, S.C.: Language documentation twenty-five years on. Language 94(4), e324–e345 (2018)
35. SIL: Fieldworks Language Explorer (FLEx) 8.3 (2018). Retrieved from 15 October 2021. https://software.sil.org/fieldworks/
36. Taylor, J., Thieberger, N. (eds.): Working together in Vanuatu: Research Histories, Collaborations, Projects and Reflections. ANU Press, Canberra (2011)
37. The Language Archive: ELAN (Version 5.2) [Computer software]. Nijmegen: Max Planck Institute for Psycholinguistics (2018). Retrieved from 15 October 2021. https://tla.mpi.nl/tools/tla-tools/elan/
38. Thieberger, N.: A grammar of South Efate: An Oceanic Language of Vanuatu. University of Hawai'i Press, Honolulu (2006)
39. Thieberger, N.: A South Efate Dictionary. University of Melbourne, Parkville (2011). https://minerva-access.unimelb.edu.au/handle/11343/28968
40. Thieberger, N.: Natrauswen nig Efat: Stories from South Efate. University of Melbourne, Parkville (2011)
41. Thieberger, N.: Guide to the Nafsan, South Efate collection (2021). Retrieved from 15 October 2021. https://www.nthieberger.net/sefate.html
42. Thieberger, Nicholas with Members of the Erakor Community: A Dictionary of Nafsan, South Efate, Vanuatu: M̃pet Nafsan ni Erakor. Oceanic Linguistics Special Publications No. 41, University of Hawaii Press, Honolulu (2021)
43. Tryon, D.: Ni-Vanuatu research and researchers. Oceania 70(1), 9–15 (1999)
44. Tryon, D.T.: New Hebrides Languages: An Internal Classification. Pacific Linguistics, Canberra (1976)
45. Vandeputte-Tavo, L.: New technologies and language shifting in Vanuatu. Pragmatics 23(1), 169–179 (2013)
46. Vanuatu National Statistics Office: 2016 Post-TC Pam Mini Census Report. Tech. rep., Ministry of Finance & Economic Management, Port Vila, Vanuatu (2017)

47. Wessel, P., et al.: The generic mapping tools version 6. Geochem. Geophy. Geosyst. **20**, 5556–5564 (2019). https://doi.org/10.1029/2019GC008515
48. Yamada, R.M.: Collaborative linguistic fieldwork: Practical application of the empowerment model. Lang. Documentation Conserv. **1**, 257–282 (2007)
49. Yamada, R.M.: Training in the community-collaborative context: A case study. Lang. Documentation Conserv. **8**, 326–344 (2014)

Design and Development of Pipeline of Preprocessing Tools for Kazakh Language Texts

Madina Mansurova[1]($\boxtimes$) ![iD], Vladimir B. Barakhnin[2,3] ![iD], Gulmira Madiyeva[1] ![iD],
Nurgali Kadyrbek[1] ![iD], and Bekzhan Dossanov[1] ![iD]

[1] Al-Farabi Kazakh National University, 71 al-Farabi Avenue, Almaty, Kazakhstan
madina.mansurova@kaznu.edu.kz
[2] Federal Research Center for Information and Computational Technologies, Lavrentiev
Avenue, 6, 630090 Novosibirsk, Russia
bar@ict.nsc.ru
[3] Novosibirsk State University, Pirogova Street, 1, 630090 Novosibirsk, Russia

Abstract. Nowadays, the Kazakh language belongs to the category of less-resourced languages, as there is a small number of resources developed and accessible to a wide range of users, such as text corpora, electronic dictionaries, morphological analyzers, thesauri, which allow to analyze text documents. The aim of this work is the design and development of pipeline of preprocessing tools for media-corpus of the Kazakh language. Media-corpus is hosted by al-Farabi Kazakh National University and serves linguists as an empirical basis for research in the contemporary written Kazakh language. The development of pipeline of preprocessing tools for media-corpus, the lexical and grammatical features of the Kazakh language were analyzed, on the basis of which the composition of the fundamental rules for changing the words (inflection) of the Kazakh language was determined. In the process of research, the tools for generation and lemmatization of the word forms of the Kazakh language were created. The proposed tools can be applied at the stage of morphological analysis in the systems of automatic analysis of the texts, in the creation of thesauruses and ontologies. For the case of the presence of homonymy, the template method was used, which allow to reduce the level of homonymy.

Keywords: Kazakh language · Pipeline · Lemmatization · Morphological model · Preprocessing tools

1 Introduction

The natural language processing (NLP) of documents is one of the most important tasks of modern information technologies. The results of NLP are used in a wide variety of fields: from the extraction of new information and knowledge, from the scientific publications in order to create information and analytical systems to support the scientific activities, to the analysis of the blogs in order to study the consumer ratings of a particular product. The modern approaches to the solving of NLP tasks provide for pipeline

Z. Vetulani et al. (Eds.): LTC 2019, LNAI 13212, pp. 129–142, 2022.
https://doi.org/10.1007/978-3-031-05328-3_9

document processing, which includes all the stages of work with a document – from its preprocessing in order to bring an Internet document (especially blog posts) to the norms of the literary written language, to at least its coordinate indexing and syntax analysis.

Within the framework of the Kazakh linguistics and applied linguistics, in Kazakhstan, of special interest is the study and development of the National Corpus of the Kazakh language due to insufficient development of the problems in this field. Despite the achievements in this field (the attempt to compile a corpus with the necessary markup, the presence of scientific investigations in the form of monographs, theses, text-books on the styles of the Kazakh language, works of a comparative character analyzing the differences in colloquial and literary languages, studies on its separate aspects) [1–5], limits of investigations do not go beyond the frames of traditional linguistics, restricting the attempts on development of the corpus or reducing them to mechanistic detection of lexical, phonetic and other differences of the Kazakh language. In [6] the authors proposed media-corpus of the Kazakh language which is hosted by al-Farabi Kazakh National University and serves linguists as an empirical basis for research in the contemporary written Kazakh language (http://corpus.kaznu.kz/). In this work, the authors presented design and development of preprocessing tools for this media-corpus of the Kazakh language, which will be made available to the public. The first task is the creation of automatic lemmatization tool. To solve the task, the lexical and grammatical features of the Kazakh language were analyzed, on the basis of which the composition of the fundamental rules for changing the words of Kazakh was determined. The second task is creation of automatic Kazakh word-forms generator tool. The developed algorithms can be applied at the stage of morphological analysis, in the construction of thesauri, thematic dictionaries, and ontologies. The third task is the development of morphological disambiguation tool for Kazakh texts. The idea of the method used in the study is to subordinate the structure of sentences to certain general laws. Despite the large variety of semantic parts of sentences, the structure of the sentences is limited and can be represented as a set of mathematical patterns. Consequently, when analyzing sentences, it becomes possible to access the tool, which is a mathematical pattern of the structure of the sentence. We tried to cope with the morphology of the Kazakh language by considering the structure of simple sentences. As an advantage of the method, various statistical indicators can be easily integrated as a determining parameter and show the possibility of improving the quality of results.

In this paper, we describe the Kazakh language pipeline, created by the researchers of Al-Farabi Kazakh National University (Almaty, Kazakhstan). Let's note that this analyzer works with the Cyrillic alphabet that is currently presented for the Kazakh language.

2 Morphological Model of the Kazakh Language

Kazakh belongs to the group of agglutinative languages in which the dominant type of inflection or inflection is agglutination, i.e. the formation of derived words and grammatical forms by attaching affixes (suffixes and endings) to the root or base of a word that have strictly defined grammatical and derivational meanings. For instance, a suffix is first added to nouns, then the ending of the plural, next the end of the possessive form, the case ending, and only after that the personal ending is added. In the case of adjectives, if they act as nouns, the word forms for these words are formed in the same way as for nouns, if they denote features, then they become modifiers and are immutable. In the case of verbs, if necessary, negative suffixes are added to create a negative form, then endings, that define the time of the verb, are added. Verbal forms in the form of participles can substantivize, and the inflection forms characteristic of nouns can be applied to them.

For each language, there are orthoepy and spelling norms of words, which consist of the rules of continuous, hyphenated, and separate spelling of words, rules for the use of capital letters and graphic abbreviations, etc. Within the framework of this project, orthoepy features will play a special role, since the Internet users, whose messages will be used for analysis tasks, often do not follow grammatical rules and orthoepy norms, as the speech, that is spoken spontaneously, significantly differs from, that prepared by certain features: redundancy (presence of repetitions, clarifications, explanations); saving speech utterances, ellipses.

Particular attention should be paid to such features of the Kazakh language as the phenomena of assimilation, dissimilation, lingual and labial synharmonism. It is through these processes and the linguistic laws associated with them that the formation of word forms and the selection of lemmas (the initial form of the word) are carried out. Based on the analysis of the patterns of formation of the word forms of Kazakh, the following set of basic rules was identified:

- In Kazakh, words never end with a sonorous consonant letter: voiced consonants *б, г, ғ, д, ж* (*b, g, ğ, d, j*), that are at the end of a word, are formed as a result of the replacement of pairing deaf consonants *n, к, қ, m, ш* (*p, k, q, t, ş*) by voiced ones before adding an end with a vowel. For example: *қазақ - қазағым* (qazaq - qazağım) (Kazakh – my Kazakh);
- hardness and softness of words are determined by the vowel sounds of the word, the rule of harmoniousness operates regarding the vowel sound in the last syllable for words borrowed from Arabic and Persian languages. Along with this, the form of the vowel sound *y* in the last syllable is determined depending on the vowel sound of the previous syllable. The rule of harmoniousness must be observed when affixes are added successively to the word.

3   The Structure of Pipeline of Preprocessing Tools

The general trends in the development of NLP algorithms are studied by the methods of bibliographic analysis [7]. However, the specific algorithms for the working with text in natural language is noticeably different depending on which class this particular language belongs to: analytical, inflectional, or agglutinative.

The Kazakh language, like other languages of the Turkic group, belongs to the agglutinative languages. In agglutinative languages, a word consists of a base, which is joined by the affixes expressing the various grammatical characteristics. The several formative affixes (sometimes called the endings) can be attached to the base of a word, while each such affix performs a grammatical function inherent only to it, the order of the affixes is strictly defined.

The Turkish language also refers to the Turkic languages. The Turkish NLP pipeline, which is described in [8], includes the following components (Table 1):

Table 1. The comparison of the set of components of the Turkish NLP pipeline and the Kazakh language pipeline.

Turkish NLP pipeline	Kazakh language pipeline
–	– Word generator
– Tokenizer	– Tokenizer
– Deasciifier	–
– Vowelizer	–
– Spelling corrector	– Spelling corrector
– Normalizer	– Normalizer: lemmatization
– isTurkish	–
– Morphological analyzer	– Morphological analyzer
– Morphological disambiguator	–
– Dependency parser	– Dependency parser

Let's note that not all of them are appropriate to use for NLP of the Kazakh language. Thus, the usage in the Kazakh language does not currently require the usage of a diacritic reducing agent, which in the Turkish NLP pipeline converts the ASCII characters into their proper Turkish forms. The vocalizer that restores the omitted vowels (usually in social media posts for shortening) is also not required for NLP in Kazakh, since such vowel omission is not a noticeable phenomenon in Kazakh social networks. There is no need for the analog of the Turkish component, which identifies the borrowed words through the morphological analysis, since in the Kazakh language, the borrowed words obey the rules of the Kazakh morphology. As for the elimination of morphological ambiguity (the component Morphological Disambiguator Turkish NLP pipeline), since this problem is found in NLP for the Kazakh language much less often than for Turkish, and its solution is very laborious, we attributed it to the subsequent stages of the work.

Let's briefly describe the existing researches in the field of the creation of separate NLP components for the Kazakh language. The pipeline for automated processing of texts written in Kazakh language is presented in the paper of the researchers from Nazarbayev University [9]. The modern tools developed by the authors for the morphological analysis of the Kazakh language are described in [10], the language model and tokenization in [11], the syntax analysis algorithms and creation of dependency trees in [12]. However, all of these articles describe the approaches to solving the problem at each of these stages, while the usage of the pipeline processing implies a certain unification of algorithms, the coordination of input and output data formats for each stage, and so on. Below we will outline our proposed approaches to solving these problems. The peculiarity of this study is that the proposed pipeline is freely available by the link http://nlp.corpus.kaznu.kz/. This pipeline can be extended with the new tools.

4   The Tool of Generation of the Word Forms

It was a difficult task to create the process of obtaining word forms that would generalize the entire lexicon of Kazakh by adding affixes in a linear form. The reason was that the form of the word, which is grammatically correct, could be erroneous, meaningless from a lexical point of view. In Kazakh, there is a simple declination and possessive declension. Words in the possessive form in the accusative and dative cases are inclined differently.

Taking into account the statements above, the following combinations of affixes were allocated to create word forms of Kazakh:

For animate nouns that answer the question "Who?":

1. Word + the plural form's ending + possessive ending. For example, дос+тар+ым (dos+tar+im – my friends).
2. Word + the plural form's ending + possessive ending + case ending. For example, көрші+лер+ім+мен (korshi+ler+im+men – with my neighbors).
3. Word + the plural form's ending + possessive ending + case ending + personal ending (Used in rare cases. For example, ауыл+дар+ың+нан+мын (auil+dar+in+nan+min – I am from your village).
4. Word + the plural form's ending + possessive ending + personal ending. For example, мұғалім+дер+мен+біз (mugalim+der+men+biz – we are with teachers).

For inanimate nouns that answer the question "What?":

1. Word + the plural form's ending + possessive ending. For example, сөмке+лер+і (somke+ler+i – their bags).
2. Word + the plural form's ending + possessive ending + case ending. For example, дәптер+лер+ім+ді (dapter+ler+im+di – my notebooks).
3. Word + the plural form's ending + case ending. For example, ағаш+тар+ға (agash+tar+ga – for trees).

For uncountable inanimate nouns. For example, тұман (tuman – fog), aya (aua – air), etc.

1. Word + possessive ending + case ending. For example, су+ы+нан (su+i+nan – out of his water).
2. Word + case ending. For example, жаңбыр+дың (janbir+ding – of the rain).

For the most common forms of adjectives:

1. Adjective, that changes the same way as a noun does. For example, әдемі (ademi – beautiful): әдемілермен (ademi+ler+men – with beauties), әдемінің (ademi+ning – at beauty).
2. Word + degrees of an adjective (mainly characteristic of a qualitative adjective: positive form, comparative form, superlative form). For example, әдемі (ademi –beautiful), әдемірек (ademi+rek – more beautiful), әп-әдемі (ap-ademi – very beautiful).

For the most common forms of verbs:

1. Word + mood form (indicative, imperative, conditional, desirable). For example, сен оқисың (sen okisyn – you read), сен оқы (sen oky – you read), сен оқысаң (sen okysan – if you will read), сен оқығайсың (sen okygaisyn – you should read).
2. Word + mood form + personal ending. For example, бар+са+м (bar+sa+m – if I go).
3. Word + form of verbal adverb. For example, оқы+п (oky+p – having read), оқы+ғалы (oky+galy – had read), оқы+ған+ша (oky+gan+sha – will have read).
4. Word + form of verbal adverb+personal ending (except forms *-ғалы*, *-гелі*, *-қалы*, *-келі* (-galı, -geli, -kalı, -keli)). For example, көр+іп+сің (kor+ip+sin – you have seen).
5. Word + participle form. For example, жаз+ылар (jaz+ilar – will probably be written), жаз+ылған (jazil+gan – written), жаз+ыл+атын (jaz+ilatin – will be written), жаз+ылмақ (jaz+ilmak – perhaps will be written).
6. Word + participle form (changes the same way as nouns do). For example, көр+ген (kor+gen – he saw).

5   The Tool of Lemmatization of the Word Forms

The task of automating the extraction of lemmas requires a dictionary of lemmas. Unfortunately, in the public domain, there was not found any dictionary of such kind in Kazakh. The language resources of the Russian language were analyzed. The most comprehensive dictionary of lemmas in the Russian was the Grammatical Dictionary of the Russian language [13] that contained 106 000 lemmas. Therefore, the task was to translate the dictionary of lemmas from Russian into Kazakh. The translation into Kazakh was made with the help of online translators Sozdik.kz [14] and Google translate [15], access to which is public. As a result, a dictionary of 93,000 lemmas of Kazakh was obtained, taking into account homonyms and synonyms. To create meta-information of the obtained lemmas, the morphological features of lemmas in Russian from A.A. Zaliznyak are used. However, it turned out that not all morphological features can be transferred automatically. The work on the analysis of the compliance of features and the introduction of additional features, characteristic of Kazakh, was carried out by philologists.

In the process of translation of the lemmas, it was revealed that the online translator Google translate does not always provide a correct translation. Therefore, the work on the examination and correction of the received translations is carried out by specialists-philologists.

For the automatic extraction of lemmas from word forms and a deeper analysis of the composition of words, two approaches were implemented: direct and reverse.

The direct approach is the task of normalizing words. Based on the Porter algorithm [16], according to the grammatical rules of Kazakh, affixes are sequentially cut off from the end of the word: that is how stemming is performed. To do this, the help of philologists, dictionaries of Kazakh affixes, namely, plural endings, endings of the possessive form, case and personal endings for nouns, which are also used for adjectives, in addition to specific adjectives for various adjectives and personal endings for verbs, suffixes and participles affixes are used.

Next, normalization is performed: on the basis of the results of the stemming, the synthesis of the normal form is carried out. This stage consists of the dictionary of lemmas' research, meant to find the closest lemma to the derived word basis. The system can return several options of lemmas. At the moment, there is no algorithm that would allow determine the lemma uniquely. Therefore, this work is carried out with the assistance of philologists. An online interface has been developed for convenience and productivity (http://corpus.kaznu.kz/). The reverse approach is carried out using a dictionary of word forms. Based on the use of the Kazakh grammar and the examination of philologists, the following results were obtained: up to 75 word forms can be generated for nouns, 96 forms for adjectives, 319 forms for verbs. For each lemma from the created dictionary of lemmas, various word forms were generated and stored in a database together with the corresponding meta information (morphological features of the word). Currently, the word form database consists of 4,006,833 units.

The task of extracting a lemma from a given word is carried out as follows: by the word form in the database, the source lemma of the given word is searched with the corresponding meta information. The system can return several options.

For approbation of the word form generator, a total of 70,000 news from https://aba i.kz/ and https://www.qamshy.kz/ were used. In these texts, 40 million non-unique words were revealed, of which 22 765 323 words were found in our list of word forms and lemmas. 161,317 words turned out to be unique among them, while there is information about the frequency of their occurrence in these texts. Out of 161,317 unique words, 19,565 words are lemmas and are summarized in texts 15,054,339 times, the other unique words are generated.

6 The Reduction of Homonymy

The idea of the proposed method is to subordinate one sentence to one law. The simplest element of the sentence is the idea of the initial and narrative texts, which, in turn, is limited in the structure of such sentences. Therefore, if you summarize the ways in which sentences are summed to a certain extent, then you will have access to a tool that provides a mathematical pattern for analyzing a phrase. For the purposes of this method, it is likely that information about an unrecognized token (not found in the dictionary) is very likely in the process of morphological analysis of the sentence, as well as in solving homonymy. There are few cases in the analysis when the phrase is interpreted differently, and this is a very difficult problem.

$\Psi = \{\Psi_1, \Psi_2, \ldots, \Psi_n\}$ – let's summarize the last sentence and summarize any existing phrase in the Kazakh language. Here, the lengths of the vectors Ψ_i $i = \overline{1, n}$ can be different.

$\Theta = \{\Theta_1, \Theta_2, \ldots, \Theta_m\}$ - possible interpretations of the current sentence. One or several coordinates Θ_i $i = \overline{1, m}$ may be unknown.

Different quantitative indicators can be used to compare the corresponding vectors of different arrays at the stages of the method, for example, the angle between vectors (Cosine similarity).

$i = \overline{1, m}$ and $j = \overline{1, n}$: $Max(\cos(\Psi_i, \Theta_j))$ satisfies the value of Ψ_i - the template we need, Θ_j - the structure that is considered suitable for the current offer.

This method provides a number of possibilities, in addition to predicting unspecified morphological features and homonyms. Using the method, you get a sentence syntax analysis to get meaningful words and phrases. The chosen quantitative measure of the method can be replaced by any other, it is not so difficult to insert different statistical parameters based on the documents in the application. The following digital transitions are used to convert sentences into numeric vectors based on tags:

Cases:

Noun - 10

Adjective - 20

Numeral - 30

Pronouns - 40

Verb - 50 (auxiliary verb - "−50")

Adverb - 60

Preposition - 70

Verb endings:

Verbal adverb - 1

Participle - 5

For example:

«Сыйлы қонақтар ауылға кеше келген еді» (Sıylı qonaqtar auılga keshe kelgen edi – It turns out that distinguished guests arrived in the village yesterday) analysis this sentence:

Сый-лы (siy-li – distinguished) <20> қо-нақ-тар (qo-naq-tar – guests) <10> ауыл-ға (awıl-ga – in the village) <10 + 2> ке-ше (ke-she – yesterday) <60> /кеш-е (kesh-e) <50 + 1> кел-ген (kel-gen – arrived) <50 + 5> е-ді (e-di – It turns out that) <−50>

Possible interpretations of the current sentence:

$\Theta_1 = \{20,10,12,60,55,-50\}$

$\Theta_2 = \{20,10,12,51,55,-50\}$

$\Theta_3 = \{20,12,12,60,55,-50\}$

$\Theta_4 = \{20,12,12,51,55,-50\}$

And the required template in Ψ.

$\Psi_1 = \{20,10,12,60,55,-50\}$

$\Psi_2 = \{20,10,15,60,55,-50\}$

By comparing the results from the Table 2, we can see that the Θ_1 and Ψ_1 are similar, so:

Analysis of «сый-лы <20> қо-нақ-тар <10> ауыл-ға <10 + 2> кеше <60> кел-ген <50 + 5> еді <−50>» is considered to be a good option.

Table 2. Cosine similarity measure between the template and a sentence.

	Θ_1	Θ_2	Θ_3	Θ_4
Ψ_1	1.0	0.99708	0.99979	0.9967
Ψ_2	0.99954	0.996844	0.999340	0.99649

It is important to note that this definition and the quantitative equivalents selected here are also subject to change during the course of the experiment, and at present, work is being done at an experimental level.

7 Pipeline Implementation

7.1 Word Generator

The description of Word Generator: to construct grammatical forms on the basis of a given word. Generation is the process opposite to morphological analysis, normalization. Generation is performed for nouns and verbs. The possibility of attaching an ending to a root is determined by the law of syngarmonism, that is, it depends on whether the last syllable of the root is hard or soft and what the last letter is (deaf, voiced, sonorous, vowel, consonant) (Fig. 1). Instructions for use:

1. Link to page: http://nlp.corpus.kaznu.kz/wordForm.
2. In the window for generating word forms write the source word.
3. Set the word type: noun or verb.
4. Press the button: "Generate".
5. The answer will be all the different forms of this word.

Fig. 1. Tool of word generator

7.2 Tokenizer

The description of Tokenizer: to break the raw text into small fragments. Tokenization breaks down raw text into words, sentences called tokens. These tokens help you understand the context, develop an NLP model, and interpret the meaning of the text by analyzing the word sequence (Fig. 2). Instructions for use:

1. Link to page: http://nlp.corpus.kaznu.kz/tokenizer.
2. In the window for tokenization you have to write a text.
3. Press the button: Process text.

Result: a list of tokens of the given text.

Fig. 2. Tool of tokenizer

7.3 Spelling Corrector

The automatic spelling correction system detects a spelling error and suggests a set of candidates for correction. Kukic [17], Pirinen and Linden [18] divide the whole process into three stages:

1. detection of an error;
2. generation of candidates for correction;
3. Rating of candidates for corrections.

We have implemented two steps: error detection, generation of candidates for correction and demonstration of candidates to the user with the possibility of choosing the most acceptable option. A sample of 1000 words was used to evaluate the results of the methods. From each word, 2 samples with 2 different errors were formed in it: a sample containing words with only one error, and a sample containing words with only two errors. The estimation of the method based on the Levenshtein distance [19], the LSTM seq2seq model, is carried out. The evaluation was carried out in accordance with the error detection procedure. The best results were obtained for the LSTM seq2seq machine learning method. In the future, it is planned to implement the ranking of candidate corrections (Fig. 3). Instructions for use:

1. Link to page: http://nlp.corpus.kaznu.kz/spellingCorrector.
2. In the window for spelling correction write the text.
3. Press the button: Process text.

Result: the words in which errors were detected will be shown, as well as possible candidates for correction.

Fig. 3. Tool of spelling corrector

7.4 Morphological Analyzer

The description of the morphological analyzer: identifying the components corresponding to the morphological composition of the token and identifying the grammatical characteristics of words. The result of the morphological analyzer is to find the root of the word, determine the part of speech, and identify affixes. Affixes are defined by detailing: affix and type of affix (Fig. 4). Instructions for use:

1. Link to page: http://nlp.corpus.kaznu.kz/lemma.
2. In the window for morphological analysis write the text.
3. Press the button: Process text.

Result: The sentence is divided into separate words, and for each word a morphological analysis is made in the form: lemma + part of speech + affixes.

7.5 Dependency Parser

This tool is based on the method of Creation of a dependency tree for sentences in the Kazakh language [12]. The description Dependency Parser: to determine the type of connection of words in a word combination. The word combination is considered as a unit of syntax, which performs a communicative function only as part of a sentence (Fig. 5). Instructions for use:

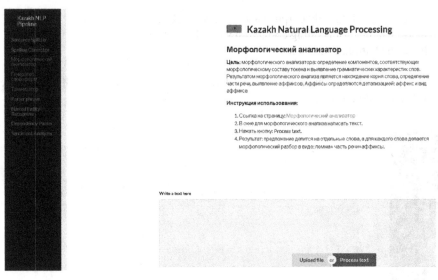

Fig. 4. Tool of morphological analyzer

1. Link to page: http://nlp.corpus.kaznu.kz/phrase.
2. In the window for determining the type of connection of words you need to write a sentence.
3. Press the button: Process text.

Result: a list of word combinations and types of word relations for these word combinations for the given sentence.

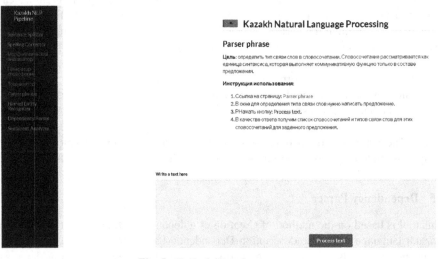

Fig. 5. Tool of dependency parser

8 Conclusion

Despite the increased activity of the Turkic languages in cyberspace as languages for accumulating and transmitting information, the presence of electronic corpuses of languages, thematic, multilingual and terminological dictionaries, natural language processing systems, almost all Turkic languages, except for Turkish, are low-resource languages. The available electronic resources are not sufficient for the creation of effective machine translation programs, machine search and natural language understanding systems, which are not only relevant, but extremely necessary both for processing the ever-increasing information on the Internet in Turkic languages, and for the development of the languages themselves. The work proposes the design and development pipeline of preprocessing tools for the Kazakh language texts. Proposed pipeline allows preprocessing the texts in the Kazakh language. The foundation that can accelerate the processing of the Kazakh language texts has been laid. The prototype of the Kazakh language pipeline is used to solve the problems of computational linguistics and is placed on http://nlp.cor pus.kaznu.kz/. Operational experience has shown the usability of the prototype, as there is a possibility of modifying the data structure.

Acknowledgments. This work was supported in part under grant of Foundation of the Ministry of Education and Science of the Republic of Kazakhstan AP09261344 "Development of methods for automatic extraction of spatial objects from heterogeneous sources for information support of geographic information systems" (2020–2022) and Erasmus+ Project "Development of the interdisciplinary master program on Computational Linguistics at Central Asian universities" (585845-EPP-1-2017-1–ES-EPPKA2-CBHE-JP).

References

1. Altenbek, G., Wang, X.-L.: Kazakh segmentation system of inflectional affixes. In: Proceedings of Joint Conference on Chinese Language Processing CIPS-SIGHAN 2010, pp. 183–190 (2010)
2. Bekmanova, G.T.: Some approaches to the problems of automatic word changes and morphological analysis in the Kazakh language. Bull. East Kazakhstan State Tech. Univ. Named D. Serikbayev, **1**, 192–197 (2009). (in Russian)
3. Sharipbaev, A.A., Bekmanova, G.T., Ergesh, B.J., Buribaeva, A.K., Karabalaeva, M.H.: The intellectual morphological analyzer based on semantic network. In: Proceedings of the International Scientific-Technical Conference «Open Semantic Technology of Intelligent Systems» OSTIS 2012, Minsk, pp. 397–400 (2012). (in Russian)
4. Zafer, H.R., Tilki, B., Kurt, A., Kara, M.: Two-level description of Kazakh morphology. In: Proceedings of the 1st International Conference on Foreign Language Teaching and Applied Linguistics (FLTAL 2011), Sarajevo, Bosnia and Herzegovina, pp. 560–564 (2011)
5. Zhubanov, A.H.: Basic principles of formalization of the Kazakh text content. Almaty (2002). (in Russian)
6. Mansurova, M., Madiyeva, G., Aubakirov, S., Yermekov, Z., Alimzhanov, Y.: Design and development of media-corpus of the Kazakh language. In: Nguyen, N.T., Papadopoulos, G.A., Jędrzejowicz, P., Trawiński, B., Vossen, G. (eds.) ICCCI 2017. LNCS (LNAI), vol. 10449, pp. 509–518. Springer, Cham (2017). https://doi.org/10.1007/978-3-319-67077-5_49

7. Barakhnin, V.B., Duisenbayeva, A.N., Kozhemyakina, O.Yu., Yergaliyev, Y.N., Muhamedyev, R.I.: The automatic processing of the texts in natural language. Some bibliometric indicators of the current state of this research area. J. Phys. Conf. Ser. **1117**, 1–9 (2018)
8. Eryiğit, G.: ITU Turkish NLP web service. In: Proceedings of the Demonstrations at the 14th Conference of the European Chapter of the Association for Computational Linguistics, pp. 1–4. Association for Computational Linguistics (2014)
9. Yessenbayev, Z., Kozhirbayev, Z., Makazhanov, A.: KazNLP: a pipeline for automated processing of texts written in Kazakh language. In: Karpov, A., Potapova, R. (eds.) SPECOM 2020. LNCS (LNAI), vol. 12335, pp. 657–666. Springer, Cham (2020). https://doi.org/10.1007/978-3-030-60276-5_63
10. Barakhnin, V.B., et al.: The software system for the study the morphology of the Kazakh language. In: The European Proceedings of Social & Behavioural Sciences, vol. XXXIII, pp. 18–27 (2017)
11. Akhmed-Zaki, D., Mansurova, M., Madiyeva, G., Kadyrbek, N., Kyrgyzbayeva, M.: Development of the information system for the Kazakh language preprocessing. Cogent Eng. **8**(1), 1896418 (2021)
12. Akhmed-Zaki, D., Mansurova, M., Kadyrbek, N., Barakhnin, V., Misebay, A.: Creation of a dependency tree for sentences in the Kazakh language. In: Nguyen, N.T., Hoang, B.H., Huynh, C.P., Hwang, D., Trawiński, B., Vossen, G. (eds.) ICCCI 2020. LNCS (LNAI), vol. 12496, pp. 709–718. Springer, Cham (2020). https://doi.org/10.1007/978-3-030-63007-2_55
13. Zaliznyak, A.A.: Grammatical Dictionary of the Russian Language: Slovoizmenenie: About 100000 Words. Russian Language, Moscow (1977).(in Russian)
14. Online Kazakh-Russian translator. https://sozdik.kz. Accessed 1 Sept 2021
15. Online translator. https://translate.google.ru. Accessed 1 Sept 2021
16. Porter, M.F.: An algorithm for suffix stripping. Program **14**(3), 130–137 (1980)
17. Kukich, K.: Techniques for automatically correcting words in text. ACM Comput. Surv. **24**, 377–439 (1992)
18. Pirinen, T.A., Lindén, K.: State-of-the-art in weighted finite-state spell-checking. In: Gelbukh, A. (ed.) CICLing 2014. LNCS, vol. 8404, pp. 519–532. Springer, Heidelberg (2014). https://doi.org/10.1007/978-3-642-54903-8_43
19. Wagner, R.A., Fischer, M.J.: The string-to-string correction problem. J. ACM **21**(1), 168–173 (1974)

Thai Named Entity Corpus Annotation Scheme and Self Verification by BiLSTM-CNN-CRF

Virach Sornlertlamvanich[1,3(✉)] [ID], Kitiya Suriyachay[2] [ID],
and Thatsanee Charoenporn[1] [ID]

[1] Faculty of Data Science, Musashino University, Tokyo, Japan
{virach,thatsanee}@ds.musashino-u.ac.jp
[2] School of ICT, Sirindhorn International Institute of Technology,
Thammasat University, Pathumthani, Thailand
m5922040075@g.siit.tu.ac.th
[3] Faculty of Engineering, Thammasat University, Pathumthani, Thailand

Abstract. Corpus is one of the essential parts of language research, especially for the low resource language. To ensure the researching result to be most effective, the corpus that has been used also requires effectiveness and accuracy. The Thai language has some special characteristics that cause difficulty in building the corpus and affect the error of those corpora. Therefore, this paper proposes an effective and efficient approach to clean up the existing Named Entity corpus before using it in any language research. The THAI-NEST corpus is adopted to verify the consistency and integrity of the data and re-design with our proposed model. The revised corpus is verified by the BiLSTM-CNN-CRF model that combined the features among word, POS, and Thai character clusters (TCCs). Experimental results show the effectiveness of the verification, which increased the accuracy by up to 12%, and the model can effectively detect and handle errors of word segmentation and NE tag consistency.

Keywords: Corpus annotation · Named entity recognition · Thai named entity · Thai corpus

1 Introduction

Methodical and accurate use of this enormous amount of data is the key to the success of many businesses. Many organizations have invested in developing information extraction and retrieval to use the data and information efficiently. Information Extraction (IE) is an automated process of extracting specific information from unstructured data, for example, the process of extracting names, addresses, and phone numbers from a web page. Named Entity Recognition (NER) is a subtask of IE that aims to recognize and classify specific entities in focused texts such as person names, locations, and organizations.

NER has gained popularity over several years. NER in the Thai language is more challenging and complex than English or other European languages. Due to there is no capitalization or special characters to identifying named entities. In addition, there is no space or word boundary in a sentence causing difficulties in word segmentation and the

Z. Vetulani et al. (Eds.): LTC 2019, LNAI 13212, pp. 143–160, 2022.
https://doi.org/10.1007/978-3-031-05328-3_10

ambiguity between common nouns and proper nouns. Moreover, incorrect word segmentation causes problems for named entity recognition and directly affects the accuracy of the NER process. The ambiguity of homographs can also provide different meanings depending on the context.

For example, "พันตำรวจเอกทวี สอดส่อง อธิบดีกรมสอบสวนคดีพิเศษ"(Colonel Tawee Sordsong Director-General of the Department of Special Investigation) the word "สอดส่อง"in this context refers to a surname of the person. However, it also means "to observe" in the general context. Thai writing styles are also a problem for NER. In general, a proper noun is written with a common noun or prefix that specifies the type of the proper noun, but Thais often use a shorter name, abbreviation, or cut its prefix. "ธรรมศาสตร์ร่วมกับเอกชนสร้างสรรค์ไอเดีย"(Thammasat collaborates with private companies to create IT ideas.), "มหาวิทยาลัยธรรมศาสตร์ก่อตั้งขึ้นในปี ค.ศ. 1934"(Thammasat University was established in 1934), and "ม.ธ. ประกาศกำหนดการในภาคการศึกษาแรก"(TU announces the schedule for the first semester.), from the examples, Thammasat University is used in three different ways including "ธรรมศาสตร์"(Thammasat) "มหาวิทยาลัยธรรมศาสตร์"(Thammasat University) and "ม.ธ."(the abbreviation of Thammasat University). Another problem of named entity in the Thai language is that named entity in different types can be the same word, such as "สุรินทร์"(Surin), which can be both a name of the person or Thailand province (location name), depending on the context of the word.

Furthermore, the consistency of named entity tags in the corpus is also an important issue because inconsistency leads to wrong name entity recognition. Corpus which is used in this paper is the THAI- NEST corpus [18]. The corpus is disjointedly generated into seven files, and each file is exclusively for each main category. In addition, the corpus contains some inconsistency of named entity tags due to the word segmentation process on the corpus. To solve these problems, we propose a method to clean up and verify the existing corpus for the Thai NER. We also performed cross annotation among the separate seven named entity tagged files to enhance the number of NE tags and to prepare for further developing the NER model.

The previous related researches are explained in Sect. 2. Section 3 describes the corpus that has been used in this research. The methodology to improve the consistency and correctness of the corpus is showed in Sect. 4. Then, in Sects. 5 and 6, we presented the result of our experiment and combined corpus approach, respectively. The conclusion is described in Sect. 7.

2 Related Work

Research on NER is widely popular in many languages, so the NER tools have been gaining attention and continually improving. However, in many languages, a small number of the corpus is used for NLP tasks, while the study of named entities is quite limited. Therefore, having a small number of the corpus is not enough [4]. Several approaches can be used to solve these problems, but one of the most influential and popular approaches is based on machine learning techniques [5].

Nevertheless, there is a pretty limited Thai NE corpus. The famous Thai corpus is the Orchid corpus created in 1996 by collecting Thai text for more than two million words and splitting all of the words with their part of speech. However, Thai did not have any clear boundary or punctuation. To build the Orchid corpus, they need to separate the paragraph into sentences and then from sentences to words before tagging each word with POS using trigram. Their trigram model will consider the word segmentation and POS tagging together within the model [15].

Many previous research papers show an Error Correction for many languages, such as Chinese, use a transformation-based error-driven machine learning technique to found error positions and produce error repairing rules [20]. The dictionary-based approach is easy, fast, and widely used, but this approach can only be used with known and unambiguous words. In Vietnamese [10], they use two entropy-based methods to detect error and inconsistency in the Vietnamese word-segmented and POS-tagged corpus. The first method is to rank the order of error candidates using a scoring function that depends on conditional entropy. The second method uses beam search to find a subset of error candidates that has been changed its label and finally leads to the decreasing of confidential entropy.

Some traditional machine learning techniques have been applied for NLP tasks over the past year. [1] introduced the NER model of the Hindi language by using Hidden Markov Model (HMM). [3] presented the Support Vector Machine (SVM) with word shape, and POS is used as features to recognize named entities in Biomedical Text such as genes, DNA, and protein.

[12] proposed Malay Named Entity Recognition using the CRF model. Some characteristics of Malay are employed for training models such as capitalization, lowercase, previous and neighboring word, word suffix, digit, word shape, and POS. [6] introduced the CRF model for Chinese electronic medical records recognition with bag-of-characters, part-of-speech, dictionary feature, and word clustering as features.

Deep Learning architectures have recently made significant advances in various fields. BiLSTM is a type of deep learning model used in many research studies. [11] conducted the NER model for recognizing Indonesian information on Twitter using Bi-LSTM with word embedding and POS tag and showed that both features provided the most F1-score. [19] presented the Deep Learning model for Chinese telecommunication information recognition using character embedding instead of word embedding. [16] applied word with POS as an input of the Bi-LSTM model to recognize the named entity in the Thai language.

Furthermore, there are several researches on NER that use a hybrid approach. [2] proposed a Bi-LSTM and CRF model for Chinese NER based on character and word embedding. [7] also presented Bi-LSTM-CNN-CRF model, which achieved performance on NER and POS tagging and successfully employed CNN to extract more useful character-level features. More recent work used the CRF as the last layer of the pipeline to handle the classification and provided the satisfactory results [17].

3   Corpus

The original corpus used in our research is the THAI-NEST corpus, which is collected from Thai online news published on the Internet including politics, economic, foreign, crime, sports, entertainment, education, and technology news [18]. The THAI-NEST corpus is already tokenized text into words and punctuation. The corpus is separated into seven files based on named entity categories, consisting of date (DAT), time (TIM), measurement (MEA), name (NAM), location (LOC), person (PER), and organization (ORG). The first three characters abbreviate each category. The number of words and named entity tags in each file is listed in Table 1.

Table 1. Number of sentences, words, and named entity tags in each file

	No. of sentence	No. of word	No. of NE tag
DAT	2,784	214,467	14,334
LOC	8,585	569,292	33,596
MEA	1,969	157,788	17,371
NAM	7,553	547,489	40,537
ORG	20,399	1,386,824	95,566
PER	33,233	2,705,218	222,075
TIM	419	41,493	3,362

3.1 Structure of the Corpus

In this experiment, the THAI-NEST corpus was designed and restructured based on the structural format of the Orchid corpus [15] as shown in Fig. 1. There are two types of mark-ups to differentiate text information line and numbering line in the corpus. Text information line beginning with "%" symbol, which is used to describe the additional information of the corpus as shown in Table 2. The numbering line begins with a "#" symbol, which is used to order the sequence of text in the corpus as shown in Table 3, and also there are four special mark-up characters as shown in Table 4.

Table 2. Mark-up for text information line

Mark-up	Description
%Title:	Title of the corpus
%Description:	Detail of the corpus or reference
%Number of sentence:	Total number of sentences in the file
%Number of word:	Total number of words in the file
%Number of NE tag:	Total number of named entity tags in the file
%Date:	Date of creating the corpus
%Creator:	Name of the creator (s)
%Email:	Email Address (es) of the creator (s)
%Affiliation:	Affiliation (s) of the creators

Table 3. Mark-up for numbering line

Mark-up	Description
#P[number]	Paragraph number of the text. The number in the bracket presents the sequence of paragraph within a text
#S[number]	Sentence number of the paragraph. The number in the bracket presents the sequence of sentence within a paragraph

Table 4. Special mark-up characters

Mark-up	Description
\\	Line break symbol for the long text
//	End of sentence
/[POS]	Tag marker for POS annotation of a word
/[NE]	Tag marker for NE annotation of a word

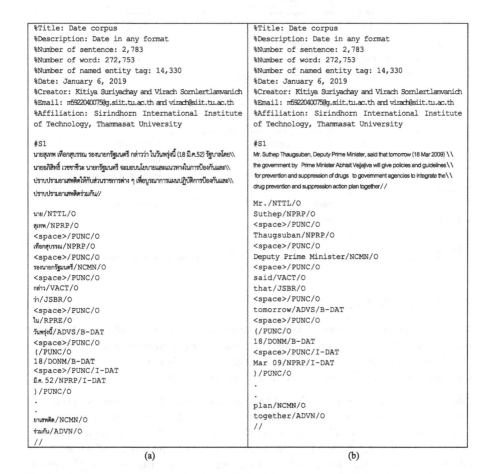

(a)   (b)

Fig. 1. Example of Data corpus file (a) in Thai original text and (b) in English translated text

For the named entity tags, BIO annotation format is used for all seven categories of the named entity as shown in Table 5, and we use 47 types of part-of-speech (POS) as defined in the Orchid corpus.

Table 5. NE tags in each corpus file

Category	Format	Description	Example
Date	B-DAT	Beginning of a date	วันที่ (Date)
	I-DAT	Inside of a date	14 กุมภาพันธ์ (Feb, 14)
Location	B-LOC	Beginning of a location name	เมือง (City)
	I-LOC	Inside of a location name	นิวยอร์ค (New York)
Measurement	B-MEA	Beginning of a measurement unit	ห้า (Five)
	I-LOC	Inside of a measurement unit	เล่ม (Books)
Name	B-NAM	Beginning of any proper name except location, person, and organization names, e.g., name of competition, name of position, etc.	ศึก (League)
	I-NAM	Inside of any proper name	ลาลีกา (La Liga)
Organization	B-ORG	Beginning of an organization name	บริษัท (Corp.)
	I-ORG	Inside of an organization name	โตโยต้า มอเตอร์ (Toyota Motor)
Person	B-PER	Beginning of a person name	นาย (Mister)
	I-PER	Inside of a person name	ณัฐวุฒิ (Natthawut)
Time	B-TIM	Beginning of a time	สิบ (Ten)
	I-TIM	Inside of a time	นาฬิกา (O'clock)
Other	O	Word does not belong to any type of entity	

3.2   Corpus Challenges

As we mentioned above, some Thai language problems need to be solved as they cause some defects and limitations in the corpus. Thus, this paper aims to solve the difficulties of the existing corpus to improve the consistency and efficiency of the NE corpus in subsequent research. The defects of the corpus are as follows:

The Error of Word Segmentation. The major challenge of this corpus is the error of word segmentation. Since the Thai language has no clear word boundary or space between words, it is challenging to segment words. If the word segmentation is incorrect, it will affect the following process, especially in the process of named entity labeling. In Fig. 2, each picture represents an example of mistakes in word segmentation. The errors of segmenting words also occur to the abbreviation such as "มี.ค."(Mar) this word means the abbreviation of March in the Thai language, but it was cut separately into different tokens as can be seen in Fig. 2(a). These wrong word segmentations also result in incorrect POS and named entity tagging.

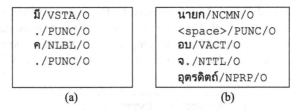

Fig. 2. Example of mistakes word segmentation in different corpus file (a) Date corpus file and (b) Name corpus file

The Inconsistency of NE Tagging. The problem of inconsistency also happens with the named entity tag. Some words that can only belong to one category are labeled as the main category at some places, while they are labeled as "Other" in some other sentences in the same file. For example, "ประเทศ"is labeled as a location name (B-LOC) and it is labeled as other (O) in another place as shown in Fig. 3. Figure 4 also present an example of inconsistency NE tag labeling, "พรีเมียร์ลีก อังกฤษ"(Premier League) is annotated both name of football league (NAM) and other (O) in the same file.

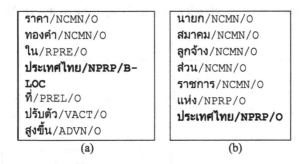

Fig. 3. Inconsistency of named entity tagging in Location corpus file

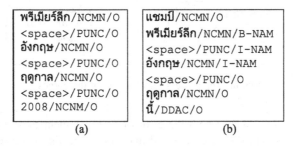

Fig. 4. Inconsistency of named entity tagging in Name corpus file

The Error of Named Entity Tag Assignment. The mistake in word segmentation will affect the POS of the word. Moreover, the incorrect word segmentation and POS annotation lead to wrong named entity tagging. Figure 5 shows incorrect name entity tags of the person's surname due to incorrect word segmentation and POS assignment.

ร้อยตำรวจเอก / NTTL/B-PER
เฉลิม/NPRP/I-PER
<space>/PUNC/O
อยู่/XVAE/O
บำรุง/VACT/O

Fig. 5. False NE tagging of surname in Person corpus file

4 Corpus Revision Methodology

4.1 Approaches to Clean the Corpus

There are three main steps for cleaning and verifying the existing THAI-NEST corpus shown in Fig. 6.

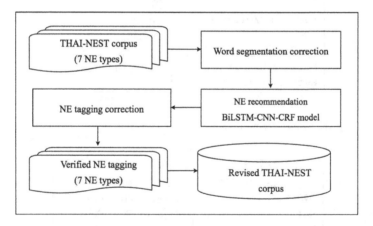

Fig. 6. The process for cleaning up the corpus

For the first step, we deal with the error of word segmentation in the corpus by searching incorrect named entities tags and manually correcting word segmenting and its POS. Next, NER models were trained for each of the main category files. Our proposed model is described in the following paragraph. Each of these models was trained separately using its own training set and validation set from each file and use these seven models to classify the entity of words. The most appropriate NE tag of each word is predicted and selected by the model. In the final step, we proceed with the cross annotation among the disjoint seven NE tagged files.

4.2 Named Entity Recognition Model

In this section, we describe the components of our model for improving and correcting the named entity tagging of the THAI-NEST corpus. The Thai NER model is presented, which was inspired by the research of [7]. Our proposed model consists of five important layers: Word Embedding, Character-level Representation, Part of Speech Embedding, Bi- LSTM layer, and the last is CRF layer. The architecture of the model is shown in Fig. 7.

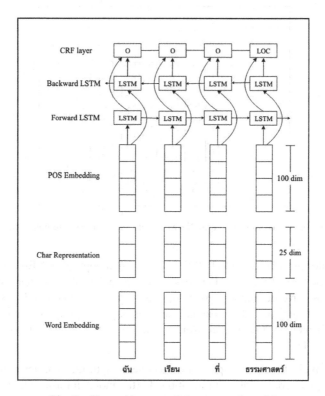

Fig. 7. The architecture of the proposed model

Word Embedding Layer. In our experiment, the Word2Vec tool of the Gensim library was used to pre-train word embedding by using the skip-gram model with 300 dimensions and window size of three words, three words before and three words after.

Thai Character Cluster (TCC)-Level Representation. Character-level representation can extract morphological information from the word and is extremely helpful in particular for languages with complicated structures or a rich morphological language, i.e., Hindi [8], Korean [9], and Thai. However, the Thai language has various characters such as vowels, consonants, tones, and special characters. In addition, a tone mark and a vowel sign cannot stand alone and they must be placed with the character only. Hence, if we use only word embedding may not significantly improve the performance of our NER model. For this reason, we used the Thai Character Cluster (TCC) technique in character-level representation, which is an unambiguous unit that is smaller than a word and larger than a character and cannot be further divided based on Thai rules to group these characters [13, 14]. For example, "ป|ระ|เท|ศ|ไท|ย"(Thailand) or "นา|ย|ก|รัฐ|ม|น|ตรี"(The prime minister). In addition, CNN has good performance for NLP and character embedding. So, the CNN layer is applied to create the character vectors of the model. Detail of the CNN layer is as shown in Fig. 8.

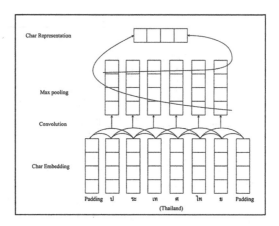

Fig. 8. Convolution Neural Network for character-level representation

POS Embedding: Part-of-speech (POS) can help to optimize NER model performance because most Thai named entities tend to be adjacent to or attached to part-of-speech such as verbs or prepositions. Several prior studies supported the use of POS in the NER model, such as Indonesian and Chinese language, so POS is introduced to be used with our model. POS of each word is encoded into the one-hot vector format in the embedding layer.

Bi-LSTM Layer. LSTM is capable of effectively capturing long-term dependencies and can retrieving rich global information. Furthermore, information from previous words is useful for prediction, but information from words coming after is also helpful. This can be done by having a second LSTM running backward, and this pair of forwarding and backward LSTMs is referred to as a bidirectional LSTM.

Conditional Random Field (CRF): The CRF is widely used model to predict the sequence of labels with the most likely tendency that corresponds to the sequence of given input sentences. The CRF model will take advantage of the neighbor tag information and consider the previous context in predicting current tags. We thus consider that the CRF is the last layer to predict the NE of each word and followed [14] to create our linear-chain CRF layer. The linear-chain CRF will find the highest scoring path through an input sequence and gives the best tags and final score.

Each layer is combined to construct the Bi-LSTM-CNN-CRF model for predicting named entity tags. Word, POS, and character-cluster vectors from each embedding layer are concatenated before being fed into the Bi-LSTM layer. Then, the outputs of Bi-LSTM are transferred to the CRF layer and decoded by the Viterbi algorithm (part of the CRF layer) to determine the most possible entity tags. The model has been able to enhance the capacity to predict target words from the vector of the surrounding context.

5 Experiment

5.1 Pre-processing

As mentioned above in Sect. 3, the format of the named entity tag in this corpus is BIO format. However, the Thai writing system usually does not have a prefix or an indication of a name. It means that we cannot measure the score of B-tag and I-tag separately because some words in the corpus do not have a prefix. Therefore, the format of the NE tag needs to change from BIO to IO format instead to solve this problem.

5.2 Experiment Setup

Each NE tagged file is divided into three parts: 80% of all sentences for the training set, 10% for the validation set, and the last 10% for the testing set.

For our neural network model, a list of all parameters needs to be set. The various parameters are adjusted on the development set to get the most suitable final parameters. All parameters and settings are displayed in Table 6.

Table 6. Parameters for all experiments

Parameter	Setting
Char_dim	30
Character-level CNN filters	30
Character-level CNN window size	3
Word_dim	100
Word_LSTM_dim	200
Word_bidirection	TRUE
POS_dim	100
Dropout_rate	0.5
Batch size	10
Learning rate (initial)	0.01
Decay rate	0.5
Gradient clipping	5.0
Learning method	SGD
Training epoch	60

6   Result and Discussion

In this experiment, the performances of models are compared based on F1-score. Seven models predict only their own testing set and are measured their F1-score separately. Table 7 shows the results of each corpus before and after the segmenting correction. Due to the prediction of the NER model, we can also manually resolve inconsistencies and invalid NE tags problems based on the results from the model. Each revised NE tagged file was retrained by using the proposed model. All F1-score after correcting the NE tags is higher than one before correcting up to an average of 12%, and the results are listed in Table 8.

Table 7. Comparison results between before and after correcting word segmentation

NE	F1-score	
	Before	After
DAT	85.04	89.14
LOC	69.27	73.68
MEA	77.58	80.45
NAM	42.76	46.91
ORG	70.19	75.03
PER	81.71	85.64
TIM	84.53	88.55

Table 8. The results after correcting NE tags in each corpus file

NE	F1-score
DAT	93.21
LOC	88.93
MEA	86.52
NAM	84.96
ORG	87.31
PER	88.90
TIM	94.76

As shown in Table 7, the f1-score of each category after solving word segmentation is higher than before correcting. The correction of word segmentation errors improves F1-score and affects the named entity prediction of the model. Once words are resolved, their entity is also correctly predicted. For example, words in the sentence "วันที่ 1 มี .ค.2551"(March 1, 2008), the NE tag is labeled as "Other" before correcting the error of segmentation but the model can correctly predict the named entity tags after correction as shown in Fig. 9. The fourth column is the predicted NE tag from the model.

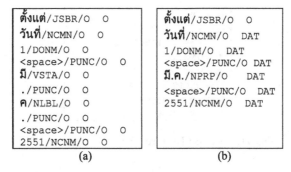

```
ตั้งแต่/JSBR/O   O          ตั้งแต่/JSBR/O   O
วันที่/NCMN/O   O           วันที่/NCMN/O   DAT
1/DONM/O   O               1/DONM/O   DAT
<space>/PUNC/O   O          <space>/PUNC/O DAT
มี/VSTA/O   O               มี.ค./NPRP/O   DAT
./PUNC/O   O                <space>/PUNC/O DAT
ค/NLBL/O   O                2551/NCNM/O   DAT
./PUNC/O   O
<space>/PUNC/O   O
2551/NCNM/O   O
       (a)                         (b)
```

Fig. 9. The predicted named entity tag (a) before and (b) after editing word segmentation in Date corpus file

Furthermore, the use of POS can improve the NER model efficiency. For instance, the word "ใน"(in) is a preposition, and "ไป"(go) is a verb, once these words occur before a noun, the noun will be a named entity as shown in the following example. In Fig. 10(a), "ความน่าเชื่อถือของธนาคารในประเทศไทย"(The reliability of the banks in Thailand"), the word "ใน"(in) is placed before the word "ประเทศไทย"(Thailand), therefore, "ประเทศไทย"(Thailand) is location (LOC) not person (PER). The sentence "นายกรัฐมนตรีจะเดินทางไปเมืองปุตราจายา"(The prime minister will go to Putrajaya), the word "ไป"(go) preceding the word "เมืองปุตราจายา"(Putrajaya) which is a city in Malaysia, "เมืองปุตราจายา"is a location as shown in Fig. 10(b).

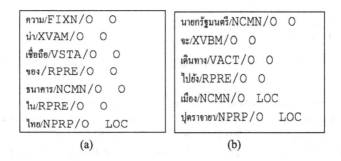

```
ความ/FIXN/O   O            นายกรัฐมนตรี/NCMN/O   O
น่า/XVAM/O   O              จะ/XVBM/O   O
เชื่อถือ/VSTA/O   O          เดินทาง/VACT/O   O
ของ/RPRE/O   O             ไปยัง/RPRE/O   O
ธนาคาร/NCMN/O   O          เมือง/NCMN/O   LOC
ใน/RPRE/O   O              ปุตราจายา/NPRP/O   LOC
ไทย/NPRP/O   LOC
       (a)                         (b)
```

Fig. 10. Examples of NE tags that close to (a) preposition, and (b) verb

However, even POS is beneficial for the named entity recognition, but there are some errors in the prediction process. As can be seen in Fig. 11, "มีกิจกรรมพิเศษ**ค้นหาอัจฉริยะไอที** เพื่อแข่งขันทักษะไอทีของนักเรียนป ระถมและมัธยมทั่วประเทศ"(There is a special event called **Search for IT genius** to compete for IT skills of a nationwide elementary and high school student.), "ค้นหาอัจฉริยะไอที"(Search for IT genius) is an event name, but the word "ค้นหา"(search) in the event name is a verb, so "ค้นหาอัจฉริยะไอที"(Search for IT genius) seems like an activity not the name of the event. Thus, the model cannot predict the named entity correctly.

```
กิจกรรม/NCMN/O  O
พิเศษ/VATT/O  O
<space>/PUNC/O  O
"/PUNC/O  O
ค้นหา/VACT/B-NAM  O
อัจฉริยะ/NCMN/I-NAM  O
ไอที/NCMN/I-NAM  O
"/PUNC/O  O
```

Fig. 11. Incorrect NE prediction in Name corpus file

In the sentence as shown in Fig. 12, "ความรู้การพัฒนาซอฟต์แวร์มาตรฐานสากลCMMI จากมหาวิทยาลัยซอฟต์แวร์ในประเทศสหรัฐอเมริกา"(CMMI International Software Development Knowledge from software university in the United States), "มหาวิทยาลัยซอฟต์แวร์"(software university) mean a university that offers software discipline instruction. This word should be predicted as other but because the word "จาก"(from) is a verb and it makes the model predict inaccurately, so "มหาวิทยาลัยซอฟต์แวร์"(software university) is incorrectly predicted as a location.

```
มาตรฐานสากล/NCMN/O  O
<space>/PUNC/O  O
CMMI/NPRP/O  O
<space>/PUNC/O  O
(/PUNC/O  O
Capability<space>Maturity<space>Model<space>Integration/NPRP/O  O
)/PUNC/O  O
<space>/PUNC/O  O
จาก/RPRE/O  O
มหาวิทยาลัย/NCMN/O  LOC
ซอฟต์แวร์/NCMN/O  LOC
<space>/PUNC/O  O
ประเทศสหรัฐอเมริกา/NPRP/B-LOC  LOC
```

Fig. 12. Incorrect NE prediction in Location corpus file

We also performed experiments with other baseline models using the same revised NE tagged file to prove the effectiveness of consistency verifying and named entity selection of the proposed models and show the importance of using POS and TCC in dealing with wrong word segmentation and NE assignment. Table 9 presents the comparison result of the performance of each model.

According to the result shown in Table 9, our Bi-LSTM-CNN-CRF model outperforms other baseline models, especially a Date and Time category file in which the F1-score was around 94% and the Time category file in which the F1-score was about 93%.

Table 9. Performance of our model and other baseline models

NE	F1-score		
	BiLSTM (Word)	BiLSTM-CNN-CRF (Word+TCC)	BiLSTM-CNN-CRF (Word+POS+Char)
DAT	79.20	84.65	92.43
LOC	75.33	79.26	87.07
MEA	72.41	77.53	85.66
NAM	67.18	73.84	82.25
ORG	73.56	77.90	85.79
PER	74.29	81.72	87.32
TIM	82.77	86.05	93.88
NE	F1-score		
	BiLSTM (Word+POS)	BiLSTM-CNN (Word+POS+TCC)	BiLSTM-CNN-CRF (Word+POS+TCC)
DAT	86.14	89.72	93.21
LOC	83.67	86.24	88.93
MEA	80.22	82.67	86.52
NAM	75.48	80.13	84.92
ORG	81.17	84.75	87.31
PER	82.35	85.07	88.90
TIM	88.12	91.36	94.76

7   Combined Corpus

Another major issue of this corpus is that it is disjointedly generated into seven files, and each file is exclusively for each main category. Due to this structure, the corpus cannot use directly to create such a model for named entity recognition which can classify all categories. We thus conduct cross annotation among the seven NE tagged files and combine all named entity types into the same file. The brief idea of this approach is to use the seven trained models obtained from our proposed model training, applying cross annotation of every named entity category for labeling named entity tags in one category file until all seven original category files are complete. Table 10 shows an example of a combined corpus.

Table 10. Named entity tags in the combined corpus (a) in Thai original text, and (b) in English translated text

```
%Title: BKD19-1 (Thai NE Corpus)            %Title: BKD19-1 (Thai NE Corpus)
%Description: Based on THAINEST corpus      %Description: Based on THAINEST corpus
%Number of sentence: 2,783                  %Number of sentence: 2,783
%Number of word: 272,753                    %Number of word: 272,753
%Date: March 17, 2019                       %Date: March 17, 2019
%Creator: Kitiya Suriyachay and Virach Sornlertlamvanich    %Creator: Kitiya Suriyachay and Virach Sornlertlamvanich
%Email: m59220400750@g.siit.tu.ac.th and virach@siit.tu.ac.th    %Email: m59220400750@g.siit.tu.ac.th and virach@siit.tu.ac.th
%Affiliation: Sirindhorn International Institute    %Affiliation: Sirindhorn International Institute
of Technology, Thammasat University         of Technology, Thammasat University

#S1                                         #S1
นายสุเทพ เทือกสุบรรณ รองนายกรัฐมนตรี กล่าวว่า ในวันพรุ่งนี้ (18 มี.ค.52) รัฐบาลโดย\\    Mr. Suthep Thaugsuban, Deputy Prime Minister, said that tomorrow (18 Mar 2009) \ \
นายอภิสิทธิ์ เวชชาชีวะ นายกรัฐมนตรี จะมอบนโยบายและแนวทางในการป้องกัน\\    the government by Prime Minister Abhisit Vejjajiva will give policies and guidelines \ \
และปราบปรามยาเสพติดให้กับส่วนราชการต่าง ๆ เพื่อบูรณาการแผนปฏิบัติการป้องกัน\\    for prevention and suppression of drugs to government agencies to integrate the \ \
และปราบปรามยาเสพติดร่วมกัน//              drug prevention and suppression action plan together / /

นาย/NTTL/B-PER                             Mr./NTTL/B-PER
สุเทพ/NPRP/I-PER                            Suthep/NPRP/I-PER
<space>/PUNC/I-PER                         <space>/PUNC/I-PER
เทือกสุบรรณ/NPRP/I-PER                      Thaugsuban/NPRP/I-PER
<space>/PUNC/O                             <space>/PUNC/O
รองนายกรัฐมนตรี/NCMN/O                      Deputy Prime Minister/NCMN/O
<space>/PUNC/O                             <space>/PUNC/O
กล่าว/VACT/O                               said/VACT/O
ว่า/JSBR/O                                 that/JSBR/O
<space>/PUNC/O                             <space>/PUNC/O
ใน/RPRE/O                                  tomorrow/ADVS/B-DAT
วันพรุ่งนี้/ADVS/B-DAT                      <space>/PUNC/O
<space>/PUNC/O                             (/PUNC/O
(/PUNC/O                                   18/DONM/B-DAT
18/DONM/B-DAT                              <space>/PUNC/I-DAT
<space>/PUNC/I-DAT                         Mar 09/NPRP/I-DAT
มี.ค.52/NPRP/I-DAT                         )/PUNC/O
)/PUNC/O                                   .
.                                          .
.
ยาเสพติด/NCMN/O                            plan/NCMN/O
ร่วมกัน/ADVN/O                             together/ADVN/O
//                                         //
```

(a)   (b)

Finally, we train the proposed model on the Combined corpus. The F1-score of the Combined model on each main category is listed in Table 11. As being shown, all F1-scores of Combined corpus are quite similar to the F1-score of models trained by each file. Nevertheless, using the Combined-Corpus approach provides better results, but the F1-Scores in some categories drop slightly (e.g., Location and Organization). In addition, the number of named entities in each corpus file is dramatically increasing after combining all seven named entity categories.

Table 11. Results of the combined corpus

NE	F1-score
DAT	94.02
LOC	87.15
MEA	87.36
NAM	86.17
ORG	85.84
PER	89.27
TIM	96.44

8 Conclusion

This paper presented an approach by generating a NER model to clean the existing named entity corpus in the Thai language and verify its consistency. We use the THAI-NEST corpus to verify the consistency of NE tags and re-annotate the named entities by the proposed model. Our model can deal with the named entity tag inconsistency problem, including word segmentation mistakes, and yield impressive results. In order to enhance the amount of NE tags and prepare for further NE tag context captures in NER model development, we have performed a cross-annotation technique of all the seven NE tagged files. The Bi-LSTM-CNN-CRF model with the word, part-of-speech, and TCC features is used to verify the revised NE tagged corpus. Furthermore, POS and TCC play an important role in solving the problems related to word segmentation errors and inconsistency of NE tags. The model provides the performance of the verification, which increases the accuracy up to 12%.

Acknowledgements. The project is financial support provided by Thammasat University Research fund under the TSRI, Contract No. TUFF19/2564 and TUFF24/2565, for the project of "AI Ready City Networking in RUN", based on the RUN Digital Cluster collaboration scheme.

References

1. Chopra, D., Joshi, N., Mathur, I.: Named entity recognition in Hindi using hidden Markov model. In: 2016 Second International Conference on Computational Intelligence & Communication Technology (CICT), pp. 581–586 (2016)
2. Shijia, E., Xiang, Y.: Chinese named entity recognition with character-word mixed embedding. In: Proceedings of the 2017 ACM on Conference on Information and Knowledge Management, pp. 2055–2058 (2017)
3. Ju, Z., Wang, J., Zhu, F.: Named entity recognition from biomedical text using SVM. In: 5th International Conference on Bioinformatics and Biomedical Engineering, pp. 1–4 (2011)
4. Lample, G., Ballesteros, M., Subramanian, S., Kawakami, K., Dyer, C.: Neural architectures for named entity recognition. In: Proceedings of the 2016 Conference of the North American Chapter of the Association for Computational Linguistics: Human Language Technologies, pp. 260–270 (2016)

5. Limsopathan, N., Collier, N.: Bidirectional LSTM for named entity recognition in Twitter messages. In: Proceedings of the 2nd Workshop on Noisy User-generated Text, pp. 145–152 (2016)
6. Liu, K., Hu, Q., Liu, J., Xing, C.: Named entity recognition in chinese electronic medical records based on CRF. In: 14th Web Information Systems and Applications Conference (WISA), pp. 105–110 (2017)
7. Ma, X., Hovy, E.: End-to-end sequence labeling via bi-directional LSTM-CNNs-CRF. In: Proceedings of the 54th Annual Meeting of the Association for Computational Linguistics (vol. 1: Long Papers), pp. 1064–1074 (2016)
8. Maimaiti, M., Wumaier, A., Abiderexiti, K., Yibulayin, T.: Bidirectional long short-term memory network with a conditional random field layer for Uyghur part-of-speech tagging. Information 8(4), 157 (2017)
9. Na, S., Kim, H., Min, J., Kim, K.: Improving LSTM CRFs using character-based compositions for Korean named entity recognition. Comput. Speech Lang. **54**, 106–121 (2019)
10. Nguyen, P., Le, A., Ho, T., Do, T.: Two entropy-based methods for detecting errors in POS-tagged treebank. In: 3th International Conference on Knowledge and Systems Engineering, pp. 150–156 (2011)
11. Rachman, V., Savitri, S., Augustianti, F., Mahendra, R.: Named entity recognition on Indonesian Twitter posts using long short-term memory networks. In: International Conference on Advanced Computer Science and Information Systems (ICACSIS), pp. 228–232 (2017)
12. Salleh, M.S., Asmai, S.A., Basiron, H., Ahmad, S.: A Malay named entity recognition using conditional random fields. In: 5th International Conference on Information and Communication Technology (ICoIC7), pp.1–6 (2017)
13. Sornlertlamvanich, V., Tanaka, H.: The automatic extraction of open compounds from text corpora. In: Proceedings of the 16th Conference on Computational Linguistics (COLING-1996), pp. 1143–1146 (1996)
14. Sornlertlamvanich, V., Tanaka, H.: Extracting open compounds from text corpora. In: Proceedings of the 2nd Annual Meetings of the Association for Natural Language Processing, pp. 213–216 (1996)
15. Sornlertlamvanich, V., Takahashi, N., Isahara, H.: Thai part-of-speech tagged corpus: ORCHID. In: Proceedings of Oriental COCOSDA Workshop, pp. 131–138 (1998)
16. Suriyachay, K., Sornlertlamvanich, V.: Named entity recognition modeling for the Thai language from a disjointedly labeled corpus. In: 5th International Conference on Advanced Informatics: Concept Theory and Applications (ICAICTA), pp. 30–35 (2018)
17. Suriyachay, K., Sornlertlamvanich, V., Charoenporn, T.: Thai named entity tagged corpus annotation scheme and self verification. In: Proceedings of the 9th Language & Technology Conference (LTC2019), pp.131–137 (2019)
18. Theeramunkong, T., et al.: THAI-NEST: a framework for Thai named entity tagging specification and tools. In: Proceedings of the 2nd International Conference on Corpus Linguistics (CILC10), pp. 895–908 (2010)
19. Wang, Y., Xia, B., Liu, Z., Li, Y., Li, T.: Named entity recognition for Chinese telecommunications field based on Char2Vec and Bi-LSTMs. In: 12th International Conference on Intelligent Systems and Knowledge Engineering (ISKE), pp. 1–7 (2017)
20. Yao, T., Ding, W., Erbach, G.: Repairing errors for Chinese word segmentation and part-of-speech tagging. In: Proceedings of the International Conference on Machine Learning and Cybernetics, pp. 1881–1886 (2002)

Computational Semantics

Analogies Between Short Sentences: A Semantico-Formal Approach

Yves Lepage[✉] [iD]

IPS, Waseda University, Kitakyushu, Japan
yves.lepage@waseda.jp

Abstract. The present article proposes a method to solve analogies between sentences by combining existing techniques to solve formal analogies between strings and semantic analogies between words. The method is applied on sentences from the Tatoeba corpus. Two datasets of more than five thousand semantico-formal analogies, in English and French, are released.

1 Goal of the Present Article

Formal analogies between strings are puzzles of the general type:

$$abc : abbccd :: efg : x \quad \text{(solution: } effggh) \tag{1}$$

or

$$król : królowa :: kr : x \quad \text{(solution: } krowa). \tag{2}$$

The first example is taken from [9]. As the two above examples illustrate, in such formal analogies, no meaning is attached to the strings. Techniques have been proposed to solve puzzles that involve prefixing, suffixing and parallel infixing [13,15], e.g.,

$$kataba : k\bar{a}tib :: sakana : x \quad \text{(solution: } s\bar{a}kin) \tag{3}$$

or

$$wyszedłem : wyszłaś :: poszedłem : x \quad \text{(solution: } poszłaś). \tag{4}$$

Analogy (3) constitutes a formal analogy. In addition, it makes sense in Arabic: *kataba,* he wrote, *kātib,* a writer, *sakana,* he lived (in some place), *sākin,* an inhabitant. Similarly, Analogy (4), in addition to being a formal analogy, makes sense in Polish. Its English translation is as follows:

$$I \text{ left (masc.)} : you \text{ left (fem. sg.)} :: I \text{ went (masc.)} : you \text{ went (fem. sg.)}. \tag{5}$$

Work supported by JSPS Grant 18K11447 (Kakenhi C) "Self-explainable and fast-to-train example-based machine translation using neural networks".

Z. Vetulani et al. (Eds.): LTC 2019, LNAI 13212, pp. 163–179, 2022.
https://doi.org/10.1007/978-3-031-05328-3_11

By considering sentences as being strings of words, these techniques solve analogies like (6).

$$\frac{You\ will\ see\ the}{man\ next\ week.} : \frac{I\ see\ the\ woman}{this\ week.} :: \frac{You\ will\ meet\ the}{man\ next\ month.} : x \qquad (6)$$

The solution is: $x = I\ meet\ the\ woman\ this\ month$.

Formal analogies do not care about meaning. They are different from semantic analogies, the over-repeated example of which is $man : woman :: king : queen$ [20] which can be stated in other languages, e.g. in French $homme : femme :: roi : x$ (solution: $reine$) or in Polish: $mężczyzna : kobieta :: król : x$ (solution: $królowa$). Semantic analogies became popular some years ago because vector representations of words, especially word embeddings [1,2,20,23,24], can be used to solve them, up to a certain extent.

This article proposes a way to combine the resolution of formal analogies between strings with the resolution of semantic analogies between words so as to solve (some) analogies between sentences, like the hand-crafted example in (7).

$$\frac{You\ will\ see\ this}{man\ this\ week.} : \frac{I\ saw\ the\ woman}{last\ week.} :: \frac{You\ will\ meet\ the}{King\ today.} : x \qquad (7)$$

The solution should be: $x = I\ met\ the\ Queen\ yesterday$. The reader should observe the change in tenses, from future to preterit, on irregular verbs, and the corresponding change in time from $this\ week$ to $last\ week$ expressed for days by the substitution of $today$ with $yesterday$. In addition, as expected, the male/female opposition has been reflected when solving $man : woman :: King : x \Rightarrow x = Queen$.

The goal we pursue by proposing a way to solve analogies between sentences is to augment the number of analogy resources available. Resources like Google set [20] or BATS [4] cover analogies between words. Our goal is to extend the type of such resources beyond words. We believe that datasets of analogies between sentences can serve as testbeds to evaluate the quality of vector representations of sentences too. Solving analogies by relying solely on vector representations of sentences is a next study in our work. We already started such study by making use of the datasets produced with the technique presented in this article [29].

The present article is an extension of an article presented at the 9th Language & Technology Conference, Poznań, 2019 [18]. The extension consists in three points:

- we provide a better explanation of the method with more demonstrative examples and mention parallel work we have done in related topics;
- we show how to extend the method to solve semantico-formal analogies to potentially create bigrams from unigrams and bigrams;
- in addition to the already released dataset in English, we release a new dataset of semantico-formal analogies between sentences in French.

2   Semantic Analogies

2.1   Vector Arithmetic

Let us recall the simplest technique used to solve analogies between words in word embedding spaces. It is based on vector arithmetic. If $\overrightarrow{w}$ notes the vector corresponding to the word w, the resolution of the analogical equation $A : B ::$ $C : x$, where A, B, C and x are words, is performed in two steps. First, a vector, v, is built as in (8).

$$v = \overrightarrow{B} - \overrightarrow{A} + \overrightarrow{C} \tag{8}$$

Then, the solution x is defined as the word in the embedding model which maximises the cosine similarity with v, i.e.,

$$\overrightarrow{x} = \arg\max_{w} \cos(\overrightarrow{w}, v), \tag{9}$$

where w ranges over all the words in the vector space. The validity of this linear conception has been questioned in [4] and other formulae have been proposed (called PairDirection, 3CosMul or LRCos), but we will stick with the above explanation (called 3CosAdd) in the remaining of this article.

2.2   Extension to Sets of Words

It is possible to extend the above-mentioned use of vector arithmetic by considering sets of words. From three sets of words, $\mathcal{A}$, $\mathcal{B}$ and $\mathcal{C}$, one can always form the vector v defined in Eq. (10).

$$v = \sum_{w_B \in \mathcal{B}} \overrightarrow{w_B} - \sum_{w_A \in \mathcal{A}} \overrightarrow{w_A} + \sum_{w_C \in \mathcal{C}} \overrightarrow{w_C} \tag{10}$$

The word x whose vector $\overrightarrow{x}$ maximises the cosine similarity with v, is considered the solution of the analogical equation between the given sets of words: $\mathcal{A} : \mathcal{B} ::$ $\mathcal{C} : x$. Note that, here, on the contrary to $\mathcal{A}$, $\mathcal{B}$ and $\mathcal{C}$, x is a single word.

This simple extension allows us to solve small analogies like:

$$\{will, see\} : \{saw\} :: \{will, meet\} : x \tag{11}$$

and obtain *met* as the solution. Similarly, the equation

$$\{this, week\} : \{last, week\} :: \{today\} : x \tag{12}$$

has *yesterday* as its solution.

However, this extension does not answer the case where the solution of the analogical equation is longer than one word. It does not allow us to get the expected answer to analogies like $\{tomorrow\} : \{yesterday\} :: \{next, week\} : x$ where the expected solution would be a sequence of two words, namely *last week*. And more importantly, it does not apply to the resolution of analogies between complete sentences.

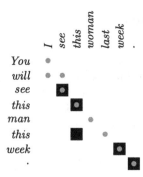

Fig. 1. Traces between two sentences using the LCS distance, indicated by dots. Because of the two dots on the second line, there are two traces. On the contrary to Fig. 3, the cells in the matrix have a value of 0 or 1, indicated by black (equality) or white cells (inequality).

3   Formal Analogies

3.1   Traces

We now review methods to solve formal, not semantic, analogies between strings. There exist two trends for solving formal analogies between strings. The first one [12,13] uses the notion of shuffle of strings to produce a solution while the second one [15,16] is based on the computation of edit distances between strings. Some extensions of these techniques are given in [17] or [25], and application to machine translation and transliteration have been introduced in [3,14,19].

The two trends share some abstract commonalities, but we will concentrate on the second one. There, the crucial notion is that of a *trace*, i.e., a sequence of edit operations, including copying, to apply so as to transform one string into another. The classical algorithms to compute an edit distance and a trace are found in [28]. Another possible algorithm to deliver a trace is given in [8].

An illustration of traces is given in Fig. 1. The two strings are actually sentences, the words of which are just symbols compared for equality. Equality is shown by black squares. The dots visualise the traces, i.e., the shortest paths linking the top left cell to the bottom right one in the matrix, which minimises the number of edit operations needed to transform the sentence on the left into the sentence at the top. There are two possible traces here, because one can go through the cell (*will, I*) or the cell (*will, saw*) to reach the first black cell when one starts from the top left of the matrix, trying to reach its right bottom.

3.2   LCS Edit Distance and Parallel Traversal of Traces

An illustration of the algorithm for the resolution of formal analogies between strings proposed in [15] (based on [10]) is given in Fig. 2. Here again, the strings are sentences with words compared for equality. The arithmetic formula $b - a + c$

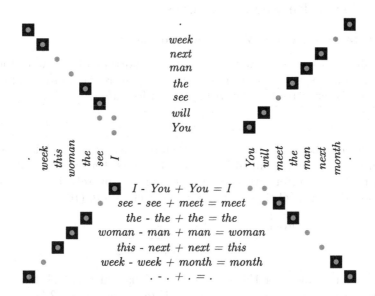

Fig. 2. Resolution of a formal analogy between sentences. Observe that the top-left sentences and their traces are the same as in Fig. 1, up to mirroring.

is applied to compute the word at hand in the solution from the words a, b and c found in A, B and C while reading the two traces, between A and B and between A and C, in parallel. Note, however, that the formula makes sense only when $a = b$ or $a = c$.

Leaving copying apart (cost of 0), the edit operations used in the computation of the trace are reduced to two: insertion and deletion, each with a cost of 1. Consequently, the substitution of a symbol with another one has a cost of 2, because this corresponds to a deletion and an insertion, each of a cost of 1.

The edit distance with insertion and deletion only is related to the similarity between strings classically defined as their longest common subsequence (LCS) through Eq. (13). For this reason, it is called the LCS distance.

$$d(A, B) = |A| + |B| - 2 \times s(A, B) \tag{13}$$

In Eq. (13), d is the LCS distance between two strings, s is their similarity, i.e., the length of their longest common subsequence (LCS), and $|s|$ denotes the length of a string s. With the additional notation that $|s|_a$ denotes the number of occurrences of character a in string s, an analogy between strings can be characterised by the system in (14).

$$A : B :: C : x \;\Rightarrow\; \begin{cases} d(x, C) = d(A, B) \\ d(x, B) = d(A, C) \\ |x|_a = |B|_a - |A|_a + |C|_a \end{cases} \tag{14}$$

4 Analogies Between Sentences

4.1 Distance Between Words and Sentences

The similarity between two words w_1 and w_2 represented by their vectors in a word embedding space of a given dimensionality n, is classically computed as the cosine of their corresponding vectors, as in Eq. (15).

$$s(w_1, w_2) = \cos(\overrightarrow{w_1}, \overrightarrow{w_2}) \tag{15}$$

The distance between two words can then be taken as the Euclidean distance between the two points pointed by the word vectors. Under the usual assumption that all word vectors are located on the unit n-sphere[1], their distance is a function of the cosine of the word vectors as given by Eq. (16).[2]

$$
\begin{aligned}
d(w_1, w_2) &= \ \|\overrightarrow{w_1} - \overrightarrow{w_2}\| \\
&= \ \sqrt{2} \times \sqrt{1 - \cos(\overrightarrow{w_1}, \overrightarrow{w_2})}
\end{aligned}
\tag{16}
$$

The maximal value for this distance is 2. This makes it look like the LCS distance, where the distance between two different characters is 2 (see Sect. 3.2). Between words, a distance of 2 would correspond to deleting a word and inserting its exact opposite word on the unit n-sphere. However, in practice, word vectors are not evenly distributed on the unit n-sphere. For instance, in the English vector space pre-trained using FastText (see Sect. 6.1), the word closest to the opposite point of *Queen* is *component.* (with a glued full stop). But we find: $\cos(\overrightarrow{Queen}, \overrightarrow{component.}) = -0.203$, which is far from -1.0. It has even been shown that some vector embeddings produce vector representations in which all vectors are all situated in an octant of the n-sphere [21].

The matrix of the distances between the words in two sentences can be visualised as in Fig. 3. Because we deal with a true mathematical distance between words, the edit distance between sequences of words is also a true mathematical distance. It can thus be computed using the standard algorithm [28]. The traces can also be computed in the standard way or directly [8]. In Fig. 3, the cells with a dot in their centre are the cells on the traces. There are two possible traces.

---

[1] This is usually the case, as normalisation is applied on raw vectors, learnt from a corpus, before any use.

[2] This is proven as follows. $^{\top}$ denotes transpose.

$$
\begin{aligned}
\|\overrightarrow{w_1} - \overrightarrow{w_2}\|^2 &= (\overrightarrow{w_1} - \overrightarrow{w_2})^{\top}(\overrightarrow{w_1} - \overrightarrow{w_2}) \\
&= \|\overrightarrow{w_1}\|^2 + \|\overrightarrow{w_2}\|^2 - 2 \times \overrightarrow{w_1}^{\top}\overrightarrow{w_2} \\
&= \|\overrightarrow{w_1}\|^2 + \|\overrightarrow{w_2}\|^2 - 2\|\overrightarrow{w_1}\|\|\overrightarrow{w_2}\| \cos(\overrightarrow{w_1}, \overrightarrow{w_2})
\end{aligned}
$$

With all norms being equal to 1, the equality is rewritten as follows, hence Eq. (16) above.

$$\|\overrightarrow{w_1} - \overrightarrow{w_2}\|^2 = 2 \times (1 - \cos(\overrightarrow{w_1}, \overrightarrow{w_2}))$$

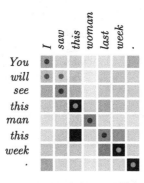

Fig. 3. Traces between two sentences using LCS distance, indicated by dots. The shades of grey for the cells reflect similarity: the darker, the more similar. The top sentence is different from the one in Fig. 1 (*saw* instead of *see*), but the traces are similar.

4.2 Analogy-Compatible Decomposition of a Quadruple of Strings

A decomposition of a pair of strings (A, B), each into two parts $(A_1.A_2, B_1.B_2)$ such that $A = A_1.A_2$, $B = B_1.B_2$ and such that at most one of the lengths of A_1, A_2, B_1 and B_2 is null, always satisfies (17).

$$d(A_1.A_2, B_1.B_2) \leq d(A_1, B_1) + d(A_2, B_2) \tag{17}$$

The equality is only reached on traces.[3] For that reason, we say that a decomposition is *trace-compatible* if it verifies Eq. (18).

$$d(A_1.A_2, B_1.B_2) = d(A_1, B_1) + d(A_2, B_2) \tag{18}$$

An *analogy-compatible* decomposition of a quadruple of strings (A, B, C, x) is a quadruple of decompositions $(A_1.A_2, B_1.B_2, C_1.C_2, x_1.x_2)$ where each individual decomposition

- $(A_1.A_2, B_1.B_2)$,
- $(A_1.A_2, C_1.C_2)$,
- $(B_1.B_2, x_1.x_2)$ and
- $(C_1.C_2, x_1.x_2)$

is trace-compatible. Given a quadruple of strings (A, B, C, x), we note $\tau(A, B, C, x)$ the set of all analogy-compatible decompositions.

4.3 Semantico-Formal Resolution of Analogies between Sentences

To solve an analogy $A : B :: C : x$ between the sentences A, B and C, we proceed as follows.

---

[3] Combined with the mirror of strings, this property is used in [8] to directly find an optimal alignment between two strings.

170   Y. Lepage

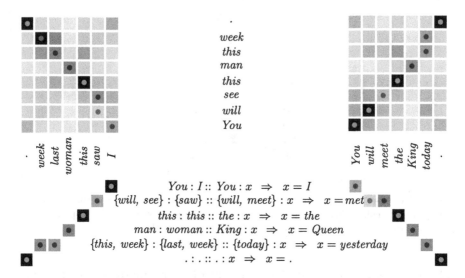

Fig. 4. Semantico-formal resolution of an analogy between sentences. Observe that the top-left matrix is the same as in Fig. 3, up to mirroring.

– We first compute the traces between A and B and between A and C, using the distance between words introduced in Sect. 4.1.
– We then explore the traces between A and B and between A and C, in parallel, in a way which is similar to the resolution of formal analogies (Sect. 3.2), so as to build the set of analogy-compatible decompositions, introduced above (Sect. 4.2). There are two cases:
  • Either $\tau(A, B, C, x)$ is empty;
    * If the length of x is 1, i.e., if x is a single word, the method presented in Sect. 2.2 can be applied to solve the analogy between the sets of words in A, B and C.
    * if the length of x is 2 and A and B are words and C is a bigram (or A and C are words and B is a bigram), we try to solve the analogy using the extension proposed below in Sect. 5.
    * in the other cases, we are unable to solve the analogy.
  • Or $\tau(A, B, C, x)$ is not empty and for any decomposition $(A_1.A_2, B_1.B_2, C_1.C_2, x_1.x_2)$, we can try and solve $A_1 : B_1 :: C_1 : x_1$ and $A_2 : B_2 :: C_2 : x_2$. The recursion will ultimately end up on the first case.

The result of the semantico-formal resolution of the example analogy between sentences given in Sect. 1 using the procedure described above is illustrated in Fig. 4. It succeeds in delivering the expected solution: *I met the Queen yesterday.*

5   Extension: Computation of Analogies Between Two Words and a Bigram

In this section, we describe an extension to solve analogies and deliver possibly more than one word out of two words and one bigram. Let us illustrate it with the following example. Suppose that we have the following analogy to solve: *yesterday* is to *tomorrow* as *previous week* is to what? The solution is *next week*.

$$yesterday : tomorrow :: last\ week : x \ \Rightarrow \ x = next\ week \qquad (19)$$

This corresponds to the case where A, B and C are such that A and B are two words and C is a bigram, or A and C are two words and B is a bigram. Remembering that $|s|$ stands for the length of string s, the following disjunction describes the case.

$$\begin{cases} |A| = 1 \\ |B| = 1 \\ |C| = 2 \end{cases} \text{ or } \begin{cases} |A| = 1 \\ |B| = 2 \\ |C| = 1 \end{cases} \qquad (20)$$

As a general remark, exchanging B with C in an analogy is always possible. This property is called the exchange of the means.

We solve this kind of analogy by selecting the best solution out of three solutions obtained from three different ways of solving it. The three different ways are explained below.

5.1   Candidate Solution Where D Is One Word

The first way is to look for a single word that would solve the puzzle. In order to make it a puzzle between vectors, we posit that the vector representing *last week* is just the average of the two word vectors for *last* and *week*.

$$\overrightarrow{last\ week} = \frac{1}{2} \times (\overrightarrow{last} + \overrightarrow{week}) \qquad (21)$$

We then solve the analogy between single vectors.

$$\overrightarrow{yesterday} : \overrightarrow{tomorrow} :: \overrightarrow{last\ week} : x \qquad (22)$$

Notice that it is not necessary to normalise the vector as we are working with cosine similarity: the norms of the vectors do not influence the value of the cosine. We call D_1 the word which is solution of Eq. (22) and $\overrightarrow{D_1}$ its vector representation.

5.2   Candidate Solution Where D Is a Bigram

The second way, and third way, of solving the analogy is to make the hypothesis that there is a word which commutes with the word *last* or the word *week* so as to compose a solution to the analogy. Said in another way, we make the hypothesis that the solution is either of the form *last <something>* or of the form *<something> week*.

Let us consider the case of *<something> week*, the other case being similar. The word *<something>* should be to *last* in the same way as *yesterday* is to *tomorrow*, however without the notion of time period. The notion of time period is what is common to *yesterday, tomorrow* and *week*. Following the main conception of word embeddings, the common meaning in these three words can be regarded as their average, i.e., it can be computed as:

$$\frac{1}{3} \times (\overrightarrow{yesterday} + \overrightarrow{tomorrow} + \overrightarrow{week}) \tag{23}$$

Consequently, the word *<something>* should be the word which is the closest to the vector given by the following computation:

$$<something> = \arg\cos_{w \in \mathcal{V}} \left(\overrightarrow{w}, \overrightarrow{tomorrow} - \overrightarrow{yesterday} + \overrightarrow{lastweek} \right.$$
$$\left. - \frac{1}{3} \times (\overrightarrow{yesterday} + \overrightarrow{tomorrow} + \overrightarrow{week}) \right) \tag{24}$$

where $\mathcal{V}$ is the vocabulary of the word embedding space. We note down $D_2 = <something>\ week$, and compute its vector representation as

$$\overrightarrow{D_2} = \frac{1}{2} \times (\overrightarrow{<something>} + \overrightarrow{week}) \tag{25}$$

In the same way, we obtain a word that commutes with *week*. We thus note $D_3 = last\ <something>$, and

$$\overrightarrow{D_3} = \frac{1}{2} \times (\overrightarrow{last} + \overrightarrow{<something>}) \tag{26}$$

5.3   Selection of the Best Candidate

Using the three possible ways explained above, one gets three words or bigrams that are candidate solutions to the analogy. The quality of the three possible analogies should thus be compared and the best answer should be selected. To this end, we compare the three solutions D_1, D_2 and D_3, and select D such that:

$$D = \arg\cos_{D \in \{D_1, D_2, D_3\}} \left(\overrightarrow{D}, \overrightarrow{tomorrow} - \overrightarrow{yesterday} + \overrightarrow{last\ week} \right) \tag{27}$$

6   Experiments and Released Datasets

We perform experiments on English and French data to confirm the fact that our technique is indeed able to solve analogies between sentences. Basically, we try to solve all analogies between all triples of sentences extracted from a resource rich in similar sentences. We of course apply restrictions.

Firstly, so as to reduce the number of possible triples, we only consider sentences shorter than 10 words.

Secondly, so as to avoid the problems mentioned at the end of Sect. 2.2, we impose that, for an analogy $A : B :: C : x,$ the length of A be equal to or greater than the length of B, and the same for B and C.

Thirdly, we impose that the number of words in common between B and A be higher than two thirds of the length of A, and the same between C and A.

Fourthly and finally, as a major restriction, so as to enforce grammaticality of the sentences generated by analogy, we impose that D, the solution of the analogy, be present in the resource. By doing this, we suppose that the sentences contained in the resource are grammatically correct. What we did is thus equivalent to retrieving possible semantico-formal analogies from the set of sentences, without allowing the generation of new sentences.

6.1   Used Datasets

We use the English and French sentences from the English-French Tatoeba corpus.[4] For English, there are 92,062 sentences shorter than 10 words with an average length of 6.9 word (standard deviation of 1.7 word), and the total number of different words is 13,813.

This corpus is well-fitted for our work as it exhibits a large number of similar sentences with simple commutations like masculine / feminine, affirmative / negative, etc., as illustrated in the following real example:

$$\frac{I\ do\ not\ know}{his\ address\ .} : \frac{I\ do\ not\ know}{her\ address\ .} :: \frac{I\ know\ his}{address\ .} : \frac{I\ know\ her}{address\ .} \quad (28)$$

For word embeddings, we use the English and French vectors trained with FastText and released at LREC 2018, among other languages [2,6].[5] In both languages, out of the 2 million words offered, we filtered out obviously spurious words by eliminating long numbers, ill-formed words, series of non-alphabetical symbols, etc. In this way, we retained 1,192,424 English words and 919,283 French words.

6.2   Released Datasets

In total with all the restrictions described above, we obtained 5,607 semantico-formal analogies in English and 5,296 semantico-formal analogies in French. Some of them are given in Tables 1 and 2. These semantico-formal analogies are released as public resources.[6]

---

[4] https://tatoeba.org/ and http://www.many-things.org/anki/.

[5] https://fasttext.cc/docs/en/pretrained-vectors.html Word embeddings for 294 languages.

[6] http://lepage-lab.ips.waseda.ac.jp/   Projects > Kakenhi Kiban C 18K11447 > Experimental Results.

6.3   Interesting Phenomena Observed

The conjugation of French is known to be much more complicated than that of English, with verbal forms depending on mood, tense, person and number. Many frequent forms are irregular. In the resource produced, we observed that, in a felicitous way, irregular forms of verbs were actually captured, making a parallel with the hand-crafted English example that we used in the first sections of this article. For instance, in the analogy between sentences shown on the first row of Table 2, the analogy between the verbs 'want' in indicative mood and 'understand' in subjunctive mood (present tense 2nd person plural vs. singular) has been correctly solved.

$$voulez : veux :: compreniez : x \;\Rightarrow\; x = comprennes \qquad (29)$$

Interestingly too, we observed the ability of our proposed method to capture the opposition between genders exhibited in pairs of synonyms. In the analogy between sentences in the second row of Table 2, *veinard* 'lucky' (masc.) is a synonym of *chanceuse* (fem.). The feminine form *amie* for the masculine noun *ami* 'friend' has been correctly proposed as a solution of the analogy between words.

$$veinard : chanceuse :: ami : x \;\Rightarrow\; x = amie \qquad (30)$$

6.4   Difficulties of Assessment

A difficulty in assessing the technique proposed in this article lies in the fact that vector arithmetic followed by the determination of the closest word always delivers a solution. Such solutions simply make no sense in the immense majority of the cases. As an example, with our data, the analogy $football : taxes :: girlfriend : x \;\Rightarrow\; x = boyfriend$ holds. Or $suit : pool :: sister : x$ yields the highly questionable solution $x = aunt$. This problem is inherited for sentences by the technique proposed in this article. It delivers a solution for a very large number of analogies between sentences, without a guarantee in meaning. For instance, in our experiments,

$$\frac{I\ do\ not\ have\ a}{suit.} : \frac{I\ do\ not\ have}{medical\ training.} :: \frac{I\ have\ not\ got\ a}{chance.} : x \qquad (31)$$

yields the solution $x = $ *I have not got medical opportunity.*

Another difficulty is the fact that the sequence of words produced may result in ungrammatical sentences (e.g. *I do not have go answers questions.*). This is the well-known problem of *over-generation*. A similar problem is that the solution sentences may contain words which actually belong to the embedding space but are misspelled words or OCR errors from the texts the word embedding space was trained on. As a example, in our experiments, we got: $x = $ *I do not recommend Engish !* (note the absence of *l*) as the solution of an analogy.

Assessing the validity of the obtained analogies would thus require a heavy and tedious work by human judges. Because of the highly subjective assessment

Table 1. Examples of semantico-formal analogies in English from the released dataset

There's hardly any coffee left in the pot.	:	There's almost no coffee left in the pot.	::	There's hardly any water in the bucket.	: x ⟹ $x =$	There's almost no water in the bucket.
I do not know what to say about that .	:	I do not know what to do now .	::	I do not know about that .	: x ⟹ $x =$	I do not think so .
You 're not from around here , are you ?	:	You 're not staying here , are you ?	::	You 're confused again , are n't you ?	: x ⟹ $x =$	You 're disappointed , are n't you ?
There is an urgent need for a new system .	:	There is an urgent need for blood donations .	::	There is an urgent need for experienced pilots .	: x ⟹ $x =$	There is an urgent need for volunteers .
I do not know his name .	:	I do not know her address .	::	I ca n't remember his name .	: x ⟹ $x =$	I ca n't remember her address .
It 's really not that interesting .	:	It 's really not that hot .	::	It 's not that bad .	: x ⟹ $x =$	It 's not that cold .

Table 2. Examples of semantico-formal analogies in French from the released dataset

Je veux ce que vous voulez .	:	Je veux ce que tu veux .	::	Je veux vraiment que vous compreniez .	: x ⟹ $x =$	Je veux vraiment que tu comprennes .
Je ne suis pas si veinard .	:	Je ne suis pas si chanceuse .	::	Je ne suis pas ton ami .	: x ⟹ $x =$	Je ne suis pas ton amie .
Je veux ce que vous voulez .	:	Je veux ce que tu veux .	::	Je sais ce que vous faisiez .	: x ⟹ $x =$	Je sais ce que tu faisais .
Nous ne sommes pas responsables .	:	Nous ne sommes pas ouvertes .	::	Nous ne sommes pas folles .	: x ⟹ $x =$	Nous ne sommes pas extravagantes .

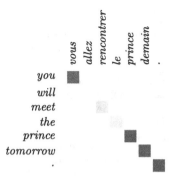

Fig. 5. Correspondences between two sentences in two different languages using the translation probabilities found in a translation table. Cells with a white colour reflect the fact that the pairs of words are not found in the translation table.

required, we do not expect any reasonable inter-judge agreement. For that reason, we propose to restrict ourselves to those analogies which deliver a sentence already present in the resource. In this way, any solution of an analogy should be a valid sentence.

7   Conclusion

In this article, we proposed an approach to solve analogies between sentences which is different from those which try to make direct use of neural networks [7, 30], or try to directly learn sentence representations [22]. Our approach combines semantic word analogies with formal string analogies.

The design of our approach highlighted some problems with word analogies. For instance, how to solve analogies like

$$last\ week : in\ two\ weeks :: yesterday : the\ day\ after\ tomorrow \qquad (32)$$

where the lengths in words do not verify the analogical arithmetic relation, i.e., $2-3 \neq 1-4$? We made a preliminary step forward in that direction by proposing an extension to our method in Sect. 5.

We discovered a pressing need for cleaning word embedding models from spurious words. In contrast to that, for languages with a much richer morphology than English, another problem will arise from the word embedding resources used: what if a declined or conjugated word form is missing from the embedding space? E.g., How to solve   *król : królów :: królowa : królowych*   (Polish king and queen in nominative singular and genitive plural) if the word *królowych* does not appear in the corpus the word embeddings have been trained from? Dynamic vector embeddings [2] dynamically construct the representation of a word from its parts or from its characters and answer the inverse problem. Here, what is needed is the generation of a new word from a given vector representation. The work presented in this article and the two datasets produced use the word as a unit. However, a different unit might allow to tackle unknown words or might allow for the generation of new words in morphologically rich languages. In a study of the density of analogies among sentences at various granularity and in various languages [5], including languages with rich morphology like German and Czech, we have shown that a unit below the word allows to discover a larger number of analogies. Adopting the current practice of segmenting using BPE or sentence pieces [11, 26] might introduce more flexibility in analogies.

The difficulties in assessing the datasets produced arise from the fact that semantic analogies between words are in fact still highly unreliable. The reliability of analogies between sentences heavily depends on the reliability of the analogies between words in the word embedding space. Datasets like the ones released in [20] (Google set) or in [4] (BATS v3.0) have indeed an insufficiently small coverage relatively to the very large possibilities opened by our technique.

Finally, to open up this conclusion to applications of the work done in this article, let us mention again the fact that one of the justifications for the present work is example-based machine translation, especially machine translation by analogy. The proposal presented here can be extended to the computation of matrices between sentences in two different languages, as illustrated in Fig. 5, and as was classically done in statistical machine translation to compute alignments. There, the similarity between words is computed based on the product of the two translation probabilities found in a translation table. This kind of matrix, in conjunction with the monolingual ones presented in this article have been exploited in an experiment in example-based machine translation, where the computation of missing matrices is performed by a neural network, before the generation of the translation sentence [27]. As a side result of that article, a dataset of large number of formal analogies (1.3 million), in two languages was produced. The work presented in the present article is a path to relax the constraint of using only formal analogies in such example-based settings.

References

1. Arora, S., Li, Y., Liang, Y., Ma, T., Risteski, A.: A latent variable model approach to PMI-based word embeddings. TACL **4**, 385–399 (2016). https://doi.org/10.1162/tacl_a_00106, https://aclanthology.org/Q16-1028/
2. Bojanowski, P., Grave, E., Joulin, A., Mikolov, T.: Enriching word vectors with subword information. TACL **5**, 135–146 (2017). https://doi.org/10.1162/tacl_a_00051, https://aclanthology.org/Q17-1010/
3. Denoual, E.: Analogical translation of unknown words in a statistical machine translation framework. In: Proceedings of the Eleventh Machine Translation Summit, MT Summit XI, pp. 135–141. Copenhagen, Denmark, September 2007. https://aclanthology.org/2007.mtsummit-papers.19.pdf
4. Drozd, A., Gladkova, A., Matsuoka, S.: Word embeddings, analogies, and machine learning: beyond king - man + woman = queen. In: COLING 2016, pp. 3519–3530 (2016). http://www.aclweb.org/anthology/C16-1332
5. Fam, R., Lepage, Y.: A study of analogical density in various corpora at various granularity. Information **12**(8), 17 (2021). https://doi.org/10.3390/info12080314, https://www.mdpi.com/2078-2489/12/8/314

6. Grave, E., Bojanowski, P., Gupta, P., Joulin, A., Mikolov, T.: Learning word vectors for 157 languages. In: LREC 2018, pp. 3482–3487 (2018). https://aclanthology.org/L18-1550/

7. He, K., Zhao, T., Lepage, Y.: Numerical methods for retrieval and adaptation in Nagao's EBMT model. In: ICACSIS 2018 (2018). https://doi.org/10.1109/ICACSIS.2018.8618226, https://ieeexplore.ieee.org/document/8618226

8. Hirschberg, D.S.: A linear space algorithm for computing maximal common subsequences. Comm. ACM **18**(6), 341–343 (1975). https://doi.org/10.1145/360825.360861

9. Hofstadter, D.: The Fluid Analogies Research Group: Fluid Concepts and Creative Analogies. Basic Books, New York (1994)

10. Itkonen, E., Haukioja, J.: A rehabilitation of analogy in syntax (and elsewhere). In: Metalinguistik im Wandel: die kognitive Wende in Wissenschaftstheorie und Linguistik, pp. 131–177. Peter Lang (1997)

11. Kudo, T.: Subword regularization: improving neural network translation models with multiple subword candidates. In: Proceedings of the 56th Annual Meeting of the Association for Computational Linguistics, vol. 1(Long Papers), pp. 66–75. Association for Computational Linguistics, Melbourne, July 2018. https://doi.org/10.18653/v1/P18-1007, https://www.aclweb.org/anthology/P18-1007

12. Langlais, P., Yvon, F.: Scaling up analogical learning. In: Coling 2008, pp. 51–54 (2008). https://aclanthology.org/C08-2013/

13. Langlais, P., Zweigenbaum, P., Yvon, F.: Improvements in analogical learning: application to translating multi-terms of the medical domain. In: EACL 2009, pp. 487–495 (2009). https://pdfs.semanticscholar.org/b7d5/1ba891a266a788bf4a8cfd791223cbb0e729.pdf

14. Langlais, P., Patry, A.: Translating unknown words by analogical learning. In: Proceedings of the 2007 Joint Conference on Empirical Methods in Natural Language Processing and Computational Natural Language Learning, EMNLP-CoNLL, pp. 877–886 (2007). https://aclanthology.org/D07-1092/

15. Lepage, Y.: Solving analogies on words: an algorithm. In: COLING-ACL 1998, vol. I, pp. 728–735 (1998). https://aclanthology.org/C98-1116/

16. Lepage, Y.: De l'analogie rendant compte de la commutation en linguistique. Habilitation thesis, Université de Grenoble (2003). https://tel.archives-ouvertes.fr/tel-00004372/

17. Lepage, Y.: Character-position arithmetic for analogy questions between word forms. In: Computational Analogy Workshop, ICCBR 2017, pp. 17–26 (2017). http://ceur-ws.org/Vol-2028/paper2.pdf

18. Lepage, Y.: Semantico-formal resolution of analogies between sentences. In: Vetulani, Z., Paroubek, P. (eds.) Proceedings of the 9th Language & Technology Conference (LTC 2019) - Human Language Technologies as a Challenge for Computer Science and Linguistics, pp. 57–61, May 2019. http://lepage-lab.ips.waseda.ac.jp/media/filer_public/32/04/32049346-75dd-4bd1-93cc-ae221e49a2e9/ltc-005-lepage.pdf

19. Lepage, Y., Denoual, E.: Purest ever example-based machine translation: detailed presentation and assessment. Mach. Transl. **19**, 251–282 (2005). https://doi.org/10.1007/s10590-006-9010-x

20. Mikolov, T., Yih, W.T., Zweig, G.: Linguistic regularities in continuous space word representations. In: NAACL-HLT 2013, pp. 746–751 (2013). http://www.aclweb.org/anthology/N13-1090

21. Mimno, D., Thompson, L.: The strange geometry of skip-gram with negative sampling. In: Proceedings of the 2017 Conference on Empirical Methods in Natural Language Processing, EMNLP 2017, pp. 2873–2878. Association for Computational Linguistics, Copenhagen, September 2017. https://doi.org/10.18653/v1/D17-1308, https://www.aclweb.org/anthology/D17-1308

22. Pagliardini, M., Gupta, P., Jaggi, M.: Unsupervised learning of sentence embeddings using compositional n-gram features. In: NAACL-HLT 2018, pp. 528–540 (2018). https://doi.org/10.18653/v1/N18-1049, https://aclanthology.org/N18-1049/

23. Pennington, J., Socher, R., Manning, C.D.: GloVe: global vectors for word representation. In: EMNLP 2014, pp. 1532–1543 (2014). http://www.aclweb.org/anthology/D14-1162

24. Peters, M.E., et al.: Deep contextualized word representations. In: NAACL-HLT 2018, pp. 2227–2237 (2018). http://aclweb.org/anthology/N18-1202

25. Rhouma, R., Langlais, P.: Fourteen light tasks for comparing analogical and phrase-based machine translation. In: Proceedings of the 25th International Conference on Computational Linguistics, COLING 2014, vol. Technical Papers, pp. 444–454. Dublin City University and Association for Computational Linguistics, Dublin, August 2014. https://aclanthology.org/C14-1043/

26. Sennrich, R., Haddow, B., Birch, A.: Neural machine translation of rare words with subword units. In: Proceedings of the 54th Annual Meeting of the Association for Computational Linguistics, vol. 1(Long Papers), pp. 1715–1725. Association for Computational Linguistics, Berlin, August 2016. https://doi.org/10.18653/v1/P16-1162, https://www.aclweb.org/anthology/P16-1162

27. Taillandier, V., Wang, L., Lepage, Y.: Réseaux de neurones pour la résolution d'analogies entre phrases en traduction automatique par l'exemple. In: Actes de la 6e conférence conjointe Journées d'Études sur la Parole (JEP, 31e édition), Traitement Automatique des Langues Naturelles (TALN, 27e édition), Rencontre des Étudiants Chercheurs en Informatique pour le Traitement Automatique des Langues (RÉCITAL, 22e édition), vol. 2. Traitement Automatique des Langues Naturelles, pp. 108–121. AFCP et ATALA (Mai 2020). https://hal.archives-ouvertes.fr/hal-02784759

28. Wagner, R.A., Fischer, M.J.: The string-to-string correction problem. J. ACM **21**(1), 168–173 (1974)

29. Wang, L., Lepage, Y.: Vector-to-sequence models for sentence analogies. In: IEEE Proceedings of the 2020 International Conference on Advanced Computer Science and Information Systems, ICACSIS 2020, pp. 441–446, October 2020. https://doi.org/10.1109/ICACSIS51025.2020.9263191, https://ieeexplore.ieee.org/document/9263191

30. Zhao, T., Lepage, Y.: Context encoder for analogies on strings. In: PACLIC 32 (2018). https://aclanthology.org/Y18-1096/

Effective Development and Deployment of Domain and Application Conceptualization in Wordnet-Based Ontologies Using Ontology Repository Tool

Jacek Marciniak[(✉)] [iD]

Adam Mickiewicz University in Poznań, Uniwersytetu Poznańskiego 4, Poznań, Poland
jacekmar@amu.edu.pl

Abstract. Ontology Repository Tool is a piece of software aimed to build wordnet-based ontologies which are an example of an information language designed to represent knowledge of various kinds, including general, domain, and application knowledge. This language extends the structure of the wordnet with new types of relations that make it possible to build synset hierarchies parallel to the wordnet structure. Ontology Repository Tool is equipped with functionalities that facilitate the development and management of polyhierarchical and polyrelational knowledge structures. Moreover, the software is intended for integration with information systems such as e-learning content repositories and information systems with multimodal data in which the wordnet-based ontology is deployed and expanded while indexing the documents stored there.

Keywords: Wordnet-based ontology · Management of knowledge of various types · Development and deployment of information languages

1 Introduction

Information languages, or indexing languages, are artificial languages designed to describe content and basic formal features of documents. Such languages are used primarily to describe the subject matter of documents stored in information systems. Depending on the language, the level of detail in describing the subject matter will vary. Most existing information languages describe the subject matter of documents at a high level of generality, because these languages are general in nature and are built to cover documents and resources with different subject matter. Examples of such languages are classification systems, subject heading lists, and some thesauri. Other information languages are domain-specific, i.e. they cover a narrow subject area. Such languages include most thesauri, glossaries and domain ontologies. There are also languages intended for building application conceptualizations, i.e. those that are built for the needs of a particular information system and are not of a general nature. Ontologies or simple glossaries are used for such purposes. A language that represents knowledge with different levels

© Springer Nature Switzerland AG 2022
Z. Vetulani et al. (Eds.): LTC 2019, LNAI 13212, pp. 180–194, 2022.
https://doi.org/10.1007/978-3-031-05328-3_12

of detail is wordnet-based ontology. It is a language designed to organize domain- and application-specific knowledge with generic knowledge derived from the wordnet.

When creating an information language, it is necessary to decide in which tool this language will be created, maintained and provided for indexing resources. The choice of tool always depends on the engineering process in which the language is created. Due to the generic nature of many languages, they are created in a process that does not take into account how the language will be used in an information system for indexing. Tools are therefore considered from two perspectives: some tools are used in the construction and development of the language, others are used when indexing resources using the language. The situation is different when an information language is created and developed during the process of indexing resources. This paper presents the process of building and developing a wordnet-based ontology, which is created in parallel with the process of indexing the resources collected in the information system. This is possible by using the Ontology Repository Tool software, which is a multi-user ontology editor and at the same time a server that can be integrated with the information system in which the ontology is used. This paper discusses the development and deployment of the PMAH ontology, which was developed while indexing content in two systems: the E-archaeology Content Repository and the Hatch system storing multimodal data.

2   Development and Deployment of Information Languages

Information languages are used to index documents and other content (e.g., multimedia content) collected in information systems. The indexing process involves determining the subject matter of a document and storing it in a concise form. In controlled systems, the indexing process is carried out by domain experts, as only they are able to select appropriate indexes describing a given resource. In uncontrolled systems, documents can be described directly by their authors.

Among information languages, there are languages based on quasi natural vocabulary, i.e. those which use terms or words coming from natural language. Examples of such languages are keyword languages, descriptor languages or thesauri such as the Art & Architecture Thesaurus [3] General Multilingual Environmental Thesaurus [4], and Geonames [5]. Another type of languages are those based on artificial vocabulary, i.e. those that use symbols that are a combination of letters and numbers. Examples of such languages are classification systems (library, library-bibliographic, and documentation classifications) such as Universal Decimal Classification, Colon Classification, Library of Congress Classification and Iconclass [8].

A more advanced information language is an ontology. Ontologies belong to the group of languages with quasi natural vocabulary. The concepts collected in an ontology can be used to index resources, although the leading application of ontologies is to build knowledge bases. The process of indexing using ontologies is efficient because during the indexing process an expert can select the optimal index by searching the ontology by referring to the relations linking the concepts. What relations will be used to link concepts depends on the specific ontology. The most commonly used relations are the subclass relation, the type of relation, and the synonymy relation. If the ontology is written in a formal form (heavy ontology) then in addition to aggregating concepts linked by relations, the ontology can also be used to derive new concepts and to validate hypotheses written as queries to the ontology [20]. Relationships between concepts collected in an information language are also encountered in thesauri [1]. A resource with a specific form are wordnets, which are called lexical databases by their creators [2]. Concepts in wordnets (synsets) are connected by relations, hence wordnets can be treated as ontologies. Because of their form, they should be treated as light ontologies.

Tools for building information languages are available for those types of languages for which it is assumed that there will be many instances of the language created by different teams. This is especially true for languages intended to create domain- or application-specific conceptualizations. Examples of such languages are glossaries, thesauri and ontologies, for which tools such as Protégé [16], and TermaTres [19] are available. Indexing language editors are also available for those languages that are generic and for which many resources are created. An example is wordnets, which are being created for many natural languages. Tools designed to create wordnets include, for example, Visdic [7], or WordnetLoom [17].

Tools used for building information languages should support teamwork, since with large languages they are created collaboratively. Moreover, they should allow for iterative improvement and completion of the created language. The basic problems that arise when creating information languages are as follows [9]:

- Deciding on the synonymy of two concepts,
- Removing terms that are too specific or too general from the language,
- Correcting the form of multi-word terms,
- Handling abbreviations and acronyms,
- Adding terms of a technical nature.

Regardless of the adopted engineering process used to create the information language, the above problems require the following steps: identifying the problem, deciding on the optimal solution, and analyzing how the postulated changes will affect the language. These steps require the work of an editorial team and should lead to the release of the next version of the information language. When the language is developed in parallel with the resource indexing process, these steps will overlap. It will then not be possible to release successive versions of the language periodically to meet the expectations of the indexers. A long period of waiting for the next version destabilizes the work of indexing teams [21].

3 General, Domain and Application Conceptualization in Wordnet-Based Ontologies

A wordnet-based ontology is an ontology whose structure is based on a wordnet lexical database (a.k.a. wordnet) and also allows for the storage of domain and application knowledge. This possibility exists by extending the wordnet structure with relations of a new type, which conceptualize concepts from a different perspective than the one expressed by the wordnet hierarchy [14]. Such an approach makes a wordnet ontology an information language in which the synset used to represent concepts is derived from the wordnet structure and the wordnet hierarchical structure is the backbone of the language thus created. According to the definition adopted for wordnets, a synset is a set of synonymous words that can be used interchangeably in a certain context. Wordnet relations between synsets are relations such as hyponymy, antonymy, holonymy, near synonymy, etc., but also relations connecting individual entities (instances) and classes – "belongs to class" relation [22]. When using a wordnet-based ontology for indexing resources, the latter relation makes it possible to introduce proper names to the ontology, which are used to record names, place names, geographic names, dates, etc. As such, wordnets are treated as generic ontologies due to the fact that they contain a Top Concept Ontology expressing generic knowledge [22]. Other generic ontologies such as the CYC [17] or SUMO [18] ontologies can also be combined with wordnets.

A wordnet-based ontology extends the underlying resource (i.e., the wordnet) with new types of relations that allow domain and application conceptualizations to be introduced into the wordnet structure [12, 14]. This is achieved through:

- Domain Categories Hierarchies,
- Assigning synsets to Domain categories,
- Linking synsets to each other using domain relations.

Domain category is a separate concept, independent of the synset. Domain categories hierarchies and connecting synsets to Domain categories allow synsets to be organized into hierarchies created by domain experts. The expert perspective can be domain-specific i.e., it can lead to an information language that can be used to index resources in more than one information system. An example of such a conceptualization is the PMAH ontology (Protection and Management of Archaeological Heritage), which covers the core concepts of archaeology and natural heritage and is generic enough to allow resources to be indexed in different content repositories [12]. The conceptualization may also be strictly application-oriented, i.e., it may cover such a narrow subject area and be so detailed that it is only applicable to a single information system. It may also happen that different conceptualizations are combined in one language, i.e., domain-specific and application-specific knowledge is represented simultaneously. This is achieved by adding multiple hierarchies built using different relations to the wordnet-based ontology.

The third of these relations introduced into the wordnet-based ontology, i.e. relations of domain character between synsets, refers to relations excluded from wordnets. According to the assumptions of the creators of WordNet, the so-called tennis problem type relations should not be included in it [2]. This term refers to the assumption that relations such as those between the words ball, net, racket, referring to the game of tennis, cannot be expressed in WordNet. This assumption stems from the approach that Word-Net, as a resource that focuses on the semantics of words, should not express relations from the level of text semantics or discourse. When wordnet is used as the backbone of an ontology used for document indexing, this type of restriction is unacceptable. Unlike wordnets, wordnet-based ontologies are precisely concerned with expressing domain and application relationships without losing the generic conceptualization provided by the wordnet. This allows the indexer to select the most appropriate indexes by referring to knowledge beyond that of a general nature.

3.1 General and Domain Conceptualizations in Wordnet-Based Ontologies

General and domain conceptualizations in wordnet ontologies are obtained through the following relations [12]:

- synsets, to express synonymy between lexical units,
- wordnet relations between synsets (hypernymy, holonymy, "belongs to class", etc.),
- calculated relations between synsets determining similarity of concepts,
- hierarchies of domain categories connected by a subclass of relationship expressing domain conceptualization,
- assignments of synsets to domain categories to specify meaning with respect to domain knowledge,
- relations between synsets of domain nature introduced by domain experts.

Synonymy is a fundamental type of relation used in information languages [1]. Unlike controlled vocabularies, the use of synsets to express synonymy between lexical units in wordnet-based ontologies excludes descriptor indication. This is not a problem since this information language is not designed to index resources in a controlled manner.

Wordnet relations are intended to organize the lexical units collected in an ontology to express relationships of a general nature (e.g. archaeology – aerial archaeology). These relations are used to calculate similarity relations between concepts (e.g. aerial archaeology is in a relation with maritime archaeology because both concepts share the same hypernym, which is archaeology). The calculated relations are used in prompting indexes during indexing and searching of the information system.

A domain hierarchy built through a subclass of relations between domain categories is a language-independent ordering of synsets coming from one domain (i.e. one subject area). The assignment of synsets to domain categories expresses the conceptualization according to the intention of the ontology's authors. In the case of the PMAH ontology, this hierarchy is treated as a basic domain-specific conceptualization. The domain relationships between synsets express the auxiliary relationships between concepts in the indexing process (e.g. archaeology – archaeological method).

The process of using wordnet-based ontologies in the indexing process boils down to the fact that the indexer uses a lexical unit derived from the ontology as an index, which is then assigned to the document (see Fig. 1). An example ordering of general and domain specific concepts from the PMAH ontology is presented in Fig. 2.

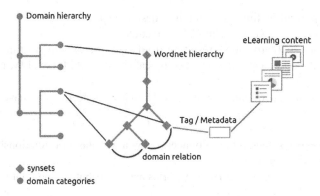

Fig. 1. Domain conceptualization embedded in wordnet-based ontology hierarchy (indexing of eLearning content case) [14]

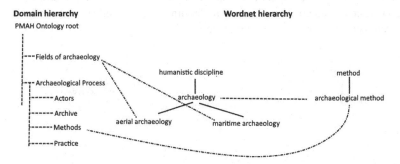

Fig. 2. General and domain conceptualizations in the PMAH ontology

3.2   Application Conceptualizations in Wordnet-Based Ontologies

In order to achieve application conceptualizations in wordnet-based ontologies, the same types of generic relations are used as in domain conceptualizations, i.e. synsets and word-net relations between synsets. In the case of application conceptualizations, however, synsets are used to store more detailed concepts than, those found in wordnets. The lack of detailed concepts in wordnets is due to their overly terminological nature. It can be assumed that terms of this type were not added there as a result of the editors' decisions, who ruled out adding terminology, either because of lexicographers' competence deficiencies and/or lack of this type of vocabulary in the corpora used to create the wordnets, or simply because of their arbitrary decisions. Among the specific lexical units introduced in the application conceptualizations in the PMAH ontology, we distinguish [14]:

- terminological concepts, i.e. astragali, animal bone, crane ulna, abandonment deposit, zoomorfic, barley seeds, feasting deposit, post retrieval pit, multi-roomed construction,

- concepts representing time as required to describe specific types of resources in the repository, i.e. "3–12 years – child", "20 + adult",
- chronology in qualitative units resulting from the representation of time adopted to describe artifacts at a particular archaeological site i.e. TP M, Level IV, North I.

In applications, the hierarchies created by Domain categories also have a different function:

- domain categories hierarchies connected by a subclass of relationship express application conceptualizations,
- assignments of synsets to domain categories act as attribute-value relations.

Domain categories hierarchies are used to organize concepts that are used in a specific application to index resources within specific data structures (attribute), which are interpreted by assigning a specific domain category. With this interpretation, assigning synsets to domain categories boils down to defining what synsets (value) can be assigned to a given data structure in the system (see Fig. 3).

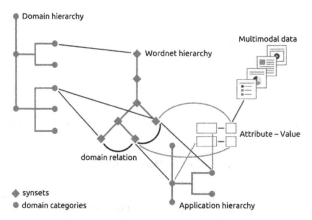

Fig. 3. Domain and application conceptualizations embedded in wordnet-based ontology hierarchy (indexing of multimodal data case) [14]

With this approach, the process of indexing a document within a particular attribute comes down to selecting one index from a limited list of indexes assigned to the domain category. For example, in the Hatch system, only values assigned to the domain category "Figurine material" can be assigned to the "Material" attribute in the "Figurine" tab (see Fig. 4 and Fig. 9). In practice, this solution comes down to defining a list of possible values that can be used in the indexing process.

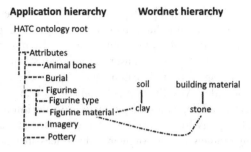

Fig. 4. General and application conceptualizations in the PMAH ontology

4   Ontology Repository Tool in Development and Deployment of Wordnet-Based Ontologies

Ontology Repository Tool is a tool designed to build wordnet-type ontologies (see Fig. 5).

Fig. 5. Ontology Repository Tool

It has an architecture that allows the tool to be used in two ways:

- As an ontology editor, where the ontology can be created and modified,
- As an ontology repository, which can be integrated into an external information system, where the wordnet-based ontology is used to index resources and where it can be extended during the indexing process.

As an ontology editor, the Ontology Repository Tool allows multiple users with different privilege levels to work simultaneously. For example, it is possible to perform operations only on a selected ontology hierarchy, and it is also possible to limit the privileges to actions of a specific type (e.g. a user can add terms, but cannot delete them). A tool treated as an ontology repository can be integrated with an external information system through the provided API. Such integration allows domain experts to perform ontology operations (e.g., adding new terms) directly from the information system while indexing resources. Ontology Repository Tool is a web browser-accessible tool, implemented in Java Spring Framework and Angular, and Services are made available through REST web services.

4.1 Ontology Editor

Among the basic functions of the tool, the ability to add, remove and modify concepts should be pointed out. The tool can simultaneously process concepts of different types. For example, in the PMAH ontology, there are 9 types of concepts, i.e. synsets for 6 language versions and 3 types of domain categories intended for building three different conceptualizations, one of domain character and two of application character. The tool allows for handling relations of any type. For the purpose of building the PMAH ontology, relations taken from wordnet such as hypernymy/hyponymy and holonymy/meronymy were used. Domain categories are hierarchized using up_category/down_category relations. Synsets are pinned to categories via domain_category/domain_representant relations. Domain relations between synsets are expressed using is_linked_to relations. A set of calculated relations between concepts is also available (e.g. "hypernym-hypernym", and "has-same-holonym"). These relations are calculated according to the algorithm defined for wordnet-based ontologies [12].

In a wordnet-based ontology, the same concept (i.e., the same synset) can occur in parallel in several hierarchies. For example, in the PMAH ontology, the term archaeology occurs in two hierarchies: the wordnet hierarchy (hypernymy relation with humanistic discipline) and the domain hierarchy (domain_category relation with Domain category Fields of archaeology) (see Fig. 2). Similarly, the term clay occurs in the wordnet hierarchy (hypernymy with soil) and the application hierarchy built for the Hatch system (hatc_domain_hiearchy relationship with Domain category Figurine material) (see Fig. 4). The presentation of the hierarchies in which the synset is embedded is implemented in the Ontology Repository Tool through Indexes (see Fig. 6). They make it possible to change the perspective (i.e. hierarchy) and provide full insight into how many conceptualizations a given concept appears in.

The tool is also equipped with a number of functions supporting the maintenance of coherence of the constructed resource, e.g. it is possible to search for orphans and tops and bottoms of the hierarchy with respect to the relation indicated by the user.

Fig. 6. Presentation of a wordnet-based hierarchy through an Index

4.2 Ontology Repository

The Ontology Repository Tool can also be used as an ontology repository. This means that the ontology can be accessed through the API by an external information system in which the ontology is used. What operations will be performed on the ontology from the external system depends on the specific implementation. First of all, it will be the use of the ontology in the process of indexing resources (selection of appropriate indexes by the indexer), but also making changes to it during the execution of the indexing process by the domain expert. The use of the tool as a repository integrated with the information system significantly improves the ontology development process. By combining the process of ontology development and usage in one tool, there is no need to release subsequent versions of the ontology before uploading it to the target system. All changes and additions are made to the ontology by domain experts when indexing the resources.

5   Deployment of Wordnet-Based Ontologies

As indicated above, the process of deployment of an wordnet-based ontology can be linked to the process of its development through an appropriate architecture of the information system under construction. Depending on the assumptions made, the scope of changes to the ontology that are made directly from the information system may vary. The Ontology Repository Tool has been used as an ontology repository in, for example, the E-archaeology Content Repository and the Hatch system. In these systems it is possible to perform the following operations on the ontology:

- Retrieving information about lexical units collected in the ontology with information about which synset they come from and the hierarchies in which the synset is located,
- Inserting new synsets into the ontology for lexical units that have been used to index documents that were not previously present in the ontology,

- Introducing synonyms to synsets that exist in the ontology for lexical units that have been used for document indexing and that previously did not exist in the ontology,
- Assigning newly added synsets to domain categories that form domain-specific conceptualizations,
- Assigning newly added synsets to domain categories creating conceptualizations of an application nature,
- Linking synsets for lexical units used to index resources using domain relationships.

In each information system, the list of operations that can be performed on the ontology varies. The differences are due to the different functions of the systems. Of course, in addition to performing operations directly from the information system, editing activities can be performed on the ontology directly in the Ontology Repository Tool. However, this was reserved for editors who take care of the consistency of the built ontology and who periodically verified the quality of the created resource.

5.1 Development and Deployment of Wordnet-Based Ontology in Repository of e-learning Content

The E-archaeology Content Repository contains e-learning materials on the protection of archaeological heritage and the management and protection of cultural and natural heritage. More than 6,200 learning objects in 9 languages and more than 30 training programs are stored in the repository. Initially, the repository contained content on the protection of archaeological heritage [11], which was later supplemented with content on the management and protection of cultural and natural heritage [13]. The indexing of e-learning materials consists in assigning lexical units selected by the indexer to the metadata keywords. This process is carried out using the PHAH ontology.

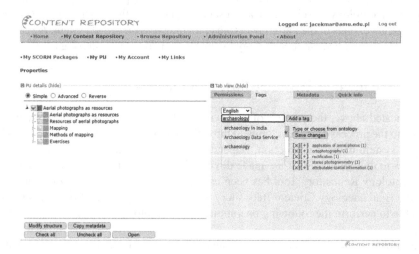

Fig. 7. Lexical units from the PMAH ontology suggested in the E-archaeology Content Repository

The ontology is used to suggest to the indexer the words and phrases that can be used as indexes (see Fig. 7). These words can also be selected from the domain hierarchy presented to the indexer. If there is no index corresponding to the domain expert in the ontology, it can create one and assign it to the e-learning content. In this case, the created lexical unit will also be immediately inserted into the ontology as a new synset. During this activity, the indexer also has the option to plug the new lexical unit into a domain category selected from the Domain categories hierarchy. Optionally, it can also add a domain relationship with other synsets found in the ontology (see Fig. 8). Because the materials are multilingual, the indexer can also link the introduced lexical unit (synset) to a synset in another language.

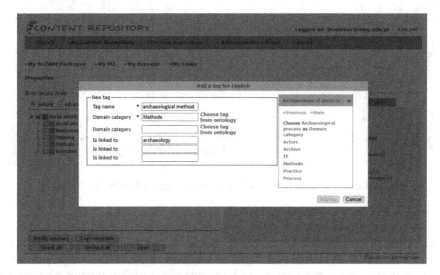

Fig. 8. Connecting a new lexical unit with a Domain category and a synset

5.2 Development and Deployment of Wordnet-Based Ontology in Hatch System

Another system in which the ontology is developed and deployed during the resource indexing process is the Hatch system. Hatch (House at Çatalhöyük) is an advanced system designed to create and maintain a digital collection [6, 10]. It is aimed at presenting a wide range of multimodal data about the Neolithic settlement at Çatalhöyük in a multiscalar and interactive form linking information of different character (types of artifacts, their attributes, relations among them) with different form of their presentation (text, photographs, graphics, maps, GIS localizations and multiscalar chronology of artifacts). It has didactic objectives and is designed to meet the needs and expectations of both professionals and general public interested in the human past.

The system was built when the excavation works at the site were at an advanced stage. The expert team had a large amount of data of different types at their disposal, such as textual descriptions, photographic material, maps, a GIS database, and an artifact chronology. This data was not structured to be made available in a system targeted at a general public that lacked expertise on the Çatalhöyük site. Due to the nature of the site, there was no indexing language that could be used to describe the resources collected in the system. Therefore, a solution was adopted in which PMAH ontology was extended to include topics related to the Neolithic site with consideration of the character of the Çatalhöyük site. This was natural since it collected lexical units from the field of archaeology in PMAH.

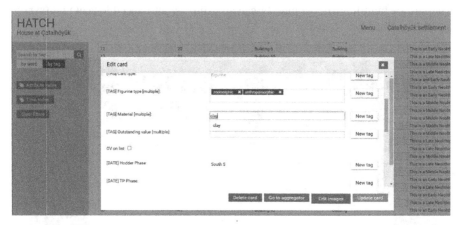

Fig. 9. Lexical units from the PMAH ontology suggested in the Hatch system

As in the previous case, an ontology is used to suggest to the indexer words and phrases that can be used as indexes (see Fig. 9). Unlike the e-learning content repository, where lexical units are assigned as tags to a single metadata, in the Hatch system the description structure is more extensive. Each object is described through a structured form with multiple fields. In this form, values for some fields (numeric values, text) are stored in the system database, others can only be assigned values (lexical units) from a limited dictionary stored in the ontology (see Fig. 9). This is accomplished by using an application conceptualization in the ontology and it boils down to mapping the fields in the Hatch system to Domain categories in the PMAH ontology. Thanks to this, the user who is supposed to assign a value to such a field in a form is prompted only with lexical units (synsets) assigned to the category associated with this field (see Fig. 9). If the expected value does not exist in the ontology, the indexer can add it. During such operation, a synset will be created for the lexical unit in the ontology, which will be immediately linked to the Domain category associated with the field (see Fig. 10). The adopted solution supports the maintenance of uniformity of indexing resources by many indexers by imposing on them a finite list of possible indexes in a given field. The new index is added only when the expected index is missing from the list. However, errors, e.g., typos, use of abbreviations, occurred in such a process [14]. These were corrected by editors working directly in the Ontology Repository Tool editor.

Fig. 10. Adding a new lexical unit and linking it with the domain category associated with the Figurine material field

6 Conclusions

The process by which an information language is built and maintained determines its quality. If the language is built in a laboratory setting by a small, fully controlled group of editors, the implementation of professional quality procedures is relatively straight-forward. If the language is built in parallel with the indexing process in the production information system, the quality-related measures must be supported by appropriate tools. The solution presented in this paper based on Ontology Repository Tool supports the work of multi-author teams who build the information language in parallel with the pro-cess of indexing resources using it. Thanks to such a combination, the domain expert, while describing the subject matter of a particular artefact introduced into the informa-tion system, has the possibility to decide with which concept the artefact will be best described. As a result, the information language is immediately provided with the con-cepts that are the most appropriate from the point of view of the needs of a particular domain. Thanks to the functions of the presented IT solution, these concepts will also be immediately connected to the hierarchical structure of the constructed information language.

References

1. Dextre Clarke, S.G., Lei Zeng, M.: From ISO 2788 to ISO 25964. The Evolution of Thesaurus Standards towards Interoperability and Data Modeling, ISO Information Standards Quarterly, Winter 2012, vol. 24 (2012)
2. Fellbaum, Ch. (ed.): WordNet: An Electronic Lexical Database, MIT Press, Cambridge (1998)
3. Getty AAT Homepage. https://www.getty.edu/research/tools/vocabularies/aat/. Accessed 20 Dec 2021
4. Gemet Homepage. http://www.eionet.europa.eu/gemet/en/about/. Accessed 20 Dec 2021
5. The GeoNames geographical database Homepage. http://www.geonames.org. Accessed 20 Dec 2021

6. Hatch Homepage. https://hatch.e-archaeology.org. Accessed 20 Dec 2021
7. Horak A., et al.: DEB Platform Tools for effective development of WordNets in application to PolNet. In: Vetulani, Z. (ed.) Proceedings of 3rd Language and Technology Conference, Poznań, Poland, 5–7 October 2007 (2007)
8. Iconclass RKD Homepage. www.iconclass.nl. Accessed 20 Dec 2021
9. Joudrey, D.N., Taylor, A.G., Wisser, K.M.: The Organization of Information, 4th edn. Libraries Unlimited, Santa Barbara (2018)
10. Marciniak, A., Marciniak, J., Filipowicz, P., Harabasz, K., Hordecki, J.: Engaging with the Çatalhöyük database. House at Çatalhöyük (HATCH) and other applications. Near Eastern Archaeol. **83**, 2 (2020)
11. Marciniak, J.: Building E-learning content repositories to support content reusability. Int. J. Emerg. Technol. Learn. (iJET) **9**(3), 45–52 (2014)
12. Marciniak, J.: Building wordnet based ontologies with expert knowledge. In: Vetulani, Z., Uszkoreit, H., Kubis, M. (eds.) LTC 2013. LNCS (LNAI), vol. 9561, pp. 243–254. Springer, Cham (2016). https://doi.org/10.1007/978-3-319-43808-5_19
13. Marciniak, J.: Methods and tools for centers of integrated teaching excellence providing training in complementary fields. In: Proceedings of 11th International Conference on Computer Supported Education (CSEDU 2019), vol. 2, pp. 527–534 (2019)
14. Marciniak, J.: Wordnet as a backbone of domain and application conceptualizations in systems with multimodal data. In: Proceedings of the LREC 2020 Workshop on Multimodal Wordnets (MMW 2020), Marseille, France, pp. 25–32 (2020)
15. Naskręt, T., Dziob, A., Piasecki, M., Saedi, C., Branco, A.: WordnetLoom - a multilingual wordnet editing system focused on graph-based presentation. In: Proceedings of the 9th Global Wordnet Conference, Singapore, 8–12 January 2018 (2018)
16. Protégé Homepage. http://protege.stanford.edu/. Accessed 20 Dec 2021
17. Ramachandran, D., Reagan, P., Goolsbey, K.: First-orderized researchCyc: expressivity and efficiency in a common-sense ontology In: Papers from the AAAI Workshop on Contexts and Ontologies: Theory, Practice and Applications, Pittsburgh, PA (2005)
18. SUMO Homepage. http://www.ontologyportal.org/. Accessed 20 Dec 2021
19. TemaTres Homepage. http://www.vocabularyserver.com. Accessed 20 Dec 2021
20. Uschold, M., Gruninger, M.: Ontologies: principles, methods, applications. Knowl. Eng. Rev. **11**(2), 93–136 (1996)
21. Weda, R.U.: Update on the Dutch AAT work. https://www.getty.edu/research/tools/vocabularies/weda_zelfde_aat_dutch_2016.pdf. Accessed 20 Dec 2021
22. Vossen, P. (ed.): Euro WordNet General Document, Version 3. University of Amsterdam (2002)

Emotions, Decisions and Opinions

Speech Prosody Extraction for Ibibio Emotions Analysis and Classification

Moses E. Ekpenyong$^{(\boxtimes)}$ [ID], Aniekan J. Ananga, EmemObong O. Udoh [ID], and Nnamso M. Umoh

University of Uyo, P.M.B. 1017, Uyo 520003, Nigeria
{mosesekpenyong, ememobongudoh}@uniuyo.edu.ng

Abstract. This paper investigates basic prosodic features of speech (duration, pitch/F0 and intensity) for the analysis and classification of Ibibio (New Benue Congo, Nigeria) emotions, at the suprasegmental (sentence, word, and syllable) level. We begin by proposing generic hypothesis/baselines, representing a cache of research works documented over the years on emotion effects of neutral speech on western languages, and adopt the circumplex model for the effective representation of emotions. Our methodology uses standard approach to speech processing and exploits machine learning for the classification of seven emotions (anger, fear, joy, normal, pride, sadness, and surprise) obtained from male and female speakers. Analysis of feature-emotion correlates reveal that syllable (duration) units yield the least standard deviation across selected emotions for both genders, compared to other units. Also, there appear to be consistency for word and syllable units – as both genders show same duration correlate patterns. For F0 and intensity features, our findings agree with the literature, as high activation effects tend to produce higher F0 and intensity values, compared to low activation effects, but neutral and low activation effects produce the lowest pitch/F0 and intensity values (for both genders). A classification of the emotions yields interesting results, as classification accuracies and errors remarkably improved in emotion-F0 and emotion-intensity classification for support vector machine (SVM) and decision tree (DT) classifiers, but the highest classification accuracies were produced by the three classifiers at the sentence unit/level for fear emotion, with the k-nearest neighbour classifier (k-NN) leading (DT: 90%, SVM: 90%, k-NN: 92.40%).

Keywords: Emotion recognition · Ibibio affects · Machine learning · Prosodic features · Speech processing · Tone languages

1 Introduction

Speech contains two distinctive channels: the linguistic and paralinguistic channels. These channels are not only important, but also mutually informative for driving efficient communication. The linguistic (or explicit) channel of speech transmits the contents of a conversation, while the paralinguistic (or implicit) channel of speech conveys cues to revealing the emotional state of a speaker as well as the implicit meaning of the communicated message, and which information is heavily dominated by prosody. Hence,

© Springer Nature Switzerland AG 2022
Z. Vetulani et al. (Eds.): LTC 2019, LNAI 13212, pp. 197–213, 2022.
https://doi.org/10.1007/978-3-031-05328-3_13

the ability to predict speakers' emotional states and their communication styles is a useful precursor to effective communication agents that can ensure suitable response to users' requests and for making conversation more natural [1]. Emotion recognition – a key step of affective computing, is the process of decoding embedded emotional message(s) from human communication signals (e.g., visual, audio, and/or other physiological cues). The communication signal of interest investigated in this paper is the audio signal.

Prosodic features (sometimes known as suprasegmental phonology) are those aspects of speech which go beyond phonemes and deal with the auditory qualities of sound. Suprasegmentals are contrastive elements of speech that cannot be easily analysed as distinct segments but rather belong to a syllable or word. One approach to describing prosodic expressions (through speech) is to select a set of basic features and develop a profile of each emotion based on the features it possesses. Hence, this paper considers the following prosodic features: duration, pitch, and intensity, for the study of Ibibio affects/emotions. Duration (or speech rate) indicates the length of phonetic segments, which transmits a wide range of meanings such as speakers' emotions [2]. Pitch measures the intonation and is represented by the fundamental frequency (F0). Pitch tends to increase for anger, joy, and fear, but decreases for sadness. It has also been found to become more variable for anger and joy [3]. Intensity represents the amplitude of vocal signals and measures the speech's loudness. Intensity has been found to increase for anger, joy, and fear, but decreases for sadness [2]. A cache of research works over the years have associated emotion effects (perceived changes) in relation to neutral speech of western languages, and this forms our general hypothesis or baselines. A summary of the common emotion effects is presented in Table 1.

Table 1. Emotions speech feature correlates for western languages [4–7]

Feature	Anger	Disgust	Fear	Joy	Sadness	Surprise
Speech rate	>	<<<	>>	> or <	<	>
Average pitch	>>>	<<<	>>>	>>	<	>
Pitch range	>>	>	>>	>>	<	>
Pitch changes	abrupt (sudden on stressed)	fall (terminal)	normal	smooth, fluid (upward inflection)	fall (downward inflection)	rise (upward inflection)
Intensity	>	<	=	>	<	>
Speech quality	breathy (chest)	groaningly	breathy (irregular voicing)	breathy (blaring tone)	deep (resonant)	breathy (blaring tone)
Articulation	tense	normal	precise	normal	slurred	precise

The analysis of speech emotions is very complex in that vocal expression is evolutionarily old nonverbal affect coded in a distinctive and continuous fashion and contains a mixture of emotions and verbal messages that are arbitrarily and categorical coded. Debates on the extent to which verbal and nonverbal aspects can be neatly detached are ongoing, but a trivial argument proving some degree of independence is illustrated by the fact that people can perceive mixed messages in speech utterances, where words may convey a uniquely different thing from nonverbal cues. In Fig. 1, the circumplex model – currently the widely accepted paradigm for the representation of affect is presented.

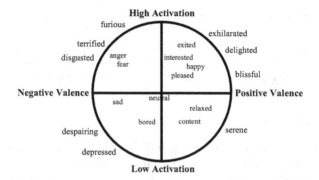

Fig. 1. Two dimensional circumplex model of emotion (Adapted from [8]).

Whereas most of the notable stress languages such as English and Spanish have been widely researched and developed, this cannot be said of tone languages. The latter are called low-resourced languages and are characterised by the lack of linguistic resources such as training corpora [9]. The language used to demonstrate our experiment is Ibibio – A tone language of the New Benue Congo language family, spoken by about 4 million speakers in the Southeast Costal region of Nigeria, West Africa. This paper is therefore pioneering because to the best of our knowledge there is no sufficient evidence for emotion technology research in African tone languages (ATLs).

2 Ibibio Vowel, Consonant and Tone Systems

2.1 Vowel System

Ibibio has a 10-vowel system (α, ɛ, ι, ɪ, o, ǫ, ʋ, ʋ̣, ʌ, ə), of which six also occur as long vowels (αα, ɛɛ, ιι, oo, ǫǫ, ʋʋ), summing up to 16. The language has great dialectal variety in its vowel inventory, especially with the central (or centralised) vowels: /ι, ə, ʌ, ʋ/. However, except for [10] who posited a six-vowel system /ι, ɛ, α, o, ʋ, ǫ/ and [11], seven: / ʋ̣, ɛ, α, o, ʌ, ʋ, ǫ/, most other analyses give Ibibio a ten-vowel system. This ten-vowel system (maximal contrast) is based largely on a pan-dialectal survey, as it is possible to find a few dialects showing a less than ten-phonemic contrast. Specifically, it is possible that some dialects have no phonemic difference between /ə/ and /o/, / ʋ/ and / ʋ̣/ as well as between /ι/ and / ɪ/; but Asanting Ibiono and Use Abat dialects do have. Also, / ʋ̣/ may be used in place of / ʋ/; just as /ι/ can alternate with / ɪ/ [12]. The phonemic vowel system of Ibibio is shown in Fig. 2.

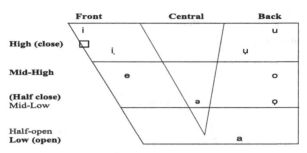

Fig. 2. Phonemic vowel (quadrilateral) chart of Ibibio.

2.2 Consonant System

Ibibio has 23 consonants [13]. These consist of six oral stops (π, β, τ, δ, κ, κπ), 6 nasal stops (μ, μν̄, ν, νψ, ν̄, ν̄w), 3 syllabic nasals (ν, μ, ν̄), 1 trill (ɾ), 1 tap (ʁ), 3 fricatives (β, φ, σ) and 3 approximants (φ, ω, ρ). Consonants have effects on pitch in both tone and non-tone languages. Certain consonants have some effect on tone realisation – usually, the prevocalic consonants appear to have a more pronounced effect on tone than postvocalic ones. Voiced and unvoiced consonants, especially the obstruent, may have a depressing and raising effect on tones, respectively. Implosives are reported not to have a depressing effect on pitch/F0 [14].

2.3 Tone System

The Ibibio language has two contrastive tones, which are High (H) and Low (L), plus a downstepped high (!H) feature. A downstepped tone may occur after a preceding H tone. High-Low (HL) or falling and Low-High (LH) or rising contour tones are other tonal realisations. Ibibio, like most African tone languages, basically is a register tone system, and has three kinds of tonal functions: lexical, morphosyntactic and syntactic. The Syllable phonotactics (syllables may be long or short, open, or closed, strong, or weak, stressed/accented or unstressed/unaccented, tonal or non-tonal – depending on the language) of most African languages like Ibibio is complex [13] when compared to Indo-European languages like English, thus dictating (or limiting) the size of corpus gathered in speech recognition studies. Also, in terms of inflectional morphotactics, African languages are agglutinating [13], with chains of suffixes, each of which has a simple grammatical meaning while English tends towards the isolating type, with little inflection, thus making speech (and by extension emotion) recognition studies AI-complete, and linguistically more complex in the former than in the latter.

3 Ibibio and Emotions

A message predominantly aimed at describing an emotion state, situation, object, or mental state connotes the referential function of a language. This function includes descriptive statements with definite descriptions as well as utterances with deictic terms privileged to the speaker [15]. A focus on the speaker, however, brings into prominence, the *emotive* (also called 'expressive' or more recently 'affective') function. In Ibibio, a classical example is the use of interjections and ideophones such as o!, ìψá!, áàνκ, and ùωá-o, and certain modifications of linguistic sounds that do not change the denotative meaning of an expression, but adds information about a particular attitude or stance that the speaker is taking. Interjections such as ìψá!and ùωá-o, form an interesting and much understudied area of spoken language. Among other features, they allow for the adoption of sounds that are not otherwise part of the Ibibio linguistic systems. As such, a study of the sequential aspects of conversations reveals that both the emotive and connotative functions are usually at play, although they might be more- or less-dominant. For instance, even when people express imprecations after an adverse happening and produce expletives such as áàνκ, ùωá-o, a certain level of recipient design is at work. This is made evident by the speakers' ability to monitor the quality and manner of articulations of such imprecations, which can range from whispers to loud cries. Essien [16] also observed that interjectional words like *â* and ìψá!are usually utilised in expressing emotions like disapproval, surprise, and happiness, and is shown in the following examples:

1. (a). àμédíá ùκèèð νδìδìà áðò áμà?'so you have finished the whole meal?' (normal)
    (b). ìψá! àμédíá ùκèèð νδìδìà áðò áμà?'so you have finished the whole meal?' (surprise)

The expression in (1) looks like a rhetorical question but 1 (a) is just a neutral expression where the speaker does not show any surprise, but just acknowledges the fact that the addressee has unexpectedly finished a whooping meal. In (1 (b)), the difference from (1 (a)) is just the introduction of one word, and the utilisation of this interjection ìψá! changes the emotion from normal to that of surprise. Just as Ameka [17] rightly observed, interjections are common words that can stand alone as an utterance to express emotional or mental states. In that case, interjections in Ibibio do not necessarily need to combine with other word classes to form a construction. While we can have the expression in (1 (b)) where the interjection is used with other word classes to drive home an emotion, this same expression can still be used alone to express surprise. Therefore, *â* and ìψá! could respectively be used to indicate disapproval and surprise on their own. At the same time, an expression like *hmm/mmm* could be used on its own to communicate lamentation. This same word can also be combined with other words to express the same emotion. On the prosodic side, the pitch of the voice could be used to show emotion or differentiate one emotion from the other. For instance, in sentences (1 (a)) and (1 (b)), a show of surprise, even with the introduction of the interjection may require raising the voice higher than the neutral case. In this case, a higher pitch (and even tempo) during a stretch of utterance would signal a different emotion in Ibibio from a normal pitch and tempo. As regards intonation, the affective or attitudinal approach to intonation study correlates strongly with mood/emotion [18]. In the performance of attitudinal functions, the rising and falling tunes are normally combined in different ways to indicate different emotions. For instance, the rising tune (also known as tune 1) is used for statements, commands and wh-questions while tune 2 (the falling tune) is used for polar questions, requests, some commands, and echo questions [19]. The pitch of voice and word class combination notwithstanding, the combination of this intonation patterns really show the emotional disposition of the speaker at a particular time. However, the rising intonational tune depicts surprise, command, and anger in Ibibio, while the falling pattern shows statements, and neutral expressions. Emotive overtones may also occur depending on the speaker's mood and can be created in several ways. One way is to use phonological devices – a process referred to as phono stylistics, and involves the use of intonation and tonal modification to add additional shades of meaning to words and sentences as in:

2. (a). èté m̀mì ákpà-o 'my father has died'
    (b). v̀dìtọ̀ èβé á!kpà-o 'children of a widow'
    (c). m̀βọ̀k túá ànψé m̀βọ́m-ọ 'please have pity on him/her'

The interplay of emotion and energy or loudness reveals another dimension to the study, as sentences that portray annoyance and pride are mostly uttered with a raise of the voice. But sentences meant to attract sympathy always carry low energy (2 (b) and (c)). This paper analyses different emotion expressions in Ibibio with a view to ascertaining how the sound patterns correlate with emotions. Emotion is investigated in this paper because it is yet to be deeply researched for spoken ATLs, and no technology resource exists for African language emotions. Further, the nonlinearity of a semantic concept such as emotion requires more precise data representation techniques to ensure effective and semantically meaningful visualisations, as a suitable methodology to achieving this is demonstrated in this paper.

4   Methods

Our methodology begins with building of the speech corpus from recorded emotions. This step involves:

(i)   Speaker selection and speech utterance recordings of the selected emotions.
(ii)   Utterance (audio) annotation.
(iii) Extraction of desired speech features and feature normalisation.

The extracted corpus dataset is categorised and projected graphically, to visualise the acoustic correlates of emotions. Using machine learning (ML) techniques, these features are trained to discover how best the selected emotions patterns influence the classification error. The ML classifiers used in this paper include decision tree (DT), support vector machine (SVM), and k-nearest neighbour (k-NN). Seven types of emotions (anger, fear, joy, normal, pride, sadness, and surprise) were considered, and two sentences constructed for each emotion class. To diffuse the 'observer's paradox' effect [20], speakers were cued to speak in a 'spontaneous' manner and not at their own convenience. Figure 3 shows a sample annotated corpus that reveals the duration, pitch, and intensity of the anger emotion:

v̀sínám̀ ψàk àσúénnè èkà m̀mì? dakka ke υσάῦ ψak m̀βóíψò. "Why have you disgraced my mother? Make way (leave the way) for me to pass".

A catalogue of the emotion types investigated in this study, sentences, and English gloss, is given in Table 2. A total of twenty speakers (ten males and ten females) were selected for the study. To ensure efficient speech processing without loss of information,

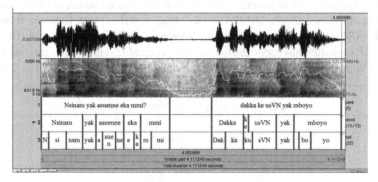

Fig. 3. Annotated emotion corpus revealing pitch, intensity and duration

Table 2. Recorded emotions used in the experiment

Emotion	Emotion sentence	English gloss
Anger	(i) νσίνάμ ψὰκ ἀσύέννὲ ἐκ ἀ μμὶ?	(i) Why have you disgraced my mother?
	(ii) δάκκά κέ ύσΛ̅ν ψάκ μβόί ψὸ	(ii) Make way (leave the way) for me to pass
Fear	(i) κὰσέ ύρλκ ίκὸτ ἀδΛκ ύφὸ κ μμὶ	(i) Watch out! (or any other exclamatory start) ... A snake has entered my house
	(ii) σὰί! ἰνό όδὸάτὸόν̅ὸ ἀδί ψ ἀὰν̅ά ἀκὸμ ύφὸκ	(ii)(exclamation) ... that thief has started removing the roofing sheets
Joy	(i) άν̅ωάάν μμὶ άμὰν έ- ψέν άωὸδὲὲν	(i) My wife has given birth to a baby boy
Emotion	Emotion sentence	English gloss
	(ii) έψέν μμὶ ἀψά ἀδάκκάάβὶ ὸ μβὰκάρά ν̅κπόν̅	(ii)My child will leave for overseas tomorrow
Normal	(i) ν̅ψά κὰά ύφὸκ ν̅ωὲδ ν̅φΙv	(i) I will go to school today
	(ii) έψέν έκὰ μμὶ ἀμέ- σὶὲρέ v≅δὲ	(ii) My brother (or sister), good morning to you too
Pride	(i) νσὲ νδὶὰ ν̅κπό δάν̅ν̅ά- μμάμα	(i) I do eat as freely as I want to
	(ii) ἐτέ μμὶ ἀνίέ έφάάκ άμὶ	(ii) My father owns this street
Sadness	(i) μμμ ἐτέ μμὶ άμὰ άκ- πά ν̅κπόν̅	(i) mmm ... my father just died yesterday
	(ii) μμμ ίδέμ ἐκὰ μμὶ ίσόν̅ν̅ό	(ii)mmm ... my mother is not feeling well
Surprise	(i) ίφά! άκέ δάμμά ἰδά- ηάάκὲ?	(i) (exclamation) ... when did s/he turn mad?
	(ii) ὺωό! ἀφὸ κέ έκέτόπκέ ίκ άν̅όδό?	(ii)(exclamation) ... were you the victim of that gunshot?

the Ibibio emotions dataset was coded using the Speech Assessment Method for Phonetic Alphabet (SAMPA) notations and tailored to suit the ergonomic needs of the language, ultimately christened Ibibio SAMPA – a machine-readable phonetic script using 7-bit printable ASCII characters based on the International Phonetic Alphabet (IPA) [21]. The SAMPA notation has become a universally acceptable standard of encoding the IPA symbols. A metadata documenting the Ibibio emotion corpus is presented in Table 3.

Table 3. Metadata of Ibibio emotion corpus

Language investigated	Ibibio
Number of speakers	20 speakers: 10 male and 10 female
Recording device	Zoom handy H4n sound recorder
Recording rate	44.1 MHz
Channel	Stereo mode
Recording environment	Semi-controlled
Number of emotions	7
Emotion recorded	Anger, Fear, Joy, Normal, Pride, Sadness, Surprise
Total files recorded	(10 + 10) * 7 = 140 files
Number of takes	2
Extracted features	Duration and F0
Type of emotions	Acted
Text processing format	SAMPA
Purpose	Classification
Publicly available	No

4.1 Classification Approach

A cross validation approach is used in this paper to accomplish the classification process. Cross validations are techniques for measuring generalisation capacity of any regression against overfitting or other limitations, by comparing several statistical models, which may further be used for better regression through averaging. 10-folds (divisions) cross validation was chosen to partition the emotion dataset. Each fold is held out in turn for testing, and a model trained for each fold using all data outside the fold. Each model performance is trained inside the fold, and the average test error over all folds computed. This training method indeed yields good estimate of the predictive accuracy of the final model trained with all data and is recommended for small data. Hence the reason for choosing the cross-validation approach over other approaches is to obtain satisfactory performance using our small dataset.

4.2 Adopted Classifiers

Support Vector Machine (SVM): SVM is a discriminative classifier capable of deciphering subtle patterns in noisy and complex datasets, i.e., given labelled training data, SVM outputs an optimal hyperplane which categorises new exemplars. In this paper, a decision boundary that maximises the "margin" separating the positive from the negative training data points is finally made by minimising $\frac{1}{2}||\overrightarrow{w}||^2$ subject to the constraints: $y_i(\overrightarrow{w}.\overrightarrow{x}_i+b) \geq 1$. The resulting Lagrange multiplier equation that requires optimisation is:

$$L = \frac{1}{2}||\overrightarrow{w}||^2 - \sum_i \alpha_i(y_i(\overrightarrow{w}.\overrightarrow{x}_i + b) - 1). \tag{1}$$

Solving the Lagrangian optimisation problem in Eq. (1) yields, w, b, and α_i, parameters that determines a unique maximal margin solution. Our SVM uses the fine Gaussian – places finely-detailed distinctions between classes, using Gaussian kernel, with kernel scale set to $\frac{\sqrt{P}}{4}$, where P represents the number of predictors.

Decision Tree: Decision trees are a simple and widely used classification technique [22]. Suppose a dataset is induced on the following scheme: $A_1, \ldots, A_m, C$, where A_j are attributes, and C is the target class. All candidate splits are generated and evaluated by the splitting criterion, and splits on continuous and categorical attributes are generated as described above. The selection of the best split is usually carried out by impurity measures. The impurity of the parent node has to be decreased by the split. Now, let $(E_1, E_2, \ldots, E_k)$ be a split induced on the set of records, E, then, a splitting criterion that makes use of the impurity measure $I(\cdot)$ is:

$$\Delta = I(E) - \sum_{i=1}^k \frac{|E_i|}{|E|} I(E_i) \tag{2}$$

Standard impurity measures are the *Shannon* entropy or the *Gini* index. Classification and regression tree (CART) uses the Gini index that is defined for a set E as follows: Let p_j be the fraction of samples in E of class c_j:

$$p_j = \frac{|\{t \in E : t|C| = c_j\}|}{|E|} \tag{3}$$

then,

$$Gini(E) = 1 - \sum_{j=1}^Q p_j^2 \tag{4}$$

where Q is the number of classes. In this paper we use the simple decision tree classifier.

k-Nearest Neighbour (KNN): k-NN classification [23] is one of the most fundamental and simple classification methods – often preferred when there is little or no prior knowledge about the distribution of data. k-NN learns the input samples and predicts the response for a new sample by analysing a certain number (k) of the nearest neighbours of the sample using voting, weight computation, and more. It is commonly based on the

Euclidean distance between a test sample and the specific training samples. The k-NN algorithm is first implemented by introducing some notations, thus: Let $S = (x_i, y_i)$: $i = 1, 2, \ldots, n$, be the training set, where x_i is a d-dimensional feature vector, and $y_i \in \{+1, -1\}$ is associated with the observed class labels. For simplicity, we consider a binary classification as this relates to the classification task at hand. Suppose that all training data are *iid* samples of random variables with unknown distribution. With previously labelled samples as the training set S, the k-NN algorithm constructs a local sub-region $R(x) \subseteq R^d$ of the input space, which is situated at the estimation point x. The predicting region $R(x)$ contains the closest training points to x, expressed as follows:

$$R(x) = \{\hat{x}|D(x, \hat{x}) \leq d_{(k)}\} \tag{5}$$

where $d_{(k)}$ is the k th order statistic of $\{D(x, \hat{x})\}_1^N$, and $D(x, \hat{x})$ is the distance metric. The k-NN algorithm is statistically designed for the estimation of posterior probability $p(y|x)$ of the observation point x:

$$p(y|x) = \frac{p(x|y)p(y)}{p(x)} \cong \frac{k[y]}{k}. \tag{6}$$

where $k[y]$ denotes the number of samples in region, which is labelled. Hence, for a given observation, x, the decision $g(x)$ is formulated by evaluating the values of $k[y]$ and selecting the class with the highest value:

$$g(x) = \begin{cases} 1, k[y = 1] \geq k[y = -1], \\ -1, k[y = -1] \geq k[y = 1]. \end{cases} \tag{7}$$

Finally, the decision that maximises the associated posterior probability is used in the k-NN algorithm. For a binary classification problem where $y_i \in \{+1, -1\}$, the KNN algorithm produces the following decision rule:

$$g(x) = \text{sgn}\left(\text{ave}_{x_i \in R(x)} y_i\right). \tag{8}$$

We adopt a k-NN classifier with the distance weighting kernel.

5   Results

5.1   Acoustic-Emotion Correlates

In this section, we document our findings as regards acoustic-emotion correlates in male and female speakers, for sentence, word, and syllable units. Tables 4 and 5 show mean feature extractions and standard deviations (SDs) of selected emotions, for male and female speakers, respectively.

We found that in male and female speakers, syllable duration units gave the least standard deviations (SD) across the various emotions. For F0 and intensity features, the SD was variable [3], and indicates that these features are non-linear across emotions, and varies according to the degree (or state) of activation-valence of the intended emotion. This is opposed to duration feature, which show consistency and is strictly linear

Table 4. Mean features of ibibio emotions (male speakers)

Speech feature	Unit	Anger		Fear		Joy		Normal		Pride		Sadness		Surprise	
		Mean	SD	Mean	SD	Mean	SD	Mean	SD	Mean	SD	Mean	SD	Mean	SD
Duration	Sentence	3.2213	0.3789	6.0257	0.7434	4.8530	0.9527	3.5851	0.7041	4.2283	0.9644	5.4397	1.3362	4.9118	0.8807
	Word	0.3023	0.1307	0.3311	0.1229	0.3202	0.1518	0.3065	0.1514	0.3697	0.1293	0.3888	0.1951	0.3331	0.1519
	Syllable	0.1458	0.0674	0.1651	0.1059	0.1511	0.0859	0.1470	0.0692	0.1748	0.0864	0.2046	0.1500	0.1628	0.1091
Pitch/F0	Sentence	173.9753	20.1374	186.2643	41.3152	150.6446	28.1163	129.2312	18.0015	145.2936	39.4561	126.2795	13.8291	215.1276	38.3312
	Word	173.9896	46.2745	184.4767	63.5429	155.5201	47.2494	131.3191	32.3447	146.2194	44.1161	127.7192	31.1052	218.2324	56.8124
	Syllable	171.4385	48.6454	183.1234	65.4774	151.2131	46.5778	130.4647	33.3057	148.0819	48.9123	127.2735	32.6847	218.5435	60.6849
Intensity	Sentence	71.3681	11.1641	64.5904	10.4888	69.2853	12.0354	64.6719	10.2274	68.8917	11.7138	61.1909	8.5930	67.0621	11.3581
	Word	75.7141	11.2432	69.9034	10.0698	72.9568	12.2833	68.0253	11.1658	71.6732	11.6424	65.0394	9.5469	73.7602	11.4105
	Syllable	75.3624	11.6382	70.0722	10.4966	73.0488	12.2704	68.0007	11.4948	71.7716	12.0546	65.2062	9.8026	73.9887	11.8463

Table 5. Mean features of ibibio emotions (female speakers)

Speech feature	Unit	Anger		Fear		Joy		Normal		Pride		Sadness		Surprise	
		Mean	SD	Mean	SD	Mean	SD	Mean	SD	Mean	SD	Mean	SD	Mean	SD
Duration	Sentence	3.1175	0.5202	5.3178	0.9232	4.1826	0.6340	3.5632	0.5068	3.7172	0.5244	5.6711	1.5592	4.5508	0.5510
	Word	0.3099	0.1545	0.3217	0.1652	0.3134	0.1639	0.3217	0.1493	0.3384	0.1148	0.4348	0.2742	0.3513	0.1958
	Syllable	0.1468	0.0823	0.1612	0.1365	0.1374	0.0889	0.1553	0.0734	0.1618	0.0746	0.2288	0.2284	0.1673	0.1272
Pitch/F0	Sentence	251.6670	42.7610	270.3883	29.9262	239.6105	51.8792	201.5260	16.0259	221.8083	33.6151	192.8358	12.8627	311.1050	51.0308
	Word	251.0401	80.4923	275.6471	70.9275	250.9130	77.2443	204.7604	46.2096	224.2902	49.3449	200.4537	45.7644	315.7004	73.0713
	Syllable	247.6529	84.7050	271.8920	73.9995	246.2229	78.7392	203.9337	49.4454	223.2533	55.6428	202.0533	48.7692	314.8463	78.0613
Intensity	Sentence	75.9353	8.0478	71.7731	7.8474	75.5319	7.7686	70.2761	7.8238	73.4468	7.0124	67.7867	6.9861	73.3965	7.2994
	Word	79.0610	9.1639	76.3378	8.2537	79.1918	9.5895	72.2230	8.1623	75.8363	7.6530	70.8944	8.2407	77.7358	8.8299
	Syllable	78.9374	9.5458	76.7780	8.8347	78.9652	9.9422	72.2966	8.6039	75.8725	8.1755	70.8651	8.9345	78.2435	9.1233

across emotions. Figures 4, 5, and 6, are visualisations of acoustic correlates of selected emotions, for sentence, word, and syllable units, respectively. We found that for duration feature (Fig. 4), as regards sentence units, female speakers, express emotions with shorter durations compared to male speakers, sadness emotion exempt. As regards word and syllable units, male speakers communicate anger, normal, sadness, and surprise emotions with shorter durations than their female counterparts, while female speakers express fear, joy, and pride emotions with shorter durations. As regards syllable features female speakers, express emotions with shorter durations for fear, joy, and pride emotions only, while male speakers, express anger, normal, sadness and surprise emotions with shorter durations.

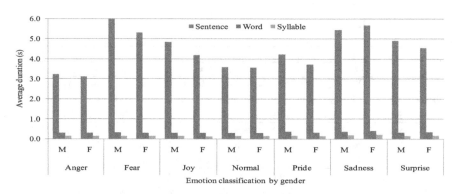

Fig. 4. Effect of speakers' emotion on average duration

For F0 feature (Fig. 5), female speakers dominate the F0 space, and have high pitch values across the various emotions compared to their male counterparts, with surprise emotion having the highest F0 value, followed chronologically by fear, anger, joy, pride, normal and sadness emotions (for both genders).

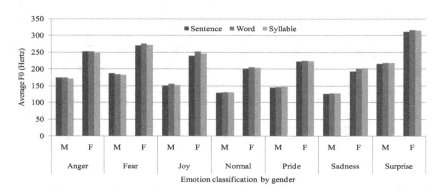

Fig. 5. Effect of speakers' emotion on average F0

Concerning intensity features, female speakers still dominate the intensity space, and were more expressive in terms of loudness/energy, compared to their male counterparts. The emotion with the highest average intensity value is joy, followed chronologically by anger, surprise, fear, pride, normal and sadness.

The results indicate that high activation effects tend to produce higher F0 and intensity values, compared to low activation effects, but neutral and low activation effects produce the lowest pitch/F0 and intensity values (for both genders). These findings agree with the literature [2] and support our generic hypothesis.

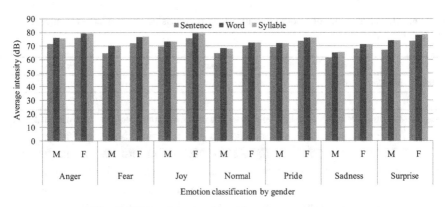

Fig. 6. Effect of speakers' emotion on average intensity

5.2 Classification Results

Figure 7 shows classification accuracies of emotion-duration correlates, categorised by gender – for sentence, word, and syllable units. We observe that syllable units present the best classification accuracies for the selected emotions, excepting fear and joy emotions – which give poor classification performance. The k-NN classifier performed dismally for male and female speakers (on the average), while the SVM and decision tree classifiers (on the average) perform well for word and syllable units, compared to sentence units.

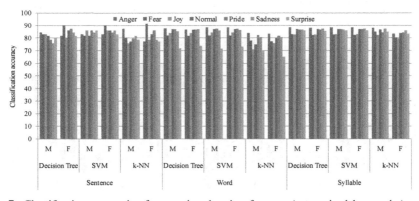

Fig. 7. Classification accuracies for emotion-duration features (categorised by gender) – for sentence, word and syllable units

The corresponding classification errors are found in Table 6, with the most significant errors highlighted. Here the k-NN classifier presented the most significant errors, compared to other classifiers (DT and SVM), which classification errors were low for other emotions excepting fear, joy, and surprise emotions (for word units), and fear and joy (for syllable units). Surprisingly, the highest classification accuracies (or least classification errors) are from the three classifiers, with k-NN leading (DT: 90%, SVM: 90%, k-NN: 92.40%), for fear emotion (sentence unit).

Table 6. Mean Error rates of emotion-duration features (classified by gender) – at sentence, word and syllable levels

Emotion	Sentence						Word						Syllable					
	Decision Tree		SVM		k-NN		Decision Tree		SVM		k-NN		Decision Tree		SVM		k-NN	
	M	F	M	F	M	F	M	F	M	F	M	F	M	F	M	F	M	F
Anger	15.70	18.60	17.10	17.10	12.90	22.90	12.40	13.50	11.70	11.70	16.30	17.00	11.90	12.10	11.90	11.90	12.10	17.00
Fear	17.10	10.00	18.60	10.00	20.00	8.60	18.10	18.60	18.10	18.10	22.10	22.90	17.20	18.00	17.50	17.70	15.20	19.90
Joy	17.10	20.00	14.30	14.30	24.30	21.40	16.30	16.10	15.70	15.70	29.40	24.10	17.50	17.50	17.30	17.30	17.70	20.60
Normal	18.60	14.30	18.60	14.30	22.90	17.10	13.10	13.40	13.00	13.00	25.40	20.20	13.20	13.20	13.00	13.00	13.50	16.40
Pride	21.40	12.70	14.30	15.70	20.70	14.30	13.10	13.50	12.90	12.90	17.80	18.60	13.40	13.30	13.00	13.00	15.80	15.80
Sadness	24.30	15.70	15.70	14.30	18.60	21.40	14.70	13.10	14.30	13.80	19.60	19.40	13.30	12.80	13.50	12.80	12.90	14.30
Surprise	20.00	18.60	14.30	17.10	21.40	22.90	28.30	26.30	28.60	26.80	30.40	35.00	14.20	14.40	14.20	14.20	15.00	16.70

The classification accuracies and errors remarkably improved for emotion-F0 (Fig. 8 and Table 7) and emotion-intensity (Fig. 9 and Table 8) classifications, as the classification at the sentence unit level exhibit improved performance for SVM classifier, compared to other classifiers. Overall performance shows that word and syllable units still maintained good performance, but with slight variations in classification at the sentence (unit) level.

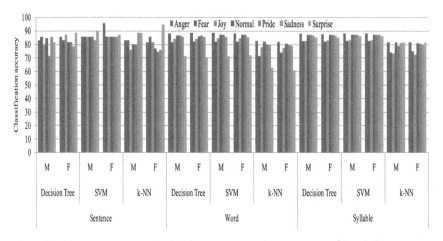

Fig. 8. SVM accuracies for emotion-F0 features (classified by gender), at sentence, word and syllable levels

Table 7. Mean Error rates of emotion-F0 features (classified by gender) – at sentence, word and syllable levels

Emotion	Sentence						Word						Syllable					
	Decision Tree		SVM		k-NN		Decision Tree		SVM		k-NN		Decision Tree		SVM		k-NN	
	M	F	M	F	M	F	M	F	M	F	M	F	M	F	M	F	M	F
Anger	17.10	14.30	14.30	4.30	17.10	18.60	11.80	11.70	11.70	11.80	17.80	18.10	12.10	12.40	11.90	11.90	18.50	18.50
Fear	14.30	17.10	14.30	14.30	17.10	14.30	18.60	18.20	18.10	18.10	28.70	26.00	17.50	17.90	17.60	17.50	26.10	25.30
Joy	20.00	12.90	14.30	14.30	24.30	18.60	16.10	16.30	15.70	15.70	22.10	22.40	17.70	17.30	17.30	17.30	26.50	27.50
Normal	15.70	18.60	14.30	14.30	20.00	22.90	13.50	13.90	13.00	13.00	17.90	19.50	13.10	13.10	13.00	13.00	18.70	18.90
Pride	28.90	18.60	14.30	14.30	20.00	25.70	13.70	13.50	12.90	12.90	20.30	20.40	13.10	13.00	13.00	13.00	20.90	19.70
Sadness	14.30	21.40	17.10	14.30	11.40	24.30	14.40	14.60	14.30	14.40	20.40	21.20	13.30	13.30	13.10	13.10	19.00	20.60
Surprise	18.60	11.40	10.00	12.90	11.40	5.70	28.50	29.60	28.70	28.30	37.10	39.40	15.10	15.00	14.10	14.00	18.50	18.70

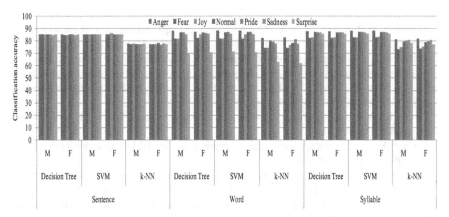

Fig. 9. SVM accuracies for emotion-intensity features (classified by gender) – at sentence, word and syllable levels

Table 8. Mean error rates of emotion-intensity features (classified by gender), at sentence, word and syllable levels

Emotion	Sentence						Word						Syllable					
	Decision Tree		SVM		k-NN		Decision Tree		SVM		k-NN		Decision Tree		SVM		k-NN	
	M	F	M	F	M	F	M	F	M	F	M	F	M	F	M	F	M	F
Anger	15.00	15.00	14.80	14.80	22.40	22.70	11.60	12.60	11.70	11.70	17.70	17.20	12.30	12.30	11.90	11.90	18.90	18.50
Fear	15.00	15.10	14.80	14.80	22.80	22.70	18.20	17.80	18.10	18.10	25.70	26.00	17.60	17.60	17.50	17.50	26.70	26.10
Joy	15.00	15.10	14.80	13.80	22.30	22.70	18.30	14.80	18.10	14.70	26.00	23.10	17.30	17.10	17.50	16.80	25.10	24.90
Normal	14.90	14.80	14.80	14.80	22.60	21.90	13.50	13.30	13.10	13.00	19.90	21.70	13.00	13.10	13.00	13.00	20.60	21.40
Pride	15.00	15.00	14.80	14.80	22.60	23.10	13.40	13.70	12.90	12.90	20.80	18.60	13.10	13.10	13.00	13.00	19.90	20.30
Sadness	15.10	15.30	14.80	14.80	23.00	22.40	15.00	14.30	14.30	14.70	22.50	22.20	13.20	13.50	13.10	13.10	19.50	19.20
Surprise	15.00	15.00	14.80	14.80	22.50	22.90	29.50	29.40	28.60	29.10	37.30	38.40	14.30	14.30	14.20	14.20	21.70	22.70

6   Conclusion and Research Perspective

Emotions are self-induced activity bursts of brief duration that does not last long. Most physiological correlates of emotions focus on the major experimental approaches currently applied to the study of emotion and its physiological or behavioural parameters. Hence, research on emotion dynamics is important from a clinical perspective as disturbances in emotion dynamics are symptomatic of several mental disorders such as depression. This paper has begun a pioneering work on discovering emotion-correlates using Ibibio speech features. A future perspective is to exploit machine learning in the discovery of hidden patterns for the recognition of symptomatic dynamics of emotions – a very essential application in the health and medical sectors; and to investigate if the tendencies in prosodic features of affect are language-specific or to some extent universal.

References

1. Moridis, C.N., Economides, A.A.: Affective learning: empathetic agents with emotional facial and tone of voice expressions. IEEE Trans. Affect. Comput. **3**(3), 260–272 (2012)
2. Sbattella, L., Colombo, L., Rinaldi, C., Tedesco, R., Matteucci, M., Trivilini, A.: Extracting emotions and communication styles from prosody. In: da Silva, H., Holzinger, A., Fairclough, S., Majoe, D. (eds.) PhyCS 2014. LNCS, vol. 8908. Springer, Heidelberg (2014). https://doi.org/10.1007/978-3-662-45686-6_2
3. Cowie, R., et al.: Emotion recognition in human-computer interaction. IEEE Signal Process. Mag. **18**(1), 32–80 (2001)
4. Scherer, K.R.: Vocal affect expression: a review and a model for future research. Psychol. Bull. **99**(2), 143–165 (1986)
5. Murray, I.R., Arnott, J.L.: Toward the simulation of emotion in synthetic speech: a review of the literature on human vocal emotion. J. Acoust. Soc. Am. **93**(2), 1097–1108 (1993)
6. Nwe, T., Foo, S., de Silva, L.: Speech emotion recognition using hidden Markov models. Speech Commun. **41**, 603–623 (2003)
7. Väyrynen, E.: Emotion recognition from speech using prosodic features. Ph.D. Thesis, University of Oulu, Oulu (2014)
8. Cowie, R., Douglas-Cowie, E., Savvidou, S., McMahon, E., Sawey, M., Schröder, M.: FEEL-TRACE: an instrument for recording perceived emotion in real time. In: ISCA tutorial and research workshop (ITRW) on speech and emotion, Newcastle, Northern Ireland, UK (2000)
9. Koleva, M.: Towards adaptation of NLP tools for closely-related Bantu languages: building a Part-of-Speech Tagger for Zulu. PhD Thesis, Saarland University, Germany (2012)
10. Kaufman, E.M.: Ibibio Grammar. Ph.D. Dissertation, California, Berkeley (1968)
11. Urua, E.A.: Aspects of Ibibio phonology and morphology. Ph.D. Thesis, University of Ibadan (1990)
12. Noah, P.: On the classification of /a/ in the Ibibio language in Nigeria. Am. J. Soc. Issues Hum. **2**(5), 322–328 (2012)
13. Urua, E., Watts, O., King, S., Yamagishi, J.: Statistical parametric speech synthesis for Ibibio. Speech Commun. **56**, 243–251 (2014)
14. Urua, E., Gibbon, D., Gut, U.: A computational model of low tones in Ibibio. In: Proceedings of 15th International Congress on Phonetic Sciences, Barcelona (2003)
15. Gumperz, J.J.: Rethinking Context: Language as an Interactive Phenomenon. Cambridge University Press, Cambridge (1992)

16. Essien, O.: A Grammar of the Ibibio Language. University Press, Ibadan (1990)
17. Ameka, F.: Interjections: the universal yet neglected part of speech. J. Pragmat. **18**, 101–118 (1992)
18. Osisanwu, W.: The aural perception of verbal expression of mood in English as a second language situation. In: Ajiboye, T., Osisanwu, W. (eds.) Mood in Language and Literature, pp. 15–23. Femolus-Fetop Publishers, Lagos (2006)
19. Crombie, W.: Intonation in English: a systematic perspective. Paper presented at the Applied Linguistics Group seminar, the Hatfield Polytechnic, Hatfield, England (1988)
20. Campbell, N.: Recording techniques for capturing natural everyday speech. In: Proceedings of 3rd International Conference on Language Resources and Evaluation, Spain (2002)
21. Gibbon, D., Urua, E.-A.: Data creation for Ibibio speech synthesis. In: Proceedings of Third Partnership International Workshop on Speech Technology for Minority Languages, pp. 1–23. Local Language Speech Technology Initiative (LLSTI) Publication (2004). http://www.llsti.org/pubs/ibibio_data.pdf
22. Tan, P.-N., Steinbach, M., Kumar, V.: Introduction to Data Mining. Pearson Addison Wesley, Boston (2006)
23. Friel, N., Pettitt, A.N.: Classification using distance nearest neighbours. Stat. Comput. **21**(3), 431–437 (2011)

Multilingual and Language-Agnostic Recognition of Emotions, Valence and Arousal in Large-Scale Multi-domain Text Reviews

Jan Kocoń[1](✉) , Piotr Miłkowski[1] , Małgorzata Wierzba[2] ,
Barbara Konat[3] , Katarzyna Klessa[3] , Arkadiusz Janz[1] , Monika Riegel[2] ,
Konrad Juszczyk[3] , Damian Grimling[4], Artur Marchewka[2] ,
and Maciej Piasecki[1]

[1] Wroclaw University of Science and Technology, Wrocław, Poland
{jan.kocon,piotr.milkowski,arkadiusz.janz,maciej.piasecki}@pwr.edu.pl
[2] Laboratory of Brain Imaging, Nencki Institute of Experimental Biology,
Polish Academy of Sciences, Warsaw, Poland
{m.wierzba,m.riegel,a.marchewka}@nencki.edu.pl
[3] Adam Mickiewicz University, Poznań, Poland
{bkonat,klessa,juszczyk}@amu.edu.pl
[4] Sentimenti Sp. z o.o., Poznań, Poland
damian@sentimenti.pl

Abstract. In this article we present extended results obtained on the multidomain dataset of Polish text reviews collected within the Sentimenti project. We present preliminary results of classification models trained and tested on 7,000 texts annotated by over 20,000 individuals using valence, arousal, and eight basic emotions from Plutchik's model. Additionally, we present an extended evaluation using deep neural multilingual models and language-agnostic regressors on the translation of the original collection into 11 languages.

Keywords: NLP · Text classification · Text regression · Deep learning · Emotions · Valence · Arousal · Multilingual · Language-agnostic

1 Introduction

Emotions are a crucial part of natural human communication, conveyed by both what we say and how we say it. In this study, we focus on the emotions attributed

This work was financed by (1) the National Science Centre, Poland, project no. 2019/33/B/HS2/02814; (2) the Polish Ministry of Education and Science, CLARIN-PL; (3) the European Regional Development Fund as a part of the 2014-2020 Smart Growth Operational Programme, CLARIN – Common Language Resources and Technology Infrastructure, project no. POIR.04.02.00-00C002/19; (4) the National Centre for Research and Development, Poland, grant no. POIR.0 1.01.01-00-0472/16 – *Sentimenti* (https://sentimenti.com).

© Springer Nature Switzerland AG 2022
Z. Vetulani et al. (Eds.): LTC 2019, LNAI 13212, pp. 214–231, 2022.
https://doi.org/10.1007/978-3-031-05328-3_14

by Polish native speakers to written Polish texts. Here, we used extensive empirical data together with a variety of machine learning (ML) techniques to automatically detect emotions expressed in natural language.

Introduction of ML to the area of text mining resulted in the rapid growth of the field in recent years. However, an automatic emotion recognition with ML remains a challenging task due to the scarcity of high-quality and large-scale data sources. Numerous approaches were attempted to annotate words concerning their emotions and valence for various languages [34, 41]. Such datasets, however, are limited in size, typically consisting of several thousands of words, while natural language lexicons are known to be much bigger.[1] This constrains their usage in natural language processing (NLP).

In the psychological tradition of emotion research, words are usually characterised according to two dominant theoretical approaches to the nature of emotion: dimensional account and categorical account. According to the first account proposed in [35], each emotion state can be represented by its location in a multidimensional space, with valence (negativity/positivity) and arousal (low/high) explaining most of the observed variance. In the competing accounts, several basic or elementary emotion states are distinguished, with more complex, subtle emotion states emerging as their combination [7,32]. The resulting categories are often referred to as *basic emotions*, and include: *trust, joy, anticipation, surprise, disgust, fear, sadness*, and *anger*. The concept of basic emotions itself has been interpreted in various ways, and thus different theories posit different numbers of categories of emotion, with [7] and [32] gaining most recognition in the scientific community. In the presented study, we include both dimensional (valence and arousal) and categorical (eight Plutchik's basic emotions) approaches to ensure a coherent and thorough description of verbally expressed emotions.

In NLP, as well as in the applied usages of emotion annotation, the dominant approach is sentiment analysis, which typically takes into account only valence (negativity/positivity). It is understandable since the emotion annotation of textual data faces difficulties in the two conventional approaches to annotation. In the first approach, a small number (usually 2 to 5) of trained annotators are engaged, which may lead to differences in individual annotations and poor inter-annotator agreement [12]. The other approach, based on crowd annotations on platforms such as Amazon Turk [29] relies on data collected from a large group of individuals, which is then used to obtain mean annotations. However, this is typically effortful and time-consuming.

In the present study, we applied an approach that proved useful in previous experiments [34]. Thus, our annotation schema follows the account of Russel and Mehrabian [35], as well as those proposed by Ekman [7] or Plutchik [32]. Finally, by combining simple annotation schema with crowd annotation, we were able to effectively acquire large amount of data, while at the same time preserving high quality of the data. Our approach to sentiment analysis opens up new possibilities of studying people's attitudes towards brands, products and their

---

[1] The largest dictionary of English, Oxford English Dictionary, for example, contains around 600,000 words in its online version https://public.oed.com/about.

features, political views, movie or music choices, or financial decisions, including stock exchange activities.

2 Data Annotation

The data used in the present work was collected within the *Sentimenti*, a large research and development project (https://sentimenti.com). In this project, a total of over 20,000 unique respondents (with approximately equal number of male and female participants) were sampled from Polish population (sex, age, native language, place of residence, education level, marital status, employment status, political beliefs and income were controlled, among other factors). To collect the data, a combined approach of different methodologies was used, namely: Computer Assisted Personal Interview (CAPI) and Computer Assisted Web Interview (CAWI).

The annotation schema was based on the procedures most widely used in previous studies aiming to create the first datasets of Polish words annotated in terms of emotion (NAWL [34]; NAWL BE [40]; plWordNet-emo [13,43]). Thus, we collected extensive annotations of valence, arousal, as well as eight emotion categories: *joy, sadness, anger, disgust, fear, surprise, anticipation,* and *trust*.

A total number of over 30,000 word meanings from Polish WordNet (plWord-Net) [30] was annotated, with each meaning assessed by at least 50 individuals on each scale. The selection of word meanings was based on the results of the plWordNet-emo [43] project, in which linguists annotated over 87K word meanings with over 178K annotations containing information about emotions and valence (statistics from May 2019). At the time when the selection was made (July 2017), 84K annotations were covering 54K word meanings and 41K synsets. We observed that 27% of all annotations (23K) were not neutral. The number of synsets having word meanings with valence different than neutral was 9K. We have adopted the following assumptions for the selection procedure:

- Word meanings that were not neutral were treated as more important,
- valence sign of the synset was the valence sign of word meanings within the synset (valid in 96% of cases),
- the maximum number of selected word meanings from the same synset was 3,
- the degree of synsets (treated as nodes in the plWordNet graph) which were sources of selected word meanings should be in range [3, 6].

Word meanings were presented to respondents in the context of short phrases, manually prepared by linguists. Subjects were instructed to assess the words for their particular meaning, as well as encouraged to indicate their immediate, spontaneous reactions. Participants had unlimited time to complete the task, and they were able to quit the assessment session at any time and resume their work later on.

Moreover, in a follow-up study, a total number of over 7,000 texts were annotated in the same way, with each text assessed by at least 25 individuals on each scale. The text paragraphs for annotation were acquired from Web reviews of two distinct domains: *medicine*[2] (2000 reviews) and *hotels*[3] (2000 reviews). Due to the scarcity of neutral reviews in these data sources, we decided to acquire yet another sample from potentially neutral Web sources being thematically consistent with the selected domains, i.e. medical information sources[4] (500 paragraphs) and hotel industry news[5] (500 paragraphs). The phrases for annotation were extracted using *lexico-semantic-syntactic patterns* (LSS) manually created by linguists to capture one of the four effects affecting sentiment: *increase, decrease, transition, drift.* Most of these phrases belong to the previously mentioned thematic domains. The source for the remaining phrases were Polish WordNet glosses and usage examples [30].

3   Data Transformation

For the first part of the experiments, we decided to carry out the recognition of specific dimensions as a classification task. Eight basic emotions were annotated by respondents on a scale of integers from the range $[0, 4]$ and the same scale was also used for the arousal dimension. For the valence dimension, a scale of integers from the range $[-3, 3]$ was used. We divided the valence scores into two groups: positive (valence_p) and negative (valence_n). This division results from the fact that there were texts that received scores for both positive and negative valence. We wanted to keep that distribution (see Algorithm 1). For the rest of the dimensions, we assigned the average value of all scores (normalised to the range $[0, 1]$) to the text.

Algorithm 1. Estimating the average value of positive and negative valence for a single review.

Require: V: list of all valence scores;
   $m = 3$: the maximum absolute value of valence;
Ensure: Pair (p, n) where p is average positive valence, and n is average negative valence;
1: $(p, n) = (0, 0)$
2: **for** $v \in V$ **do**
3:     **if** $v < 0$ **then** $n = n + |v|$ **else** $p = p + v$
   return $(p \div (|V| \cdot m), n \div (|V| \cdot m))$

---
[2] www.znanylekarz.pl.
[3] pl.tripadvisor.com.
[4] naukawpolsce.pap.pl/zdrowie.
[5] hotelarstwo.net, www.e-hotelarstwo.com.

3.1   Scores Distribution

As a part of this study, a collection of 7004 texts was annotated. To investigate
the underlying empirical distribution of emotive scores, we analysed our data
concerning each dimension separately. We performed two statistical tests to verify the multimodality of the score distribution in our sample for each dimension.
The main purpose of this analysis was to identify if there exists a specific decision boundary splitting our data into distinct clusters, to separate the examples
sharing the same property (e.g., positive texts) from the examples that do not
share this property (e.g. nonpositive texts). The first test was Hartigans' dip
test. It uses the maximum difference for all averaged scores between the empirical distribution function and the unimodal distribution function that minimises
the maximum difference [10]. There are the unimodal null hypothesis and a multimodal alternative. The second one is Silverman's mode estimation test, which
uses the kernel density estimation method to examine the number of modes in
a sample [37]. If the null hypothesis of unimodality ($k = 1$) was rejected, we
also tested if there are two modes ($k = 2$) or more [28]. We used *locmodes* R
package to perform statistical testing [1] with Hartigans' and Silverman's tests
on our annotation data. For all dimensions we could not reject the null hypothesis of bimodality and only in 2 cases (arousal, disgust) we could reject the null
hypothesis of unimodality by the result of both tests (see Table 1).

Table 1. p–values for Silverman's test with $k = 1$ (SI_mod1), $k = 2$ (SI_mod2) and
Hartigans' dip test (HH).

Dimension	SI_mod1	SI_mod2	HH
Trust	0.034	0.736	0.226
Joy	0.000	0.842	0.000
Anticipation	0.522	0.321	0.000
Surprise	0.288	0.360	0.000
Disgust	0.784	0.500	0.178
Fear	0.892	0.674	0.032
Sadness	0.000	0.424	0.000
Anger	0.000	0.630	0.000
Arousal	0.340	0.606	0.118
Valence_n	0.000	0.812	0.000
Valence_p	0.000	0.460	0.000

The distributions of averaged scores for all texts are presented in Fig. 1. We
decided to partition all scores for each dimension into two clusters using k-means
clustering [11]. Clusters are represented in Fig. 1 with different colours. We assign
a label (corresponding to the dimension) if the score for the dimension is higher
than the threshold determined by k-means. Each review may be described with
multiple labels.

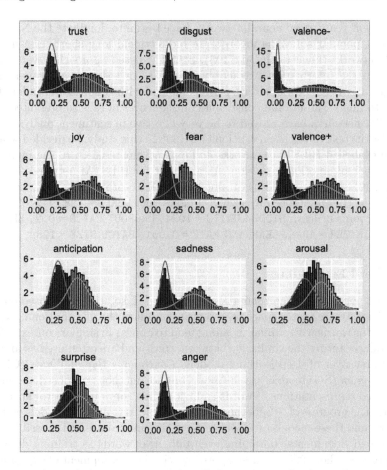

Fig. 1. Distribution of average scores for all dimensions. The first column contains positive emotions and the second column contains negative emotions. The third column contains the *valence-*, *valence+*, and *arousal* distributions. On the X-axis are the normalized emotion rating values, and on the Y-axis is the number of people who rated texts with values from a given bucket.

4   Experiments

In our experimental part, we decided to use a popular baseline model based on fastText algorithm [5,14] as a reference method for the evaluation. fastText's supervised models were used in many NLP tasks, especially in the area of sentiment analysis, e.g., for hate speech detection [3], emotion and sarcasm recognition [8] or aspect-based sentiment analysis in social media [42]. The unsupervised fastText models were also used to prepare word embeddings for Polish (see Sect. 4.1). In our experiments, we used the supervised fastText model as a simple multilabel text classifier for sentiment and emotion recognition. We used *one-versus-all cross-entropy* loss and 250 training epochs with KGR10 pretrained word vectors [20] (described in Sect. 4.1) for all evaluation cases.

In recent years, deep neural networks have begun to dominate the NLP field. The most popular solutions incorporate bidirectional long short-term memory neural networks (henceforth BiLSTM). BiLSTM-based approaches were mainly applied in the information extraction area, e.g., in the task of proper names recognition, where the models are often combined with conditional random fields (CRF) to impose additional constraints on sequences of tags as presented in [9].

LSTM networks have proved to be very effective in sentiment analysis, especially for the task of valence detection [4, 24, 39]. In this study, we decided to adopt the multilabelled BiLSTM networks and expand our research to the more challenging task of emotion detection. As an input for BiLSTM networks, we used pretrained fastText embeddings trained on KGR10 corpus [20]. The parameters used for the training procedure were as follows: MAX_WORDS = 128 (94% of reviews have 128 words or less), HIDDEN_UNITS = 1024, DROPOUT_RATIO = 0.2, EPOCHS = 250, OPTIMIZER = ADAM, LEARNING_RATE = 0.001, BATCH_SIZE = 128.

4.1 Word Embeddings

The most popular text representations in recent ML solutions are based on *word embeddings*. Dense vector space representations follow the distributional hypothesis that the words with similar meaning tend to appear in similar contexts. Word embeddings capture the similarity between words and are often used as an input for the first layer of deep learning models. *Continuous Bag-of-Words* (CBOW) and *Skip-gram* (SG) models are the most common methods proposed to generate distributed representations of words embedded in a continuous vector space [26].

With the progress of ML methods, it is possible to train such models on larger datasets, and these models often outperform the simple ones. It is possible to use a set of text documents containing even billions of words as training data. The work presented here follows these approaches in the development of experiments, and we are researching texts from similar domains. Both architectures (CBOW and SG) describe how the neural network learns the vector representations for each word. In CBOW architecture, the task is *predicting the word given its context*, and in SG the task is *predicting the context given the word*.

Numerous methods have been developed to prepare vector space representations of words, phrases, sentences, or even full texts. The quality of vector space models depends on the quality and the size of the training corpus used to prepare the embeddings. Hence, there is a strong need for proper evaluation metrics, both intrinsic and extrinsic (task-based evaluation), to evaluate the quality of vector space representations including word embeddings [31, 36]. Pretrained word embeddings built on various corpora are already available for many languages, including the most representative group of models built for English [23] language.

In [20] we introduced multiple variants of word embeddings for Polish built on KGR10 corpora. We used the implementation of CBOW and Skip-gram methods provided with fastText tool [5]. These models are available under an open license in the CLARIN-PL project repository[6]. With these embeddings, we obtained a favourable results in two NLP tasks: recognition of temporal expressions [20] and recognition of named entities [25]. For this reason, the same model of word embeddings was used for this work, which is *EC1* [20] (`kgr10.plain.skipgram.dim300.neg10.bin`).

4.2   Multilingual and Language-Agnostic Text Representations

The aforementioned content representation methods also have variants trained on corpora in other commonly used languages. However, this approach requires training dataset in the same language. A translator DeepL[7] was used to solve this problem. The study dataset was translated using machine translation into the following languages: Dutch, English, French, German, Italian, Russian, Portuguese and Spanish.

An alternative approach that does not require data translation is to move away from language dependencies. As a result of the success of word representation using embeddings in machine learning and the growing popularity of methods such as transfer learning, work has also been undertaken to develop solutions for representation of content in a language-independent manner. This removes the problem of translation error and alteration of the meaning of statements. Such methods are referred to as language-agnostic.

Widely used solution to implement this approach is Language-Agnostic SEntence Representations (LASER) [2]. LASER uses the same BiLSTM encoder to encode all languages. It has been trained on 93 languages, which include all European languages, most of Asian and Indian languages, and many others.

4.3   Evaluation Procedure

In the case of monolingual classification on the Polish part of the set, we prepared three evaluation scenarios to test the performance of fastText and BiLSTM baseline models. The most straightforward scenario is a single domain setting (SD) where the classifier is trained and tested on the data representing the same thematic domain. In a more realistic scenario, the thematic domain of training data differs from the application domain. This means that there may exist a discrepancy between the feature spaces of training and testing data, which leads to a significant decrease of the classifier's performance in the application domain.

---

[6] https://clarin-pl.eu/dspace/handle/11321/606.
[7] https://www.deepl.com/.

To test the classifier's ability to bridge the gap between source and target domains, we propose a second evaluation scenario called 1-Domain-Out (DO). This scenario is closely related to the task of unsupervised domain adaptation (UDA), where we focus on transferring the knowledge from labelled training data to unlabelled testing data. The last evaluation scenario is a multidomain setting where we merge all available labelled data representing different thematic domains into a single training dataset (MD).

- *Single Domain, SD* – train/dev/test sets are from the same domain (3 settings, metric: F1-score).
- *1-Domain-Out, DO* – train/dev sets are from two domains, the test set is from the third domain (3 settings, metric: F1-score).
- *Mixed Domains, MD* – train/dev/test sets are randomly selected from all domains (1 setting, metrics: precision, recall, F1-score, ROC_AUC).

We prepared seven evaluation settings with a different domain-based split of the initial set of texts. The final division is presented in Table 2.

Table 2. The number of texts in the evaluation settings.

Type	Setting	Train	Dev	Test	SUM
SD	Hotels	2504	313	313	3130
	Medicine	2352	293	293	2938
	Other	750	93	93	936
DO	Hotels-other	3660	406	–	4066
	Hotels-medicine	5462	606	–	6068
	Medicine-other	3487	387	–	3874
MD	All	5604	700	700	7004

To tune our baseline method, we decided to use a *dev* set. We calculated the optimal decision threshold for each dimension using receiver operating characteristic (ROC) curve, taking the threshold which produces the point on ROC closest to $(FPR, TPR) = (0, 1)$.

In the case of regression on a multilingual set (direct prediction of the level of intensity of emotions) was prepared based on the same BiLSTM model using linear activation on the output layer. Tests were run on the same train/dev/test sets for Mixed Domains (MD) only. Two measures were used to evaluate this approach:

- Mean squared error (MSE) represents the residual error which is nothing but the sum of the squared difference between actual values and the predicted/estimated values.
- R-Squared (R^2) represents the fraction of response variance captured by the regression model.

All experiments on the regression task were repeated 30 times so that strong tests could be performed. This removed the amount of uncertainty caused by the randomness of the neural network model learning process. In addition to the deviations due to the large number of results, the significance of the difference between the results and their distribution was counted. If the difference between results was under $p < 0.05$, they were treated as insignificantly different.

5 Monolingual Classification Results (Polish)

Table 3 shows the results for SD evaluation. There are 11 results for each of the 3 domains. BiLSTM classifier outperformed fastText in 27 out of 33 cases. Table 4 shows the results for DO evaluation. Here BiLSTM classifier provided better quality for 31 out of 33 cases. The last MD evaluation results are in Table 6 (P, R, F1-score) and Fig. 2 (ROC). BiLSTM outperformed fastText in 31 out of 36 cases (Table 6). ROC_AUC is the same for both classifiers in 4 cases (2 of them are micro-and macro-average ROC). For the rest of the curves, BiLSTM outperformed fastText in 7 out of 9 cases. The most interesting phenomenon can be observed in Table 4 where the differences are the greatest. This may indicate that the deep neural network is able to capture domain-independent features (pivots), which is an important capability for domain adaption tasks.

6 Multilingual and Language-Agnostic Regression Results

Table 6 and 7 show the results for the regression tasks for MSE and R^2 measures, respectively. LASER outperformed fastText in **all** cases. Figure 3 presents distribution of the aforementioned measures for fastText-based and LASER-based models. It is clearly visible that language-agnostic models provide answers almost without any out-liners from box-plot charts. That means that the general precision of the system is far better than provided by language-dedicated models.

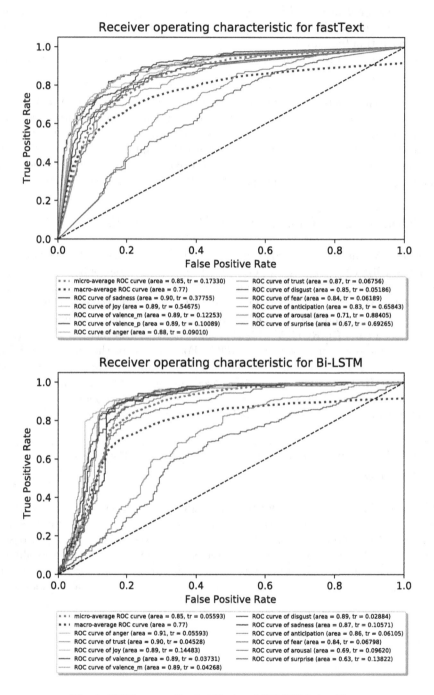

Fig. 2. ROC curves for fastText and BiLSTM classifiers.

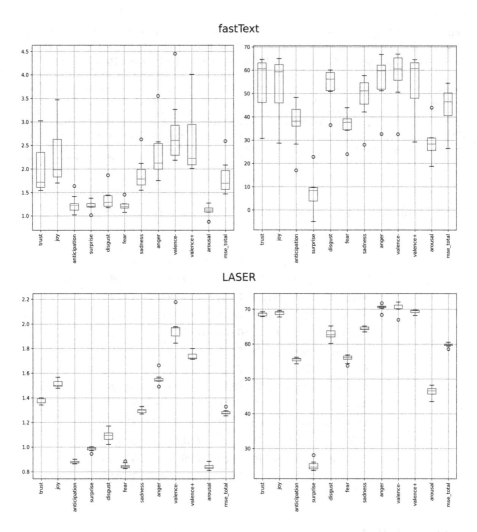

Fig. 3. Distribution of MSE (left) and R^2 (right) for fastText-based (top) and LASER-based (bottom) models evaluated for all languages from the dataset combined on all dimensions.

Table 3. F1-scores for *Single Domain* evaluation. (Train, Dev, Test) sets for settings are the same as in Table 2, rows 1–3.

Setting	Classifier	Trust	Joy	Anticipation	Surprise	Disgust	Fear	Sadness	Anger	Valence−	Valence+	Arousal
1. Hotels	fastText	83.41	89.08	77.91	62.63	81.86	65.81	86.04	88.33	88.43	90.53	66.67
	BiLSTM	80.83	86.84	82.11	46.62	82.76	63.53	88.46	89.54	89.54	89.74	67.66
2. Medicine	fastText	74.14	75.00	75.49	62.00	73.20	45.90	64.32	59.09	56.18	75.37	61.54
	BiLSTM	82.47	84.15	80.31	64.38	74.04	83.04	86.33	85.23	82.40	82.18	65.31
3. Other	fastText	45.28	62.86	51.52	48.57	46.51	45.28	77.27	48.15	66.67	66.67	62.34
	BiLSTM	70.37	80.49	64.71	33.90	62.75	68.66	79.52	65.52	75.95	80.52	65.17

Table 4. F1-scores for *1-Domain-Out* evaluation. (Train/Dev, Test) sets (see Table 2) for these settings are: 4. (Hotels-Other.Train/Dev, Medicine.Test), 5. (Hotels-Medicine.Train/Dev, Other.Test), 6. (Medicine-Other.Train/Dev, Hotels.Test).

Setting	Classifier	Trust	Joy	Anticipation	Surprise	Disgust	Fear	Sadness	Anger	Valence−	Valence+	Arousal
4. Hotels-Other vs Medicine	fastText	65.54	61.73	58.10	59.03	63.20	71.33	75.27	71.97	72.79	61.44	63.08
	BiLSTM	67.52	71.25	70.32	62.62	69.80	74.03	80.40	73.97	76.61	74.56	66.00
5. Hotels-Medicine vs Other	fastText	42.42	65.96	45.95	20.51	48.65	17.65	25.45	05.71	39.29	61.05	37.50
	BiLSTM	56.67	75.00	60.53	51.52	48.39	57.69	61.90	43.48	56.34	73.17	35.29
6. Medicine-Other vs Hotels	fastText	73.45	71.86	73.25	56.32	71.21	50.96	72.96	76.60	78.26	73.93	35.18
	BiLSTM	82.79	87.07	84.76	62.07	82.57	63.44	86.14	87.14	87.07	88.89	51.88

Table 5. Precision, recall and F1-score for *Mixed Domains* evaluation.

Dim.	fastText			BiLSTM		
	P	R	F	P	R	F
Trust	65.32	79.02	71.52	73.91	82.93	78.16
Joy	70.61	81.14	75.51	77.51	84.65	80.92
Anticip.	72.28	77.66	74.78	79.66	81.91	80.77
Surprise	65.07	64.31	64.69	67.67	59.88	63.54
Disgust	66.80	77.73	71.85	71.71	84.09	77.41
Fear	69.20	77.78	73.24	68.84	81.20	74.51
Sadness	81.73	82.55	82.14	83.88	85.57	84.72
Anger	80.92	78.52	79.70	82.03	89.63	85.66
Valence$_n$	75.79	87.00	81.01	81.31	89.53	85.22
Valence$_p$	73.41	77.41	75.36	77.61	84.10	80.72
Arousal	67.48	69.16	68.31	67.09	66.04	66.56
Avg.	71.69	77.48	74.38	75.57	80.87	78.02

Table 6. *MSE* of models recognising level of emotions evaluated for all languages from the dataset. The best model for a specified language (rows) is marked in **bold**.

Embeddings	Language	Trust	Joy	Anticipation	Surprise	Disgust	Fear	Sadness	Anger	Valence-	Valence+	Arousal	MSE_{total}
fastText	Dutch	2.4 ± 0.1	2.84 ± 0.12	1.41 ± 0.06	1.38 ± 0.06	1.45 ± 0.06	1.25 ± 0.05	2.12 ± 0.09	2.57 ± 0.11	3.14 ± 0.13	3.27 ± 0.14	1.08 ± 0.05	2.08 ± 0.09
LASER	Dutch	**1.36 ± 0.04**	**1.48 ± 0.04**	**0.87 ± 0.03**	**1.0 ± 0.03**	**1.1 ± 0.03**	**0.84 ± 0.02**	**1.28 ± 0.04**	**1.55 ± 0.05**	**1.72 ± 0.05**	**1.95 ± 0.06**	**0.83 ± 0.02**	**1.27 ± 0.04**
fastText	English	1.61 ± 0.07	1.77 ± 0.07	1.12 ± 0.05	1.19 ± 0.05	1.25 ± 0.05	1.17 ± 0.05	1.66 ± 0.07	2.0 ± 0.08	2.09 ± 0.09	2.29 ± 0.1	1.08 ± 0.05	1.57 ± 0.07
LASER	English	**1.39 ± 0.04**	**1.56 ± 0.05**	**0.88 ± 0.03**	**1.0 ± 0.03**	**1.17 ± 0.03**	**0.83 ± 0.02**	**1.31 ± 0.04**	**1.66 ± 0.05**	**1.75 ± 0.05**	**2.18 ± 0.06**	**0.89 ± 0.03**	**1.33 ± 0.04**
fastText	French	1.71 ± 0.07	1.98 ± 0.08	1.17 ± 0.05	1.25 ± 0.05	1.43 ± 0.06	1.26 ± 0.05	1.73 ± 0.07	2.12 ± 0.09	2.22 ± 0.09	2.61 ± 0.11	1.16 ± 0.05	1.69 ± 0.07
LASER	French	**1.4 ± 0.04**	**1.57 ± 0.05**	**0.9 ± 0.03**	**0.98 ± 0.03**	**1.11 ± 0.03**	**0.84 ± 0.02**	**1.27 ± 0.04**	**1.54 ± 0.05**	**1.8 ± 0.05**	**1.98 ± 0.06**	**0.85 ± 0.02**	**1.3 ± 0.04**
fastText	German	1.55 ± 0.07	1.7 ± 0.07	1.1 ± 0.05	1.19 ± 0.05	1.19 ± 0.05	1.07 ± 0.05	1.55 ± 0.07	1.79 ± 0.08	2.01 ± 0.08	2.22 ± 0.09	1.08 ± 0.05	1.5 ± 0.06
LASER	German	**1.37 ± 0.04**	**1.53 ± 0.04**	**0.87 ± 0.03**	**0.98 ± 0.03**	**1.11 ± 0.03**	**0.85 ± 0.02**	**1.31 ± 0.04**	**1.57 ± 0.05**	**1.76 ± 0.05**	**1.97 ± 0.06**	**0.84 ± 0.02**	**1.29 ± 0.04**
fastText	Italian	3.02 ± 0.13	3.47 ± 0.15	1.64 ± 0.07	1.2 ± 0.05	1.87 ± 0.08	1.46 ± 0.06	2.63 ± 0.11	3.55 ± 0.15	4.01 ± 0.17	4.45 ± 0.19	1.21 ± 0.05	2.59 ± 0.11
LASER	Italian	**1.36 ± 0.04**	**1.51 ± 0.04**	**0.87 ± 0.03**	**0.97 ± 0.03**	**1.04 ± 0.03**	**0.83 ± 0.02**	**1.28 ± 0.04**	**1.49 ± 0.04**	**1.74 ± 0.05**	**1.87 ± 0.05**	**0.83 ± 0.02**	**1.25 ± 0.04**
fastText	Polish	2.35 ± 0.1	2.63 ± 0.11	1.26 ± 0.05	1.26 ± 0.05	1.4 ± 0.06	1.19 ± 0.05	1.82 ± 0.08	2.54 ± 0.11	2.95 ± 0.12	2.93 ± 0.12	1.27 ± 0.05	1.96 ± 0.08
fastText	Polish KGR10	1.55 ± 0.07	1.6 ± 0.07	1.11 ± 0.05	1.23 ± 0.05	1.19 ± 0.05	1.02 ± 0.04	1.68 ± 0.07	2.06 ± 0.09	1.87 ± 0.08	2.61 ± 0.11	1.25 ± 0.05	1.56 ± 0.07
LASER	Polish	**1.4 ± 0.04**	**1.5 ± 0.04**	**0.88 ± 0.03**	**0.94 ± 0.03**	**1.03 ± 0.03**	**0.87 ± 0.03**	**1.33 ± 0.04**	**1.54 ± 0.04**	**1.72 ± 0.05**	**1.84 ± 0.05**	**0.81 ± 0.02**	**1.26 ± 0.04**
fastText	Portuguese	1.8 ± 0.08	2.13 ± 0.09	1.22 ± 0.05	1.19 ± 0.05	1.29 ± 0.05	1.24 ± 0.05	1.99 ± 0.08	2.34 ± 0.1	2.42 ± 0.1	2.87 ± 0.12	1.17 ± 0.05	1.79 ± 0.08
LASER	Portuguese	**1.34 ± 0.04**	**1.5 ± 0.04**	**0.89 ± 0.03**	**0.99 ± 0.03**	**1.12 ± 0.03**	**0.83 ± 0.02**	**1.27 ± 0.04**	**1.54 ± 0.04**	**1.72 ± 0.05**	**1.97 ± 0.06**	**0.83 ± 0.02**	**1.27 ± 0.04**
fastText	Russian	1.54 ± 0.06	1.83 ± 0.08	1.02 ± 0.04	1.17 ± 0.05	1.17 ± 0.05	1.07 ± 0.05	1.58 ± 0.07	1.75 ± 0.07	2.09 ± 0.09	2.18 ± 0.09	0.88 ± 0.04	1.47 ± 0.06
LASER	Russian	**1.34 ± 0.04**	**1.48 ± 0.04**	**0.86 ± 0.03**	**0.99 ± 0.03**	**1.06 ± 0.03**	**0.89 ± 0.03**	**1.32 ± 0.04**	**1.56 ± 0.05**	**1.71 ± 0.05**	**1.9 ± 0.06**	**0.85 ± 0.03**	**1.27 ± 0.04**
fastText	Spanish	1.72 ± 0.07	1.85 ± 0.08	1.24 ± 0.05	1.3 ± 0.05	1.2 ± 0.05	1.17 ± 0.05	1.79 ± 0.08	2.07 ± 0.09	2.23 ± 0.09	2.41 ± 0.1	1.12 ± 0.05	1.65 ± 0.07
LASER	Spanish	**1.36 ± 0.04**	**1.49 ± 0.04**	**0.87 ± 0.03**	**1.0 ± 0.03**	**1.09 ± 0.03**	**0.85 ± 0.02**	**1.3 ± 0.04**	**1.54 ± 0.05**	**1.72 ± 0.05**	**1.97 ± 0.06**	**0.84 ± 0.02**	**1.28 ± 0.04**

Table 7. R^2 of models recognising level of emotions evaluated for all languages from the dataset. The best model for a specified language (rows) is marked in **bold**.

Embeddings	language	trust	Joy	Anticipation	Surprise	Disgust	Fear	Sadness	Anger	Negative	Positive	Arousal	R^2_{total}
fastText	Dutch	45.0 ± 1.89	41.71 ± 1.76	28.31 ± 1.19	−4.86 ± −0.2	50.85 ± 2.14	34.53 ± 1.45	42.06 ± 1.77	51.17 ± 2.15	44.52 ± 1.87	50.54 ± 2.13	30.98 ± 1.3	37.71 ± 1.59
LASER	Dutch	**68.87 ± 2.02**	**69.55 ± 2.04**	**55.76 ± 1.63**	**24.21 ± 0.71**	**62.74 ± 1.84**	**56.09 ± 1.64**	**64.94 ± 1.9**	**70.68 ± 2.07**	**69.74 ± 2.04**	**70.42 ± 2.06**	**47.28 ± 1.39**	**60.02 ± 1.76**
fastText	English	63.13 ± 2.66	63.61 ± 2.68	43.19 ± 1.82	9.69 ± 0.41	57.49 ± 2.42	39.16 ± 1.65	54.6 ± 2.3	62.15 ± 2.62	63.13 ± 2.66	65.3 ± 2.75	30.97 ± 1.3	50.22 ± 2.11
LASER	English	**68.08 ± 1.99**	**68.03 ± 1.99**	**55.26 ± 1.62**	**23.75 ± 0.7**	**60.18 ± 1.76**	**56.86 ± 1.67**	**64.25 ± 1.88**	**68.45 ± 2.01**	**69.05 ± 2.02**	**67.02 ± 1.96**	**43.5 ± 1.27**	**58.58 ± 1.72**
fastText	French	60.9 ± 2.56	59.27 ± 2.5	40.81 ± 1.72	4.66 ± 0.2	51.25 ± 2.16	34.26 ± 1.44	52.57 ± 2.21	59.71 ± 2.51	60.81 ± 2.56	60.51 ± 2.55	26.11 ± 1.1	46.44 ± 1.96
LASER	French	**67.93 ± 1.99**	**67.8 ± 1.99**	**54.31 ± 1.59**	**25.71 ± 0.75**	**62.11 ± 1.82**	**56.08 ± 1.64**	**65.13 ± 1.91**	**70.71 ± 2.07**	**68.25 ± 2.0**	**70.02 ± 2.05**	**45.64 ± 1.34**	**59.43 ± 1.74**
fastText	German	64.43 ± 2.71	65.04 ± 2.74	44.05 ± 1.85	9.8 ± 0.41	59.58 ± 2.51	43.94 ± 1.85	57.65 ± 2.43	66.04 ± 2.78	64.53 ± 2.72	66.42 ± 2.8	30.95 ± 1.3	52.04 ± 2.19
LASER	German	**68.6 ± 2.01**	**68.49 ± 2.01**	**55.78 ± 1.63**	**25.14 ± 0.74**	**62.31 ± 1.83**	**55.62 ± 1.63**	**64.22 ± 1.88**	**70.26 ± 2.06**	**68.94 ± 2.02**	**70.18 ± 2.06**	**46.18 ± 1.35**	**59.61 ± 1.75**
fastText	Italian	30.73 ± 1.29	28.72 ± 1.21	17.07 ± 0.72	8.44 ± 0.36	36.46 ± 1.53	23.99 ± 1.01	28.04 ± 1.18	32.61 ± 1.37	29.22 ± 1.23	32.52 ± 1.37	23.02 ± 0.97	26.44 ± 1.11
LASER	Italian	**68.85 ± 2.02**	**68.98 ± 2.02**	**56.01 ± 1.64**	**26.16 ± 0.77**	**64.53 ± 1.89**	**56.56 ± 1.66**	**64.86 ± 1.9**	**71.72 ± 2.1**	**69.33 ± 2.03**	**71.73 ± 2.1**	**47.11 ± 1.38**	**60.53 ± 1.77**
fastText	Polish	46.11 ± 1.94	45.96 ± 1.94	36.02 ± 1.52	3.84 ± 0.16	52.82 ± 2.21	37.64 ± 1.58	50.27 ± 2.12	51.82 ± 2.18	47.99 ± 2.02	55.64 ± 2.34	18.82 ± 0.79	40.65 ± 1.71
fastText	Polish KGR10	64.6 ± 2.72	67.04 ± 2.82	43.81 ± 1.84	6.46 ± 0.27	59.50 ± 2.51	46.7 ± 1.97	54.06 ± 2.28	60.84 ± 2.56	67.05 ± 2.82	60.42 ± 2.54	20.3 ± 0.85	50.08 ± 2.11
LASER	Polish	**68.04 ± 1.99**	**69.23 ± 2.03**	**55.17 ± 1.62**	**28.14 ± 0.82**	**65.27 ± 1.91**	**54.41 ± 1.59**	**63.56 ± 1.86**	**70.88 ± 2.08**	**69.6 ± 2.04**	**72.08 ± 2.11**	**48.2 ± 1.41**	**60.42 ± 1.77**
fastText	Portuguese	58.77 ± 2.47	56.29 ± 2.37	38.12 ± 1.6	9.46 ± 0.4	56.14 ± 2.36	35.39 ± 1.49	45.42 ± 1.91	55.65 ± 2.34	57.35 ± 2.41	56.53 ± 2.38	25.57 ± 1.08	44.97 ± 1.89
LASER	Portuguese	**69.23 ± 2.03**	**69.17 ± 2.03**	**54.83 ± 1.61**	**24.58 ± 0.72**	**61.79 ± 1.81**	**56.59 ± 1.66**	**65.22 ± 1.91**	**70.87 ± 2.08**	**69.61 ± 2.04**	**70.14 ± 2.06**	**47.21 ± 1.38**	**59.93 ± 1.76**
fastText	Russian	64.65 ± 2.72	62.37 ± 2.63	48.28 ± 2.03	22.86 ± 0.96	60.06 ± 2.53	42.85 ± 1.8	56.83 ± 2.39	66.76 ± 2.81	63.16 ± 2.66	66.91 ± 2.82	43.98 ± 1.85	54.43 ± 2.29
LASER	Russian	**69.26 ± 2.03**	**69.62 ± 2.04**	**56.23 ± 1.65**	**24.71 ± 0.72**	**63.86 ± 1.87**	**53.78 ± 1.58**	**63.71 ± 1.87**	**70.47 ± 2.06**	**69.77 ± 2.04**	**71.17 ± 2.09**	**45.49 ± 1.33**	**59.82 ± 1.75**
fastText	Spanish	60.61 ± 2.55	62.07 ± 2.61	37.25 ± 1.57	0.83 ± 0.03	59.01 ± 2.48	38.71 ± 1.63	51.09 ± 2.15	60.79 ± 2.56	60.72 ± 2.56	63.49 ± 2.67	28.33 ± 1.19	47.54 ± 2.0
LASER	Spanish	**68.84 ± 2.02**	**69.28 ± 2.03**	**55.89 ± 1.64**	**24.17 ± 0.71**	**62.86 ± 1.84**	**55.69 ± 1.63**	**64.36 ± 1.89**	**70.76 ± 2.07**	**69.71 ± 2.04**	**70.19 ± 2.06**	**46.57 ± 1.36**	**59.85 ± 1.75**

7   Conclusions

In this preliminary study, we focused on basic neural language models to prepare and evaluate baseline approaches to recognise emotions, valence, and arousal in multidomain textual reviews. Further plans include the evaluation of hybrid approaches combining ML and lexico-syntactic rules augmented with semantic analysis of word meanings. We also plan to automatically expand the annotations of word meanings to the rest of the plWordNet using the propagation methods presented in [18,19]. We intend to test other promising methods later, such as Google BERT [6], OpenAI GPT-2 [33] and domain dictionary construction methods utilising WordNet [22].

Automatic emotion annotation has both scientific and applied value. In recent decades, the rapid growth of the Internet produced the *big data revolution* by making unprecedented amounts of data available [16]. This also includes textual data coming directly from social media and other sources. Modern business collects vast amounts of customers' opinions and attitudes associated with brands and products. In this context, monitoring opinions, reactions, and emotions present great value, as they fuel people's decisions and behaviour [38]. However, most of the existing solutions are still limited to manual annotation and simplified methods of analysis (Table 5).

The large database built in the Sentimenti project covers a wide range of Polish vocabulary and introduces an extensive emotive annotation of word meanings in terms of their valence, arousal, as well as basic emotions. The results of such research can be used in several applications – media monitoring, chatbots, stock price forecasting, search engine optimisation for advertisements, and other types of content.

We also provide a preliminary overview of ML methods for automatic analysis of people's opinions in terms of expressed emotions and their attitudes. Since our results are based on the data obtained from a wide cross-section of Polish population, we can adapt our method to specific target groups of people. This introduces the much needed human aspect to artificial intelligence and ML in natural language processing.

Language-agnostic embedding models (LASER) can not only provide multi-lingual vector representations of the text, but they can also outperform single-language dedicated models (fastText) while being used in the same monolingual manner. Even though LASER operates on the sentence level and fastText on the word level, the models based on the former were able to achieve better results each time. It not only achieves similar performance, but in many cases outperforms them by orders of magnitude, providing a more accurate measurement of the level of intensity of emotion.

8   Further Work and Data Availability

In the future, we want to explore other model architectures that are also able to incorporate human context in emotion prediction. This approach makes it possible to determine not only what emotions a text in general is likely to evoke, but more importantly, what emotions it evokes for a particular human. We have already done some work on this topic [15,17,21,27]. We are currently planning to focus on using personalized architectures based on transformers and finetuning transformers with information about the human context.

Currently, a subset of 6,000 annotated word meanings is available in open access[8] along with the accompanying publication [41]. Also, a small number of text reviews (100 documents) translated into 11 languages with full annotations were made available as part of the work [21] on the CLARIN-PL repository.[9]

References

1. Ameijeiras-Alonso, J., Crujeiras, R.M., Rodríguez-Casal, A.: Mode testing, critical bandwidth and excess mass. TEST **28**(3), 900–919 (2018). https://doi.org/10.1007/s11749-018-0611-5
2. Artetxe, M., Schwenk, H.: Massively multilingual sentence embeddings for zero-shot cross-lingual transfer and beyond. Trans. Assoc. Comput. Linguist. **7**, 597–610 (2019)
3. Badjatiya, P., Gupta, S., Gupta, M., Varma, V.: Deep learning for hate speech detection in tweets. In: Proceedings of the 26th International Conference on World Wide Web Companion, pp. 759–760. International World Wide Web Conferences Steering Committee (2017)
4. Baziotis, C., Pelekis, N., Doulkeridis, C.: Datastories at SemEval-2017 Task 4: deep LSTM with attention for message-level and topic-based sentiment analysis. In: Proceedings of the 11th International Workshop on Semantic Evaluation (SemEval-2017), pp. 747–754 (2017)
5. Bojanowski, P., Grave, E., Joulin, A., Mikolov, T.: Enriching word vectors with subword information. Trans. Assoc. Comput. Linguist. **5**, 135–146 (2017)
6. Devlin, J., Chang, M.W., Lee, K., Toutanova, K.: BERT: pre-training of deep bidirectional transformers for language understanding. arXiv preprint arXiv:1810.04805 (2018)

---

[8] https://osf.io/f79bj.

[9] https://github.com/CLARIN-PL/human-bias.

7. Ekman, P.: An argument for basic emotions. Cogn. Emot. **6**(3–4), 169–200 (1992)
8. Felbo, B., Mislove, A., Søgaard, A., Rahwan, I., Lehmann, S.: Using millions of emoji occurrences to learn any-domain representations for detecting sentiment, emotion and sarcasm. In: Proceedings of the 2017 Conference on Empirical Methods in Natural Language Processing, pp. 1615–1625 (2017)
9. Habibi, M., Weber, L., Neves, M., Wiegandt, D.L., Leser, U.: Deep learning with word embeddings improves biomedical named entity recognition. Bioinformatics **33**(14), i37–i48 (2017). https://doi.org/10.1093/bioinformatics/btx228
10. Hartigan, J.A., Hartigan, P.M., et al.: The dip test of unimodality. Ann. Stat. **13**(1), 70–84 (1985)
11. Hartigan, J.A., Wong, M.A.: Algorithm as 136: a K-means clustering algorithm. J. R. Stat. Soc. Ser. C Appl. Stat. **28**(1), 100–108 (1979)
12. Hripcsak, G., Rothschild, A.S.: Technical brief: agreement, the F-measure, and reliability in information retrieval. JAMIA **12**(3), 296–298 (2005). https://doi.org/10.1197/jamia.M1733
13. Janz, A., Kocoń, J., Piasecki, M., Zaśko-Zielińska, M.: plWordNet as a basis for large emotive lexicons of Polish. In: LTC'17 8th Language and Technology Conference. Fundacja Uniwersytetu im. Adama Mickiewicza w Poznaniu, Poznań, November 2017
14. Joulin, A., Grave, E., Bojanowski, P., Mikolov, T.: Bag of tricks for efficient text classification. In: Proceedings of the 15th Conference of the European Chapter of the Association for Computational Linguistics: Volume 2, Short Papers, pp. 427–431. Association for Computational Linguistics, Valencia, April 2017. https://www.aclweb.org/anthology/E17-2068
15. Kanclerz, K., et al.: Controversy and conformity: from generalized to personalized aggressiveness detection. In: Proceedings of the 59th Annual Meeting of the Association for Computational Linguistics and the 11th International Joint Conference on Natural Language Processing (Volume 1: Long Papers), pp. 5915–5926. Association for Computational Linguistics, August 2021. https://doi.org/10.18653/v1/2021.acl-long.460
16. Kitchin, R.: The Data Revolution: Big Data, Open Data, Data Infrastructures and Their consequences. Sage, Thousand Oaks (2014)
17. Kocoń, J., Figas, A., Gruza, M., Puchalska, D., Kajdanowicz, T., Kazienko, P.: Offensive, aggressive, and hate speech analysis: from data-centric to human-centered approach. Inf. Process. Manag. **58**(5), 102643 (2021)
18. Kocoń, J., Janz, A., Piasecki, M.: Classifier-based polarity propagation in a WordNet. In: Proceedings of the 11th International Conference on Language Resources and Evaluation (LREC 2018) (2018)
19. Kocoń, J., Janz, A., Piasecki, M.: Context-sensitive sentiment propagation in WordNet. In: Proceedings of the 9th International Global Wordnet Conference (GWC 2018) (2018)
20. Kocoń, J., Gawor, M.: Evaluating KGR10 Polish word embeddings in the recognition of temporal expressions using BiLSTM-CRF. CoRR arXiv:1904.04055 (2019)
21. Kocoń, J., et al.: Learning personal human biases and representations for subjective tasks in natural language processing. In: 2021 IEEE International Conference on Data Mining (ICDM). IEEE (2021)
22. Kocoń, J., Marcińczuk, M.: Generating of events dictionaries from polish wordnet for the recognition of events in polish documents. In: Sojka, P., Horák, A., Kopeček, I., Pala, K. (eds.) TSD 2016. LNCS (LNAI), vol. 9924, pp. 12–19. Springer, Cham (2016). https://doi.org/10.1007/978-3-319-45510-5_2

23. Kutuzov, A., Fares, M., Oepen, S., Velldal, E.: Word vectors, reuse, and replicability: towards a community repository of large-text resources. In: Proceedings of the 58th Conference on Simulation and Modelling, pp. 271–276. Linköping University Electronic Press (2017)
24. Ma, Y., Peng, H., Cambria, E.: Targeted aspect-based sentiment analysis via embedding commonsense knowledge into an attentive LSTM. In: Thirty-Second AAAI Conference on Artificial Intelligence (2018)
25. Marcińczuk, M., Kocoń, J., Gawor, M.: Recognition of named entities for polish-comparison of deep learning and conditional random fields approaches. In: Proceedings of PolEval 2018 Workshop. Institute of Computer Science, Polish Academy of Sciences, Warsaw, Poland (2018)
26. Mikolov, T., Sutskever, I., Chen, K., Corrado, G.S., Dean, J.: Distributed representations of words and phrases and their compositionality. In: Advances in Neural Information Processing Systems, pp. 3111–3119 (2013)
27. Milkowski, P., Gruza, M., Kanclerz, K., Kazienko, P., Grimling, D., Kocon, J.: Personal bias in prediction of emotions elicited by textual opinions. In: Proceedings of the 59th Annual Meeting of the Association for Computational Linguistics and the 11th International Joint Conference on Natural Language Processing: Student Research Workshop, pp. 248–259. Association for Computational Linguistics, August 2021. https://doi.org/10.18653/v1/2021.acl-srw.26
28. Neville, Z., Brownstein, N.C.: Macros to conduct tests of multimodality in SAS. J. Stat. Comput. Simul. **88**(17), 3269–3290 (2018)
29. Paolacci, G., Chandler, J.: Inside the Turk: understanding mechanical Turk as a participant pool. Curr. Dir. Psychol. Sci. **23**(3), 184–188 (2014)
30. Piasecki, M., Broda, B., Szpakowicz, S.: A WordNet from the ground up. Oficyna Wydawnicza Politechniki Wrocławskiej Wrocław (2009)
31. Piasecki, M., Czachor, G., Janz, A., Kaszewski, D., Kędzia, P.: WordNet-based evaluation of large distributional models for Polish. In: Proceedings of the 9th Global WordNet Conference (GWC 2018), pp. 232–241 (2018)
32. Plutchik, R.: A psychoevolutionary theory of emotions. Soc. Sci. Inf. **21**(4–5), 529–553 (1982). https://doi.org/10.1177/053901882021004003
33. Radford, A., Wu, J., Child, R., Luan, D., Amodei, D., Sutskever, I.: Language models are unsupervised multitask learners. OpenAI Blog, p. 8 (2019)
34. Riegel, M., et al.: Nencki Affective Word List (NAWL): the cultural adaptation of the Berlin Affective Word List–Reloaded (BAWL-R) for Polish. Behav. Res. Meth. **47**(4), 1222–1236 (2015). https://doi.org/10.3758/s13428-014-0552-1
35. Russell, J.A., Mehrabian, A.: Evidence for a three-factor theory of emotions. J. Res. Pers. **11**(3), 273–294 (1977). https://doi.org/10.1016/0092-6566(77)90037-X
36. Schnabel, T., Labutov, I., Mimno, D.M., Joachims, T.: Evaluation methods for unsupervised word embeddings. In: Proceedings of Empirical Methods in Natural Language Processing Conference (EMNLP), pp. 298–307 (2015)
37. Silverman, B.W.: Using kernel density estimates to investigate multimodality. J. Roy. Stat. Soc. Ser. B (Methodol.) **43**(1), 97–99 (1981)
38. Tversky, A., Kahneman, D.: Rational choice and the framing of decisions. In: Multiple Criteria Decision Making and Risk Analysis Using Microcomputers, pp. 81–126. Springer, Cham (1989). https://doi.org/10.1007/978-3-642-74919-3_4
39. Wang, Y., Huang, M., Zhao, L., et al.: Attention-based LSTM for aspect-level sentiment classification. In: Proceedings of the 2016 Conference on Empirical Methods in Natural Language Processing, pp. 606–615 (2016)

40. Wierzba, M., et al.: Basic emotions in the Nencki Affective Word List (NAWL BE): new method of classifying emotional stimuli. PLoS ONE **10**(7), e0132305 (2015). https://doi.org/10.1371/journal.pone.0132305

41. Wierzba, M., et al.: Emotion norms for 6,000 Polish word meanings with a direct mapping to the Polish wordnet. Behav. Res. Meth. (2021). https://doi.org/10.3758/s13428-021-01697-0, https://osf.io/f79bj/

42. Wojatzki, M., Ruppert, E., Holschneider, S., Zesch, T., Biemann, C.: Germeval 2017: shared task on aspect-based sentiment in social media customer feedback. In: Proceedings of the GermEval, pp. 1–12 (2017)

43. Zaśko-Zielińska, M., Piasecki, M., Szpakowicz, S.: A large WordNet-based sentiment lexicon for Polish. In: Proceedings of the International Conference Recent Advances in Natural Language Processing, pp. 721–730 (2015)

Construction and Evaluation of Sentiment Datasets for Low-Resource Languages: The Case of Uzbek

Elmurod Kuriyozov[1](✉) , Sanatbek Matlatipov[2] , Miguel A. Alonso[1] ,
and Carlos Gómez-Rodríguez[1]

[1] Universidade da Coruña, CITIC, Grupo LYS, Departamento de Computación,
Facultade de Informática, Campus de Elviña, 15071 A Coruña, Spain
{e.kuriyozov,miguel.alonso,carlos.gomez}@udc.es
[2] National University of Uzbekistan,
University Street 4, 100174 Tashkent, Uzbekistan
s.matlatipov@nuu.uz

Abstract. To our knowledge, the majority of human language processing technologies for low-resource languages don't have well-established linguistic resources for the development of sentiment analysis applications. Therefore, it is in dire need of such tools and resources to overcome the NLP barriers, so that, low-resource languages can deliver more benefits. In this paper, we fill that gap by providing its first annotated corpora for Uzbek language polarity classification. Our methodology considers collecting a medium-size manually annotated dataset and a larger-size dataset automatically translated from existing resources. Then, we use these datasets to train what, to our knowledge, are the first sentiment analysis models on the Uzbek language, using both traditional machine learning techniques and recent deep learning models. Both sets of techniques achieve similar accuracy (the best model on the manually annotated test set is a convolutional neural network with 88.89% accuracy, and on the translated set, a logistic regression with 89.56% accuracy); with the accuracy of the deep learning models being limited by the quality of available pre-trained word embeddings.

Keywords: Sentiment analysis · Low-resource languages · Uzbek language

This work has received funding from ERDF/MICINN-AEI (ANSWER-ASAP, TIN2017-85160-C2-1-R; SCANNER-UDC, PID2020-113230RB-C21), from Xunta de Galicia (ED431C 2020/11), and from Centro de Investigación de Galicia "CITIC", funded by Xunta de Galicia and the European Union (ERDF - Galicia 2014–2020 Program), by grant ED431G 2019/01. Elmurod Kuriyozov was funded for his PhD by El-Yurt-Umidi Foundation under the Cabinet of Ministers of the Republic of Uzbekistan.

Z. Vetulani et al. (Eds.): LTC 2019, LNAI 13212, pp. 232–243, 2022.
https://doi.org/10.1007/978-3-031-05328-3_15

1   Introduction

The Natural Language Processing (NLP) field has achieved high accuracy results, allowing the creation of useful applications that play an important role in many areas now. In particular, the adoption of deep learning models has boosted accuracy figures across a wide range of NLP tasks. As a part of this trend, sentiment classification, a prominent example of the applications of NLP, has seen substantial gains in performance by using deep learning approaches compared to its predecessor approaches [2]. However, this is the case for only about twenty high-resource languages out of seven thousand, so that, low-resource languages still lack access to those performance improvements. Neural network models, which have gained wide popularity in recent years, are generally considered as the best supervised sentiment classification technique for resource-rich languages so far [2,23,27], but they require significant amounts of annotated training data to work well.

Meanwhile, the fact that a language can be considered a low-resource language does not necessarily mean that it is spoken by a small community. For instance, the language we focus on in this paper is Uzbek, which is spoken by more than 33 million native speakers in Uzbekistan as well as elsewhere in Central Asia and a part of China.[1]

It is also important to point out that NLP tools in general, and sentiment analysis tools in particular, benefit from taking into account the particularities of the language under consideration [9,24]. Uzbek is a Turkic language that is the first official and only declared national language of Uzbekistan. The language of Uzbeks (in native language: *O'zbek tili* or *O'zbekcha*) is a null-subject, agglutinative language and has many dialects, varying widely from region to region, which introduces more difficult problems to tackle.[2]

The main contributions of this paper are:

1. The creation of the first annotated dataset for sentiment analysis in Uzbek language, obtained from reviews of the top 100 Google Play Store applications used in Uzbekistan. This manually annotated dataset contains 2500 positive and 1800 negative reviews. Furthermore, we have also built a larger dataset by automatically translating (using Google Translate API) an existing English dataset[3] of application reviews. The translated dataset has ≈10K positive and ≈10K negative app reviews, after manually eliminating the major machine translation errors by either correcting or removing them completely.
2. The definition of the baselines for sentiment analyses in Uzbek by considering both traditional machine learning methods as well as recent deep learning techniques fed with fastText pre-trained word embeddings.[4] Although all

---

[1] https://en.wikipedia.org/wiki/Uzbek_language.

[2] Little information about Uzbek languages is available in English. A good starting point for readers who are interested could be: http://aboutworldlanguages.com/uzbek.

[3] https://github.com/amitt001/Android-App-Reviews-Dataset.

[4] https://fasttext.cc.

the tested models are relatively accurate and differences between models are small, the neural network models tested do not manage to substantially out-perform traditional models. We believe that the quality of currently available pre-trained word embeddings for Uzbek is not enough to let deep learning models perform at their full potential.

3. The definition of the steps for translating an available dataset automatically to a low-resource language, analysing the quality loss in the case of English-Uzbek translation.

All the resources, including the datasets, the list of top 100 apps whose reviews were collected, the source code used to collect the reviews and the one for baseline classifiers, are publicly available at the project's repository.[5]

The remainder of this paper is organized as follows: after this Introduction, Sect. 2 describes related work that has been done so far. It is followed by a description of the methodology in Sect. 3 and continues with Sect. 4 which focuses on Experiments and Results. The final Sect. 5 concludes the paper and highlights the future work.

2   Related Work

We only know of one existing sentiment analysis resource for the Uzbek language: a multilingual collection of sentiment lexicons presented in [4] that includes Uzbek, but the Uzbek lexicon is very small and is not evaluated on an actual sentiment analysis system or dataset. To our knowledge, there are no existing annotated corpora on which it could be evaluated.

There has been some relevant work on Uzbek language so far, such as representation of Uzbek morphology [15], morpheme alignment [12], transliteration between Latin and Cyrillic scripts [14], as well as Uzbek word-embeddings [11, 13]. After completion of the work described in this paper, there has been a rise in publication of work in sentiment analysis and text classification on Uzbek language by other authors as well [19–21].

Other languages of the Turkic family such as Turkish and Kazakh have made considerable progress in the field. For example, a system for unsupervised sentiment analysis on Turkish texts is presented in [25], based on a customization of SentiStrength (Thelwall and Paltoglou) by translating its polarity lexicon to Turkish, obtaining a 76% accuracy in classifying Turkish movie reviews as positive or negative.

Sentiment analysis of Turkish political news in online media was studied in [10] using four different classifiers (Naive Bayes, Maximum Entropy, SVM, and character-based n-gram language models) with a variety of text features (frequency of polar word unigrams, bigrams, root words, adjectives and effective polar words) concluding that the Maximum Entropy and the n-gram models are more effective when compared to SVM and Naive Bayes, reporting an accuracy of 76% for binary classification.

---

[5] https://github.com/elmurod1202/uzbek-sentiment-analysis.

A sentiment analysis system for Turkish that gets a 79.06% accuracy in binary sentiment classification of movie reviews is described in [6], but it needs several linguistic resources and tools, such as a dependency parser and a WordNet annotated with sentiment information, which are not available for Uzbek.

[26] presented a rule-based sentiment analysis system for Kazakh working on a dictionary, morphological rules and an ontological model, achieving 83% binary classification accuracy for simple sentences. [16] one of the author created stemming tool for Uzbek language which is very handy for using it sentiment analysis.

A modern Deep Learning approach for solving Kazakh and Russian-language Sentiment Analysis tasks was investigated in [22]. Particularly, Long Short-Term Memory (LSTM) was used to handle long-distance dependencies,and word embeddings (word2vec, GloVe) were used as the main feature.

3   Methodology

3.1   Data Collection

When it comes to choosing an available source to collect data for low-resource languages, the usual approach for resource-rich languages, such as Twitter data [28] or movie reviews [3], may not qualify and end up being very scarce or not sufficient to work with. So one has to find out what is the most widespread web service from which a large amount of open data can be collected for a specific low-resource language. In the case of Uzbek, most of its speakers use mobile devices for accessing the Internet, and Android retains a share of more than 85% of the mobile Operating Systems market (as of February 2019).[6] This is the reason why the reviews of Google Play Store Applications have been chosen as the data source for our research.

We selected the list of top 100 applications used in Uzbekistan, retrieving for each review its text and its associated star rating (from 1 to 5 stars). In order to promote future research on the Uzbek language, the project repository that has been created to share the sources of this paper contains a file with the list of URLs for those apps and the Python script for crawling the Play Store reviews. Due to Google's anti-spam and anti-DDOS policies, there are certain limitations on harvesting data, such as that only the most relevant 40 reviews can be obtained in a single request and up to 4500 in several requests (the corresponding source code has also been included).

3.2   Pre-processing

We observed that the collection of texts (together with the associated star ratings) downloaded by the above procedure was noisy, so we performed a correction process. The comments containing only emojis, names or any other irrelevant content, such as username mentions, URLs or specific app names were removed.

---

[6] http://gs.statcounter.com/os-market-share/mobile/uzbekistan.

Tezkor o'girish: Cyrillic --> Latin

Илова зўр ясалган, шикоятим йўқ	Ilova zo'r yasalgan, shikoyatim yo'q

Fig. 1. Savodxon.uz: Online Cyrillic to Latin alphabet transformation tool, specifically for Uzbek language. (The example text in English: "The app is nicely created, I have no complaints")

Those written in languages different from Uzbek (mostly in Russian and some in English) were manually translated. There is another small inconvenience, specific to dealing with Uzbek texts: although currently the official and most-used alphabet for the language is the Latin one, some people still tend to write in the Cyrillic alphabet, which was the official alphabet decades ago and is still used in practice [7]. Those Cyrillic comments were collected and transformed to the Latin one using an available online tool.[7] A small example is shown in Fig. 1.

3.3 Annotation

This paper is intended to present only a binary classified dataset, so the main task was to label the reviews as positive or negative. A neutral class was not considered for the sake of simplicity since this is, to our knowledge, the first sentiment analysis dataset for our chosen language, so we preferred to start from the simplest setting. The annotation process was done by two native Uzbek speakers manually labeling the reviews, giving them a score of either 0 or 1, meaning that the review is either negative or positive, respectively. A third score was obtained from the dataset's rating column as follows:

- Reviews with 4- and 5-star ratings were labeled as positive (1);
- Reviews with 1- and 2-star ratings were labeled as negative (0);
- The majority of reviews with 3-star rating also turned out to have negative opinion so we labeled them as negative (0) as well, but both annotators removed the objective reviews.

Finally, the review was given a polarity according to the majority label. This process resulted into 2500 reviews annotated as positive and 1800 as negative.

3.4 Translation

In order to further extend the resources to support sentiment analysis, another larger dataset was obtained through machine translation. An available English dataset of positive and negative reviews of Android apps, containing 10000

---

[7] Online Cyrillic to Latin transformation tool: https://savodxon.uz.

reviews of each class, was automatically translated using MTRANSLATE[8]: an unofficial Google Translate API from English to Uzbek. The next step was to determine whether the translation was accurate enough to work with. Thus, we manually went through the translation results quickly and examined a random subset of the reviews, large enough to make a reasonable decision on overall accuracy. Although the translation was not clear enough to use for daily purposes, the meaning of the sentences was approximately preserved, and in particular, the sentiment polarity was kept (except for very few exceptional cases). An example of the translation can be seen in Fig. 2.

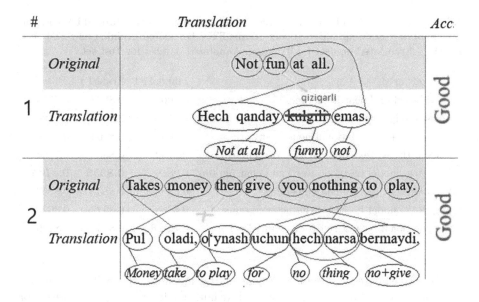

Fig. 2. An example of the translation process on two random negative reviews. As can be observed, the polarity of the comments is preserved.

As a result, we have obtained two datasets with the sizes shown in Table 1. While the translated dataset is quite balanced, the manually annotated dataset has about 3:4 ratio of negative to positive reviews. Each of the datasets has been split into a training and a test set following a 90:10 ratio, for the experiments in the next section.

4   Experiments and Results

To create the baseline models for Uzbek sentiment analysis, we chose various classifiers from different families, including different methods of Logistic Regression (LR), Support Vector Machines (SVM), and recent Deep Learning methods,

---

[8] https://github.com/mouuff/mtranslate.

Table 1. Number of reviews per dataset and polarity class

Datasets	Positive	Negative	Total
Manual dataset	2500	1800	4300
Translated dataset	9632	8853	18485

such as Recurrent Neural Networks (RNN) and Convolutional Neural Networks (CNN).

Table 2. Accuracy results with different training and test sets. **ManualTT** - Manually annotated Training and Test sets. **TransTT** - Translated Training and Test sets. **TTMT** - Translated dataset for Training, Annotated dataset for Test set.

Methods used	ManualTT	TransTT	TTMT
SVM based on linear kernel model	0.8002	0.8588	0.7756
Logistic regression model based on word ngrams	0.8547	0.8810	0.7720
Recurrent + convolutional neural network	0.8653	0.8864	0.7850
RNN with fastText pre-trained word embeddings	0.8782	0.8832	0.7996
Log. Reg. model based on word and char ngrams	0.8846	**0.8956**	**0.8145**
RNN without pre-trained embeddings	0.8868	0.8832	0.8052
Log. Reg. model based on character ngrams	0.8868	0.8945	0.8021
Convolutional Neural Network (Multichannel)	**0.8888**	0.8832	0.8120

We implemented LR and SVM models by means of the Scikit-Learn [17] machine learning library in Python with default configuration parameters. For the LR models, we implemented a variant based on word n-grams (unigrams and bigrams), and one with character n-grams (with n ranging from 1 to 4). We also tested a model combining said word and character n-gram features.

In the case of Deep Learning models, we used Keras [5] on top of TensorFlow [1]. We use as input the FastText pre-trained word embeddings of size 300 [8] for Uzbek language, that were created from Wiki pages and Common-Crawl,[9] which, to our knowledge, are the only available pre-trained word embeddings for Uzbek language so far. The source code for all the chosen baseline models is available on the project's GitHub repository.

For the CNN model, we used a multi-channel CNN with 256 filters and three parallel channels with kernel sizes of 2,3 and 5, and dropout of 0.3. The output of the hidden layer is the concatenation of the max pooling of the three channels. For RNN, we use a bidirectional network of 100 GRUs. The output of the hidden layer is the concatenation of the average and max pooling of the hidden states. For the combination of deep learning models, we stacked the CNN on top of

---

[9] http://commoncrawl.org.

the GRU. In the three cases, the final output is obtained through a sigmoid activation function applied on the previous layer. In all cases, Adam optimization algorithm, an extension of stochastic gradient descent, was chosen for training, with standard parameters: learning rate $\alpha = 0.0001$ and exponential decay rates $\beta_1 = 0.9$ and $\beta_2 = 0.999$. Binary cross-entropy was used as loss function.

As our performance metric, we use classification accuracy. This is the most intuitive performance measure for a binary classifier, and it is merely a ratio of correctly predicted observations to total observations [18].

$$\text{accuracy} = \frac{\sum \text{true positive} + \sum \text{true negative}}{\sum \text{total population}}$$

Since we have worked on relatively small dataset, other metrics, such as the runtime complexity and memory allocations were not taken into account.

Table 2 shows the accuracy obtained in three different configurations: a first one working on the manually annotated dataset (ManualTT), a second one on the translated dataset (TransTT) and a third one in which training was performed on translated dataset while testing was performed on the manually annotated dataset.

The LR based on word n-grams obtained a binary classification accuracy of 88.1% on the translated dataset, while the one based on character n-grams, with its better handling of misspelled words, improved it to 89.45%. To take advantage of both methods, we combined the two and got 89.56% accuracy, the best performance for the translated dataset obtained in this paper. The deep learning models have shown accuracies ranging from 86.53% (using RNN+CNN) to 88.88% (using Multichannel CNN) on our manually annotated dataset, the latter being the best result on this dataset, while the RNN+CNN combination performed well on the translated dataset with 88.64% average accuracy, slightly better than others (88.32% for single RNN and CNN models).

Table 3 shows per-class metrics of our best result on the translated dataset, obtained from the LR model based on word and character n-grams trained on that same dataset. Although the results obtained have been good in general terms, those obtained for deep learning models have not clearly surpassed the results obtained by other classifiers. This is mainly due to some of the complexities of Uzbek language. Indeed, Uzbek morphology [15] is highly agglutinative, and this aspect makes it harder to rely on word embeddings: a single word can have more than 200 forms generated by adding suffixes, sometimes even an entire sentence in English language can be described by one word. An example of how agglutinative the language is shown in Fig. 3.

Table 3. Performance metrics of the best result on the translated dataset.

Classes	Precision	Recall	F1-score
Negative	0.89	0.91	0.90
Positive	0.90	0.88	0.89

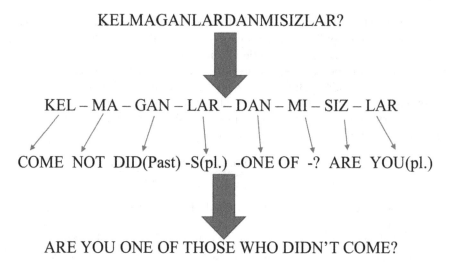

Fig. 3. An example of the agglutinative aspect of Uzbek language. Here we describe how just one Uzbek word can correspond to an entire sentence in English.

This agglutinative nature of Uzbek poses a major challenge for the definition of word embeddings. In our experiments, we could not associate a word-embedding to about 37% of words occurring in reviews. The reason for that was the noise of the reviews dataset we used, and which contained a large amount of misspelled words. Additionally, while our dataset contains only words in Latin alphabet, about the half of the word embeddings we used were in Cyrillic, decreasing the chance of the word to be found.

5   Conclusion and Future Work

In this paper, we have presented a new Sentiment Analysis dataset for Uzbek language, collected from the reviews of Top 100 Android applications in Uzbekistan in Google Play Store. This dataset contains 4300 negative and positive reviews with a 3:4 ratio between the respective classes. It was manually annotated by two annotators, also considering the star rating provided by the reviewers. We also presented another new and relatively larger (20K) dataset of the same type, but this time it was automatically translated to Uzbek using Google Translate from an existing app review dataset in English language.

From the results of the experiments presented here, one can conclude that deep learning models do not perform better in sentiment classification than classic models for a low-resource language. We achieved our best accuracy (89.56%) on the translated dataset using a logistic regression model using word and character n-grams. The modern deep learning approaches have shown very similar results, without substantially outperforming classic ones in accuracy as they tend to do when used for resource-rich languages. We believe this to be due to lack of resources to feed the deep learning models: for example, the pre-trained word embeddings need to be enhanced (trained on a larger dataset) in order to benefit from the recent methods.

Our future work will be focused on creating more fundamental resources for the Uzbek language, such as tagged corpora, pre-trained word embeddings, lexicon and treebanks allowing us to build essential NLP tools, like part-of-speech taggers and parsers, which in turn can be used to improve sentiment analysis and other NLP tasks. An alternative to improve the deep learning models tested in this work would be to use character embeddings, which should be a good fit for an agglutinative language because they can capture information about parts of words and reduce the sparsity due to the large number of different words.

References

1. Abadi, M., et al.: TensorFlow: large-scale machine learning on heterogeneous systems (2015). https://www.tensorflow.org/
2. Barnes, J., Klinger, R., Walde, S.S.I.: Assessing state-of-the-art sentiment models on state-of-the-art sentiment datasets. arXiv preprint arXiv:1709.04219 (2017)
3. Chakraborty, K., Bhattacharyya, S., Bag, R., Hassanien, A.E.: Comparative sentiment analysis on a set of movie reviews using deep learning approach. In: Hassanien, A.E., Tolba, M.F., Elhoseny, M., Mostafa, M. (eds.) AMLTA 2018. AISC, vol. 723, pp. 311–318. Springer, Cham (2018). https://doi.org/10.1007/978-3-319-74690-6_31
4. Chen, Y., Skiena, S.: Building sentiment lexicons for all major languages. In: Proceedings of the 52nd Annual Meeting of the Association for Computational Linguistics (Vol. 2: Short Papers), pp. 383–389. Association for Computational Linguistics, Baltimore, Maryland, June 2014. https://doi.org/10.3115/v1/P14-2063
5. Chollet, F., et al.: Keras (2015). https://github.com/fchollet/keras
6. Dehkharghani, R., Yanikoglu, B., Saygin, Y., Oflazer, K.: Sentiment analysis in Turkish at different granularity levels. Nat. Lang. Eng. 23(4), 535–559 (2017)
7. Dietrich, A.: Language policy and hegemony in the Turkic republics. In: Andrews, E. (ed.) Language Planning in the Post-Communist Era: The Struggles for Language Control in the New Order in Eastern Europe, Eurasia and China, pp. 145–167. Springer, Cham (2018). https://doi.org/10.1007/978-3-319-70926-0_6
8. Grave, E., Bojanowski, P., Gupta, P., Joulin, A., Mikolov, T.: Learning word vectors for 157 languages. In: Proceedings of the International Conference on Language Resources and Evaluation (LREC 2018) (2018)
9. Jang, H., Shin, H.: Language-specific sentiment analysis in morphologically rich languages. In: Coling 2010: Posters, pp. 498–506. Coling 2010 Organizing Committee, Beijing, China, August 2010. https://www.aclweb.org/anthology/C10-2057
10. Kaya, M., Fidan, G., Toroslu, I.H.: Sentiment analysis of Turkish political news. In: Proceedings of the The 2012 IEEE/WIC/ACM International Joint Conferences on Web Intelligence and Intelligent Agent Technology, WI-IAT 2012, vol. 01, pp. 174–180. IEEE Computer Society, Washington (2012). http://dl.acm.org/citation.cfm?id=2457524.2457679
11. Kuriyozov, E., Doval, Y., Gómez-Rodríguez, C.: Cross-lingual word embeddings for Turkic languages. In: Proceedings of the 12th Language Resources and Evaluation Conference (LREC 2020), pp. 4054–4062. European Language Resources Association, Marseille (2020). https://aclanthology.org/2020.lrec-1.499/

12. Li, X., Tracey, J., Grimes, S., Strassel, S.: Uzbek-English and Turkish-English morpheme alignment corpora. In: Proceedings of the Tenth International Conference on Language Resources and Evaluation (LREC 2016), pp. 2925–2930 (2016)
13. Mansurov, B., Mansurov, A.: Development of word embeddings for Uzbek language. arXiv preprint arXiv:2009.14384 (2020)
14. Mansurov, B., Mansurov, A.: Uzbek Cyrillic-Latin-Cyrillic machine transliteration. arXiv preprint arXiv:2101.05162 (2021)
15. Marciniak, M., Mykowiecka, A. (eds.): Aspects of Natural Language Processing. LNCS, vol. 5070. Springer, Heidelberg (2009). https://doi.org/10.1007/978-3-642-04735-0
16. Matlatipov, S., Tukeyev, U., Aripov, M.: Towards the Uzbek language endings as a language resource. In: Hernes, M., Wojtkiewicz, K., Szczerbicki, E. (eds.) ICCCI 2020. CCIS, vol. 1287, pp. 729–740. Springer, Cham (2020). https://doi.org/10.1007/978-3-030-63119-2_59
17. Pedregosa, F., et al.: Scikit-learn: machine learning in Python. J. Mach. Learn. Res. **12**, 2825–2830 (2011)
18. Powers, D.A.: Evaluation: from precision, recall and F-measure to ROC, informedness, markedness & correlation. J. Mach. Learn. Technol. **2**, 2229–3981 (2011). https://doi.org/10.9735/2229-3981
19. Rabbimov, I., Kobilov, S., Mporas, I.: Uzbek news categorization using word embeddings and convolutional neural networks. In: 2020 IEEE 14th International Conference on Application of Information and Communication Technologies (AICT), pp. 1–5. IEEE (2020)
20. Rabbimov, I., Kobilov, S., Mporas, I.: Opinion classification via word and emoji embedding models with LSTM. In: Karpov, A., Potapova, R. (eds.) SPECOM 2021. LNCS (LNAI), vol. 12997, pp. 589–601. Springer, Cham (2021). https://doi.org/10.1007/978-3-030-87802-3_53
21. Rabbimov, I., Mporas, I., Simaki, V., Kobilov, S.: Investigating the effect of emoji in opinion classification of Uzbek movie review comments. In: Karpov, A., Potapova, R. (eds.) SPECOM 2020. LNCS (LNAI), vol. 12335, pp. 435–445. Springer, Cham (2020). https://doi.org/10.1007/978-3-030-60276-5_42
22. Sakenovich, N.S., Zharmagambetov, A.S.: On one approach of solving sentiment analysis task for Kazakh and Russian languages using deep learning. In: Nguyen, N.-T., Manolopoulos, Y., Iliadis, L., Trawiński, B. (eds.) ICCCI 2016. LNCS (LNAI), vol. 9876, pp. 537–545. Springer, Cham (2016). https://doi.org/10.1007/978-3-319-45246-3_51
23. Socher, R., et al.: Recursive deep models for semantic compositionality over a sentiment treebank. In: Proceedings of the 2013 Conference on Empirical Methods in Natural Language Processing, Seattle, Washington, USA, pp. 1631–1642, October 2013. https://www.aclweb.org/anthology/D13-1170
24. Vilares, D., Alonso, M.A., Gómez-Rodríguez, C.: A syntactic approach for opinion mining on Spanish reviews. Nat. Lang. Eng. **21**(01), 139–163 (2015)
25. Vural, A.G., Cambazoglu, B.B., Senkul, P., Tokgoz, Z.O.: A framework for sentiment analysis in Turkish: application to polarity detection of movie reviews in Turkish. In: Computer and Information Sciences III, pp. 437–445. Springer, London, October 2012. https://doi.org/10.1007/978-1-4471-4594-3_45
26. Yergesh, B., Bekmanova, G., Sharipbay, A., Yergesh, M.: Ontology-based sentiment analysis of Kazakh sentences. In: Gervasi, O., et al. (eds.) ICCSA 2017. LNCS, vol. 10406, pp. 669–677. Springer, Cham (2017). https://doi.org/10.1007/978-3-319-62398-6_47

27. Zhang, L., Wang, S., Liu, B.: Deep learning for sentiment analysis: a survey. Wiley Interdiscip. Rev. Data Min. Knowl. Discov. **8**(4) (2018). https://doi.org/10.1002/widm.1253

28. Zimbra, D., Abbasi, A., Zeng, D., Chen, H.: The state-of-the-art in Twitter sentiment analysis: a review and benchmark evaluation. ACM Trans. Manag. Inf. Syst. (TMIS) **9**(2), 5 (2018)

Using Book Dialogues to Extract Emotions from Texts

Paweł Skórzewski[✉][iD]

Adam Mickiewicz University, Poznań, Poland
`pawel.skorzewski@amu.edu.pl`

Abstract. Detecting emotions from a text can be challenging, especially if we do not have any annotated corpus. We propose to use book dialogue lines and accompanying phrases to obtain utterances annotated with emotion vectors. We describe two different methods of achieving this goal. Then we use neural networks to train models that assign a vector representing emotions for each utterance. These solutions do not need any corpus of texts annotated explicitly with emotions because information about emotions for training data is extracted from dialogues' reporting clauses. We compare the performance of both solutions with other emotion detection algorithms.

Keywords: Emotion detection · Sentiment analysis · Text-based emotion detection

1 Introduction

1.1 Sentiment Analysis and Emotion Detection

Sentiment analysis can be described as the use of computational linguistics methods to identify, extract and analyze non-factual information. Emotion detection involves extracting and analyzing emotions from data and can be viewed as a sub-field of sentiment analysis. The notion of sentiment analysis is often used in a narrower sense: as the analysis of sentiment polarity of words and sentences (positive vs. negative vs. neutral).

1.2 Related Work

The importance of both sentiment analysis and emotion detection continuously increases because of their application in market analysis and opinion mining [30].

Most of the research in the domain of sentiment analysis focuses on classifying texts as positive or negative [8,25]. Emotion detection and analysis is a relatively new research field in natural language processing, but the study of emotions has a long history in psychology and sociology [8]. The majority of popular emotion models treat an emotional state as a combination of some basic emotions. Plutchik distinguished 8 basic emotions [26], while Ekman distinguished 6 basic emotions [11].

Z. Vetulani et al. (Eds.): LTC 2019, LNAI 13212, pp. 244–255, 2022.
https://doi.org/10.1007/978-3-031-05328-3_16

Traditional emotion classification consists of assigning one or more emotion categories to a given text fragment (an utterance, a sentence, or a document). Systems for emotion classification have been developed so far for different kinds of text, like children fairy tales [4], newspaper headlines [30], poems [6] or blog posts [5,14]. There are also studies on extracting the intensity of emotions from text [22]. Sometimes detection of emotions in texts is assisted by analyzing emotions in speech so that one can use additional non-lexical features [19].

However, most of these studies are conducted for the English language. There are several studies on detecting emotions in texts for other languages, including French (emotion lexicon – [3], Spanish (emotion detection in tweets – [13], Chinese (emotion detection in tweets – [32] or Japanese (text-based affect analysis – [27]. This paper focuses on Polish, which is not well studied in the field of emotion analysis. However, presented methods are language-independent and could also be used for other languages, provided that adequate text corpora are available.

1.3 Emotion Models: The Wheel and the Hourglass

Robert Plutchik in his "wheel of emotions" theory distinguished 8 basic emotions, arranged in 4 pairs of opposite emotions: *joy—sadness, trust—disgust, fear—anger, surprise—anticipation* [26]. He claims that every emotion can be viewed as a combination of these 8 basic emotions, e.g. *love = joy + trust* or *pessimism = sadness + anticipation*. Moreover, all emotions can occur in varying degrees of intensity, e.g. *annoyance* is weak *anger* and *terror* is strong *fear*.

The hourglass of emotions is an extension of this model [7]. It uses a 4-dimensional space to represent emotions, with 4 dimensions equivalent to Plutchik's basic emotion pairs (see Table 1).

Table 1. Four dimensions of the hourglass model of emotions.

dimension	-1 $\longleftrightarrow$ $+1$	
Pleasantness	Sadness	$\longleftrightarrow$ Joy
Attention	Surprise	$\longleftrightarrow$ Anticipation
Sensitivity	Fear	$\longleftrightarrow$ Anger
Aptitude	Disgust	$\longleftrightarrow$ Trust

In the hourglass model, every emotion can be represented as a vector $(P, At, S, Ap) \in [-1, 1]^4$, where P stands for *pleasantness*, At – *attention*, S – *sensitivity*, and Ap – *aptitude*. Such a vector is called a *sentic vector* by the authors. We will call it also simply an *emotion vector*.

The hourglass model allows us to represent not only basic emotions (e.g. *joy* is represented as a vector $(0.5, 0, 0, 0)$) and derivative emotions (e.g. *anxiety = anticipation + fear* is represented as $(0, 0.5, -0.5, 0)$) but also varying degrees of emotion intensity (e.g. *rage* = strong *anger* as $(0, 0, 1, 0)$) and the entire spectrum of mixed emotions (e.g. $(-0.9, 0.5, 0.1, -0.4)$). Conversely, every vector $(P, At, S, Ap) \in [-1, 1]^4$ can be seen as representing some emotional state.

1.4   Challenges in Analysis of Sentiment and Emotions

Many methods of emotion and sentiment analysis are lexicon-based [32]. There are various possible ways of combining individual sentiment (or emotions) to obtain the overall sentiment of an utterance [16].

However, one must keep in mind that the emotional tone of a text is mostly not a simple combination of emotions related to particular words [21]. Though of neutral sentiment itself, some expressions significantly impact the sentiment of a whole utterance (e.g., negation or modal verbs). The same word can carry different sentiments or emotions depending on the context. In addition, there are such phenomena as sarcasm, irony, humor, metaphors, or colloquial expressions. Sometimes, unambiguous determination of utterance's sentiment may be challenging even for a human recipient. Some research studies are trying to deal with the challenges mentioned above, e.g., recognizing and processing humor [10] or detecting irony [29].

On the other hand, the important challenge for machine-learning-based emotion classification of texts is the lack of reliable training data [4]. While finding annotated data for classic sentiment analysis (positive/negative) is relatively easy (many online review systems involve some rating or scoring opinions), texts annotated explicitly with emotions are virtually impossible to find in the wild. For some types of texts, like e.g. Twitter or Facebook posts, creating an annotated corpora of sentences and emotions can be made using emoticons [24, 32] or hashtags [20]. It can be implied from these examples that in order to create a useful training corpus indirectly, we should find texts containing some "metadata" that can be used as an indication of emotion. What kinds of text can serve this purpose, and how to obtain them, are crucial questions in this context.

1.5   Research Goals

Dialogues in books seem to be a natural example of such texts with "metadata". Usually, it is dialogues, not narratives, that carry an emotional charge of the story. Moreover, dialogues are often accompanied by reporting clauses and other descriptions of the manner of speaking.

The primary goal of our research was to investigate to what extent book dialogues are useful as a dataset for training text-based emotion models. We also wanted to explore which methods of obtaining emotion-related data from such texts perform best.

The availability of an annotated corpus of dialogues prepared for another study was an additional incentive for conducting this research. We thought it would be interesting to see if such data could be successfully used for learning a good emotion model.

2   Methodology

2.1   Processing Pipeline Overview

To evaluate text-based emotion models created in various ways, we prepared a multi-stage processing pipeline shown in Fig. 1. We built several neural models trained on sentences with emotion annotations obtained from the annotated dialogue corpus (see Subsect. 2.2). These models differ both in terms of their architectures (Subsect. 2.5) and the way of representing emotional value (Subsect. 2.3). For comparison, we also built several lexicon-based classifiers, which interpolate the emotional values of constituent words to the whole sentences (described in Subsect. 2.4).

We evaluated all classifiers on the separate dataset prepared via manual annotation (Subsect. 2.6). The obtained results are discussed in Sect. 3.

The individual elements of the pipeline are described in detail in the following subsections.

2.2   Datasets

Annotated Dialogues Corpus. As a source for data to create a training corpus, we used a part of the corpus created for the paper [18]. Kubis's corpus was created on the basis of the transcripts of books from the online service Wolne Lektury [2], which collects the texts of books in Polish that belong to the public domain. It contains 1.37 million utterances (23 million tokens). Even though the original books constituting the corpus's source were written in the 19th and 20th centuries, the corpus's language is modern Polish: Most of the books' texts were contemporized before publication. The remaining texts have been pre-processed with a diachronic normalizer.

Kubis's corpus contains texts that have been split into paragraphs, tokenized and lemmatized. Each token (lemma) is tagged with part of speech and morphological information. There are also annotations about which parts of text contain named entities, which parts are dialogue lines, which parts indicate speaker, and which parts indicate the manner of speaking. Annotations indicating the manner of speaking were particularly helpful for creating a training corpus annotated with emotions.

For example, the corpus contains the following utterance: – *Wszystko jak najlepiej – wykrzyknął wesoło lekarz.* (' "Everything is fine", the doctor exclaimed cheerfully.'). The particular phrases and words of this utterance are annotated as follows:

- the phrase *wszystko jak najlepiej* ('everything is fine') is annotated as "dialogue line",
- the phrase *wykrzyknął wesoło lekarz* ('the doctor exclaimed cheerfully') is annotated as "reporting clause",
- the phrase *wesoło* ('cheerfully') is annotated as "manner of speaking".

The idea is to use emotions extracted from the manner of speaking ('cheerfully'), or from the whole reporting clause ('the doctor exclaimed cheerfully'), to describe the emotional state related to the dialogue line ('everything is fine').

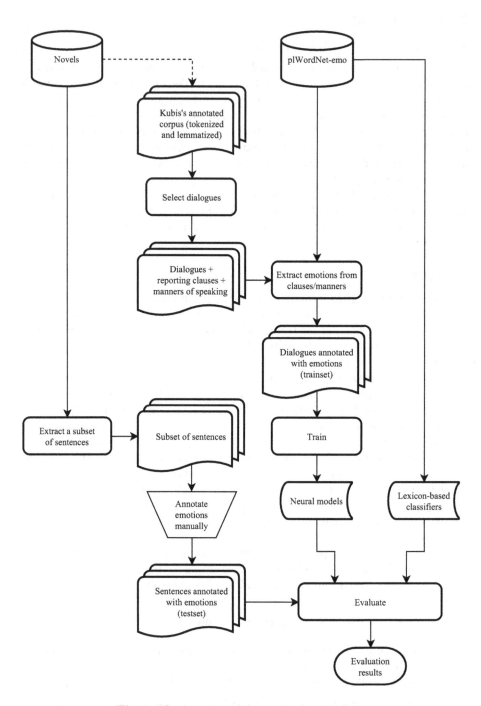

Fig. 1. The overview of the processing pipeline.

Emotions Database. For assigning emotional values to particular words, we use the plWordNet-emo sentiment lexicon for Polish [33]. It contains comprehensive emotion and sentiment annotations for 30 000 lexical units.

2.3 Emotion Representation

We use two different ways of representing emotions: simple label-based approach and a vector-based approach.

Emotion Labels. In some models, we simply use a set of labels to mark emotions. Each sentence is assigned a label from a set of Plutchik's basic emotions.

Sentic Vectors. In a vector-based approach, we use a 4-dimensional space model based on Cambria's hourglass model of emotions. We assign a 4-dimensional sentic vector (P, At, S, Ap) to each of Plutchik's basic emotions in the way shown in Table 2.

Table 2. Representation of basic emotions as vectors.

Emotion	Vector	Emotion	Vector
Joy	$(1,0,0,0)$	Sadness	$(-1,0,0,0)$
Anticipation	$(0,1,0,0)$	Surprise	$(0,-1,0,0)$
Anger	$(0,0,1,0)$	Fear	$(0,0,-1,0)$
Trust	$(0,0,0,1)$	Disgust	$(0,0,0,-1)$

This representation not only allows us to represent every basic and derivative emotion as a 4-dimensional vector but also works the other way round: every vector $(P, At, S, Ap) \in [-1,1]^4$ can be seen as a representation of some emotion, e.g., $(-0.9, 0.1, 0.5, -0.1)$ is a representation of an emotional state dominated by intense sadness and some anger.

2.4 Lexicon-Based Classifiers

A lexicon-based classifiers assigns an emotion to a given utterance by aggregating emotions related to its constituent words.

First, the input utterance is tokenized, lemmatized and POS-tagged. Then, for every word (token: lemma and POS-tag) of the utterance, we can assign an emotion using the plWordNet-emo sentiment lexicon. If no information about part-of-speech is given, we consider all lexical units from the plWordNet-emo corpus with a lemma identical to the specific word's lemma. For each of these lexical units, we calculate an emotion vector as the aggregate of the sentic vectors of basic emotion labels corresponding to this lexical unit. Subsequently, the emotion vector corresponding to the word is calculated as the aggregate of emotion vectors for its lexical units. If the part of speech of the input word is known, only lexical units of the same part of speech are taken into account.

For each lexical unit, understood as a triple: lemma + semantic variant + part of speech, a set of labels taken from the set of Plutchik's basic emotions is assigned. Sometimes there are no emotion labels assigned to a given lexical unit. In other words, zero, one or more basic emotions can be assigned to each lexical unit.

After an emotion vector for each word of the utterance is determined, we aggregate these vectors again to obtain a sentic vector for the whole utterance.

We can obtain different classifiers depending on the function used for aggregating vectors. In our study, we considered and evaluated classifiers based on the following aggregation functions::

- arithmetic mean,
- maximum in terms of absolute value (i.e., the function that chooses the number whose absolute value is the greatest),
- constant function equal zero (as a baseline).

2.5   Neural Models

We built several neural models of text-based emotions. The models differ in terms of architecture (LSTM or the Transformer), emotion representation (labels or sentic vectors), and emotion annotation source (manners or reporting clauses).

LSTM Models. As a first layer, we used pre-trained word embeddings for Polish, available under GNU General Public License [1]. The embeddings were trained on Polish National Corpus and Wikipedia with skip-gram architecture and hierarchical softmax algorithm, using Gensim library [28]. The embeddings convert input utterances to vectors of size 100.

The next layers are: a Long-Short Term Memory (LSTM) layer [15] of size 100, with dropout value of 0.5 [12] and a hyperbolic tangent activation function; and a dense layer of size 100, with dropout value of 0.5 and a hyperbolic tangent activation function.

The last layer and the loss function depends on the emotion representation. In the variant where emotions are represented as sentic vectors, we use a dense layer with 4-dimensional output and a sigmoid activation function, and cosine proximity loss. In the variant where emotions are treated as categorical labels, we use a dense layer with softmax activations, and a categorical crossentropy loss function. In both cases, we use Adam optimizer [17]. Each model was trained for several epochs, using early stopping to decide when training should be stopped to prevent overfitting.

Transformer-based Models. The Transformer is a simple network architecture based solely on attention mechanism that gained popularity in recent years thanks to its decent performance in natural-language-processing related tasks [31]. The Transformer were used to train BERT – a language model for Engilsh, trained using Masked Language Modelling (MLM) and Sentence Structural Objective (SSO) [9]. In our study, we decided to use HerBERT – a BERT-based Language Model trained on Polish corpora [23]. Then, we fine-tuned the model with our emotion data. Thus we obtained four Transformer based models - one for every combination of two parameters (manners vs. reporting clauses; labels vs. sentic vectors). Each model was trained for several epochs, using early stopping to decide when to stop training.

Emotion Annotation Source. We had no corpus annotated explicitly with emotions or sentic vectors to be our training data, so we prepared the training corpus the following way.

We used two different methods to obtain a reliable sentic vector for training utterances, both based on the properties of dialogues in books. Therefore, to create a training dataset, we selected only utterances that were dialogue lines in the books.

In the first method ("reporting clauses"), we used dialogues and corresponding reporting clauses to create a training dataset. In each dialogue line where a reporting clause was present, we took the proper dialogue line as the input utterance, and the reporting clause as the basis for calculating the sentic vector. The sentic vectors were calculated in the way described in Subsect. 2.4 but using the reporting clause as the input. This way, we obtained 72 198 training examples.

In the second method ("manners"), we used dialogues and corresponding expressions describing the manner of speaking to create a training dataset. In each dialogue line with an identifiable manner of speaking, we took the proper dialogue line as the input utterance, and the expression indicating the manner of speaking as the basis for calculating sentic vector. The sentic vectors were calculated in the way described in Subsect. 2.4 but using the expression indicating the manner of speaking as the input. This way, we obtained 11 033 training examples.

These two methods resulted in obtaining a corpus of utterance—sentic vector pairs, which was used in turn as training data for the neural network. Note that the number of obtained training examples was significantly lower than the size of the original Kubis's corpus because only a fraction of original utterances were dialogue lines and met the required conditions.

Having a corpus of utterance—sentic vector pairs, it was easy to obtain a corpus of utterance—emotion label pairs, by simply projecting each sentic vector to the nearest basic Plutchik emotion.

2.6   Evaluation Dataset

For evaluation purposes, we prepared an annotated corpus of 598 utterances and corresponding sentic vectors. The corpus was created through manual annotation. At least four independent annotators have annotated each utterance.

The annotators were volunteer students interested in machine learning. Each annotator was presented an utterance and had to point one or more of 8 Plutchik's basic emotions, or choose "neutral". Each basic emotions chosen by the annotator was converted to a sentic vector in the way shown in Table 2; "neutral" answer was converted to the sentic vector $(0,0,0,0)$. All sentic vectors corresponding to these annotations have been averaged to one sentic vector to represent the complete emotional state the annotator associates with this utterance. Then, all the sentic vectors corresponding to annotations for a given utterance have been averaged. This way, we obtain 894 pairs of utterances and corresponding emotion vectors, constituting the evaluation dataset.

3 Evaluation Results

Besides F_1-score, defined as the harmonic mean of the precision and recall, we used two other metrics for evaluation purposes:

- root-mean-square error (RMSE), defined as:

$$\text{RMSE} = \sqrt{\frac{1}{m} \sum_{i=1}^{m} \left\| v^{(i)} - v_{\text{ref}}^{(i)} \right\|^2}, \tag{1}$$

- mean cosine distance (MCosD), defined as:

$$\text{MCosD} = \frac{1}{m} \sum_{i=1}^{m} \left(1 - \frac{v^{(i)} \cdot v_{\text{ref}}^{(i)}}{\left\| v^{(i)} \right\| \cdot \left\| v_{\text{ref}}^{(i)} \right\|} \right), \tag{2}$$

where m is the total number of utterances in the testing corpus, $v^{(i)}$ is the obtained sentic vector for ith utterance, $v_{\text{ref}}^{(i)}$ is the reference sentic vector for ith utterance.

Tables 3 and 4 show evaluation results for different models.

Table 3. The evaluation results for selected neural models.

Architecture	Emotion ann. source	Emotion repr	RMSE	MCosD	F_1-score
Transformer	Manners	Vectors	2.117	0.981	0.178
Transformer	Manners	Labels	1.353	0.971	0.138
Transformer	Reporting clauses	Vectors	2.369	0.977	**0.183**
Transformer	Reporting clauses	Labels	**0.938**	1.000	0.118
LSTM	Manners	Vectors	1.463	**0.933**	0.128
LSTM	Manners	Labels	**0.938**	1.000	0.118
LSTM	Reporting clauses	Vectors	1.654	0.936	0.138
LSTM	Reporting clauses	Labels	**0.938**	1.000	0.118

Table 4. The evaluation results for lexicon-based models.

Model	RMSE	MCosD	F_1-score
Lexicon-based arithmetic mean	**0.492**	**0.880**	0.194
Lexicon-based max. w. r. t. abs	1.524	0.937	
Zero baseline	0.496	N/A	

4   Conclusion

Different neural models performed very similarly in terms of mean cosine distance. RMSE values were more diverse but still outperformed by simple averaging sentic vectors for particular words in the utterance.

The fact that models trained on dialogues' reporting clauses performed not as well as simple averaging emotion vectors of particular words may indicate that the emotional state of the utterance is more closely related to the words it consists of than to the words describing it. It may also suggest that authors tend to describe manners of speaking explicitly only if it cannot be inferred from a dialogue line itself.

There is still room for improvement of our system's performance. Collecting more training data (dialogues with reporting clauses and words determining the manner of speaking) should improve the performance of the neural models. We can also try to find ways to use other meta-information about utterances that could be more useful to determine the emotions related to these utterances.

References

1. Distributional models for Polish (2021). http://dsmodels.nlp.ipipan.waw.pl. Accessed 15 Oct 2021
2. Wolne Lektury. About the project (2021). https://wolnelektury.pl/info/oprojekcie. Accessed 15 Oct 2021
3. Abdaoui, A., Azé, J., Bringay, S., Poncelet, P.: FEEL: a French expanded emotion lexicon. Lang. Resour. Eval. **51**(3), 833–855 (2017)
4. Alm, C.O., Roth, D., Sproat, R.: Emotions from text: machine learning for text-based emotion prediction. In: Proceedings of the Human Language Technology Conference and Conference on Empirical Methods in Natural Language Processing, pp. 579–586. Association for Computational Linguistics, Vancouver (2005)
5. Aman, S., Szpakowicz, S.: Identifying expressions of emotion in text. In: Matoušek, V., Mautner, P. (eds.) TSD 2007. LNCS (LNAI), vol. 4629, pp. 196–205. Springer, Heidelberg (2007). https://doi.org/10.1007/978-3-540-74628-7_27
6. Auracher, J., Albers, S., Zhai, Y., Gareeva, G., Stavniychuk, T.: P is for happiness, N is for sadness: universals in sound iconicity to detect emotions in poetry. Discourse Process. **48**(1), 1–25 (2010)
7. Cambria, E., Livingstone, A., Hussain, A.: The hourglass of emotions. In: Esposito, A., Esposito, A.M., Vinciarelli, A., Hoffmann, R., Müller, V.C. (eds.) Cognitive Behavioural Systems. LNCS, vol. 7403, pp. 144–157. Springer, Heidelberg (2012). https://doi.org/10.1007/978-3-642-34584-5_11

8. De Bruyne, L., De Clercq, O., Hoste, V.: LT3 at SemEval-2018 task 1: a classifier chain to detect emotions in tweets. In: Proceedings of The 12th International Workshop on Semantic Evaluation (SemEval-2018), pp. 123–127. Association for Computational Linguistics (2018). http://aclweb.org/anthology/S18-1016

9. Devlin, J., Chang, M., Lee, K., Toutanova, K.: BERT: pre-training of deep bidirectional transformers for language understanding. CoRR abs/1810.04805 (2018). http://arxiv.org/abs/1810.04805

10. Dybala, P., Yatsu, M., Ptaszynski, M., Rzepka, R., Araki, K.: Towards joking, humor sense equipped and emotion aware conversational systems. In: Advances in Affective and Pleasurable Design, Advances in Intelligent Systems and Computing, vol. 483, pp. 657–669. Springer (2017). https://doi.org/10.1007/978-3-319-41661-8_64

11. Ekman, P.: An argument for basic emotions. Cognit. Emot. **6**(3–4), 169–200 (1992)

12. Gal, Y., Ghahramani, Z.: A theoretically grounded application of dropout in recurrent neural networks. In: Advances in Neural Information Processing Systems, pp. 1019–1027 (2016)

13. Gil, G.B., de Jesús, A.B., Lopéz, J.M.M.: Combining machine learning techniques and natural language processing to infer emotions using Spanish Twitter corpus. In: International Conference on Practical Applications of Agents and Multi-Agent Systems, pp. 149–157. Springer (2013)

14. Gill, A.J., French, R.M., Gergle, D., Oberlander, J.: Identifying emotional characteristics from short blog texts. In: 30th Annual Conference of the Cognitive Science Society, pp. 2237–2242. Cognitive Science Society Washington, DC (2008)

15. Hochreiter, S., Schmidhuber, J.: Long short-term memory. Neural Comput. **9**(8), 1735–1780 (1997)

16. Kim, S.M., Hovy, E.: Determining the sentiment of opinions. In: Proceedings of the 20th International Conference on Computational Linguistics, p. 1367. Association for Computational Linguistics (2004)

17. Kingma, D.P., Ba, J.: Adam: a method for stochastic optimization. In: The International Conference on Learning Representations (ICLR), San Diego (2015)

18. Kubis, M.: Quantitative analysis of character networks in Polish XIX and XX century novels. paper presented (2019)

19. Metze, F., Batliner, A., Eyben, F., Polzehl, T., Schuller, B., Steidl, S.: Emotion recognition using imperfect speech recognition. In: Eleventh Annual Conference of the International Speech Communication Association (2010)

20. Mohammad, S.: #Emotional tweets. In: *SEM 2012: The First Joint Conference on Lexical and Computational Semantics - Volume 1: Proceedings of the main conference and the shared task, and Volume 2: Proceedings of the Sixth International Workshop on Semantic Evaluation (SemEval 2012), pp. 246–255. Association for Computational Linguistics, Montréal, Canada, 7–8 June 2012. http://www.aclweb.org/anthology/S12-1033

21. Mohammad, S.M.: Sentiment analysis: detecting valence, emotions, and other affectual states from text. In: Meiselman, H. (ed.) Emotion Measurement, pp. 201–237. Elsevier (2016)

22. Mohammad, S.M., Bravo-Marquez, F.: Emotion intensities in tweets. In: Proceedings of the Sixth Joint Conference on Lexical and Computational Semantics (*Sem). Vancouver, Canada (2017)

23. Mroczkowski, R., Rybak, P., Wróblewska, A., Gawlik, I.: HerBERT: efficiently pretrained transformer-based language model for Polish. In: Proceedings of the 8th Workshop on Balto-Slavic Natural Language Processing. pp. 1–10. Association for Computational Linguistics, Kiyv, Ukraine, April 2021. https://www.aclweb.org/anthology/2021.bsnlp-1.1

24. Pak, A., Paroubek, P.: Twitter as a corpus for sentiment analysis and opinion mining. In: LREc, vol. 10, pp. 1320–1326 (2010)
25. Pang, B., Lee, L., et al.: Opinion mining and sentiment analysis. Found. Trends Inf. Retr. **2**(1–2), 1–135 (2008)
26. Plutchik, R.: A general psychoevolutionary theory of emotion. In: Theories of emotion, pp. 3–33. Elsevier (1980)
27. Ptaszynski, M., Dybala, P., Rzepka, R., Araki, K., Masui, F.: ML-Ask: Open source affect analysis software for textual input in Japanese. J. Open Res. Softw. **5**(1), 1–16 (2017)
28. Řehůřek, R., Sojka, P.: software framework for topic modelling with large corpora. In: Proceedings of the LREC 2010 Workshop on New Challenges for NLP Frameworks, pp. 45–50. ELRA, Valletta, Malta, May 2010. http://is.muni.cz/publication/884893/en
29. Reyes, A., Rosso, P., Buscaldi, D.: From humor recognition to irony detection: the figurative language of social media. Data Knowl. Eng. **74**, 1–12 (2012)
30. Strapparava, C., Mihalcea, R.: SemEval-2007 task 14: affective text. In: Proceedings of the 4th International Workshop on Semantic Evaluations (SemEval-2007), pp. 70–74. Prague, June 2007
31. Vaswani, A., et al.: Attention is all you need. In: Advances in Neural Information Processing Systems, pp. 5998–6008 (2017)
32. Yuan, Z., Purver, M.: Predicting emotion labels for Chinese microblog texts. In: Gaber, M., et al. (eds.) Advances in Social Media Analysis. Studies in Computational Intelligence, vol. 602, pp. 129–149. Springer, Cham (2015). https://doi.org/10.1007/978-3-319-18458-6_7
33. Zaśko-Zielińska, M., Piasecki, M., Szpakowicz, S.: A large wordnet-based sentiment lexicon for Polish. In: Angelova, G., Bontcheva, K., Mitkov, R. (eds.) International Conference Recent Advances in Natural Language Processing. Proceedings, pp. 721–730. Hissar, Bulgaria (2015)

Digital Humanities

NLP Tools for Lexical Structure Studies of the Literary Output of a Writer. Case Study: Literary Works of Tadeusz Boy-Żeleński and Julia Hartwig

Zygmunt Vetulani[1]([✉]) [iD], Marta Witkowska[2] [iD], and Marek Kubis[1] [iD]

[1] Adam Mickiewicz University, 4, Uniwersytetu Poznańskiego Street, 61-614, Poznań, Poland
{vetulani,mkubis}@amu.edu.pl
[2] 10, Fredry Street, 61-710 Poznań, Poland
marta.witkowska@amu.edu.pl

Abstract. In this paper we present the use of NLP tools for lexical structure studies of the literary output of a writer. We present the usage made of several tools of our own design or developed externally. In this number were POLEX and Text SubCorpora Creator (TSCC1.3.) systems developed at AMU, as well as NLTK libraries, Corpusomat (pol. Korpusomat), and others. In particular these systems and tools were used to characterize the lexical component of the linguistic instrumentarium of Tadeusz Boy-Żeleński prose and journalistic author active from 1920 to 1941 and Julia Hartwig, an outstanding Polish poet and prose author active from 1954 to 2016.

Keywords: NLP tools · Lexical structure of text · Lexical competence benchmark · Virtual lexicon · Regression models · Boy-Żeleński prose corpus · Hartwig poetry and prose corpora

1 Introduction

In our work, we are interested in the achievements of two eminent writers Tadeusz Boy-Żeleński and Julia Hartwig, whose main literary activities cover the last 100 years of Poland rebuilt after World War I. We will analyze selected elements of the writing workshop of both Boy-Żeleński (prose) and Julia Hartwig (poetry and prose) on the basis of the entirety of their writings represented by the respective corpora. We place particular emphasis on adjectives, considered by Hartwig herself to be one of the important indicators of her work (Legeżyńska 2017).

The purpose of using the NLP tools discussed in the article is to create the basis for drawing empirically substantiated conclusions about the characteristic features of the writer's literary workshop. We are primarily interested in the use of vocabulary (quantitative aspects) and the dynamics of the development of the creative workshop. We find it interesting and important to study the dynamics of this workshop development using incremental analysis.

Z. Vetulani et al. (Eds.): LTC 2019, LNAI 13212, pp. 259–276, 2022.
https://doi.org/10.1007/978-3-031-05328-3_17

The paper is organized as follows: In Sect. 2 we describe the corpora collected for the study. Section 3 presents NLP tools developed internally at AMU. External tools that we use are discussed in Sect. 4. In Sect. 5 we use the presented tools to investigate literary output of Boy-Żeleński and Hartwig. The last section contains concluding remarks.

2 Linguistic Data

We investigated the literary output of two notable Polish authors: Tadeusz Boy-Żeleński (1874–1941) and Julia Hartwig (1921–2017). Respectively, we collected two corpora (see Annex). The collected texts reported observations made by the authors at the occasion of their travels as well as various literary events. Both Boy-Żeleński and Hartwig were also outstanding translators of French literature masterpieces but this part of their literary output was not taken into account[1].

2.1 Tadeusz Boy-Żeleński Corpus

The texts of the Boy-Żeleński corpus are in the public domain and were collected from www.wolnelektury.pl and https://pl.wikisource.org (see Annex) (Table 1).

Table 1. Tadeusz Boy-Żeleński Corpus.

Tadeusz Boy-Żeleński Corpus	Size
Number of pages	1143
Number of paragraphs	11371
Number of words	607544
Number of characters (excluding spaces)	3582496
Number of characters (including spaces)	4187970

2.2 Julia Hartwig Corpus

For the purposes of her research Marta Witkowska collected two corpora of Julia Hartwig works. The first, smaller one, composed uniquely of the Author's diaries, summarized in Table 2, was used for comparative analyzes of the literary language of Boy-Żeleński and Hartwig (Vetulani et al. 2019).

---

[1] Translation work undoubtedly influences the entire literary workshop of the translator and thus his original literary work, but the work of translation itself includes, by definition, elements of the literary workshop of the translated author. For this reason, these elements should be omitted in the study of Julia Hartwig's original literary instruments.

Table 2. Julia Hartwig Corpus (prose-all).

Julia Hartwig Corpus	Size
Number of pages	1499
Number of paragraphs	12311
Number of words	717669
Number of characters (excluding spaces)	4283357
Number of characters (including spaces)	4994603

The second corpus of Julia Hartwig's works is much larger, it covers both the prose (prose-all) and poetic (poetry-all) works of Hartwig. The author's achievements are represented by the electronic reference corpus of published prose and poetry from the period from 1954 to 2016, collected for this purpose. This corpus was collected and prepared[2] by Marta Witkowska doctoral dissertation (2021). Texts of Julia Hartwig were made available by the rights holders in the electronic form (publishers: Zeszyty Literackie and Wydawnictwo Literackie) or in the form of printed text requiring OCR processing.

The research on selected elements of Julia Hartwig's literary workshop discussed in this article refers to a collection of texts covering over sixty years (1954–2016) of creative activity, which includes prose and poetry. In order to create an empirical basis for the conducted work, a representative[3] corpus of approx. 800,000 words was collected and developed, in which we distinguish two sub-bodies: for prose (prose-all) and for poetry (poetry-all). The corpus includes artistic, journalistic and para-documentary texts that constitute the core of the author's legacy. In particular, the sub-corpus of prose consists of 10 items, which are journals, reportages, literary monographs, a collection of essays and a collection of columns from 1954–2014. The subcorpus of poetry consists of 22 volumes, which are collections of poems and poetic prose published in the years 1956–2016 (see Annex) (Table 3).

Table 3. Quantitative data for Julia Hartwig's prose and poetry subcorpora.[4]

Corpus	Characters	Words
Prose-all	4 614 710	669 163
Poetry-all	985 681	151 794

---

[2] The study includes annotation and acquisition of frequency lists and concordances from the corpus.

[3] Fulfilling the requirement of representativeness for quantitative research required a lot of diligence, for example, it resulted in the necessity to exclude from the corpus extensive works constituting reissues of earlier titles.

[4] The table has been compiled for the source version of the corpus.

3 AMU Tools Used in Studies of Literary Outcome of an Author

3.1 System POLEX

The POLEX system developed at the Department of Computer Linguistics and Artificial Intelligence[5] at AMU consists of a digital morphological dictionary POLEX (Vetulani et al. 1998) and related NLP tools of a utilitary character[6].

Dictionary POLEX. The POLEX dictionary is a digital database created including the basic lexical resource of Polish, a total of over 110,000 lexemes. Based on the achievements of the classics of Polish lexicography (Doroszewski, Polański, Tokarski, Szymczak, Saloni, Mędak and others), a new approach to describing Polish morphology has been developed, focused on the needs of computer processing of Polish texts. Its characteristic feature was to propose an unambiguous set of inflectional paradigmatic classes characterized by the fact that one alpha-numeric code corresponds to exactly one (ordered) set of endings. A position code was used in which the items are significant and linguistically motivated, which facilitates the development and maintenance of the dictionary by lexicographers and thus maintaining a high quality of the resource. These features make it possible to easily expand and verify the dictionary by an expert without the need to use special tools (editor). The unambiguousness of the code guarantees the precision of description so far not achieved in the existing dictionaries[7].

The morphological unit of POLEX dictionaries has the following structure:

> BASIC_FORM + LIST_OF_STEMES + PARADIGMATIC_CODE + STEMS_DISTRIBUTION

For example, the dictionary entries for the lexemes frajer1 and frajer2[8] would look like this:

frajer; frajer,frajerz; N110; 1:1-5,9-13;2:6-8,14

frajer; frajer,frajerz; N110; 1:1-5,8-14;2:6-7

The POLEX dictionary is related to derivative tools, such as a lemmatizer or a generator of inflectional forms, used until now, also in the works discussed in this article.

---

[5] Department of Computer Linguistics and Artificial Intelligence (1993–2020) headed by Zygmunt Vetulani.

[6] POLEX was created in the years 1994–1996 as part of a grant from the State Committee for Scientific Research, led by Zygmunt Vetulani, and was the first real-size electronic dictionary for Polish designed for the NLP application for Polish language processing.

[7] A detailed description of the methodological assumptions and properties of the POLEX dictionary can be found in the book (Vetulani et al. 1998).

[8] The lexemes *frejer1* and *frajer2* have the same meaning and syntax requirements, they differ at the inflectional and pragmatic levels.

The Lemmatizer POLEX. We will use the term lemmatization to refer to the procedure of finding the basic (dictionary) form in a lexeme, i.e. in a set of inflectional forms that a word can take without losing its meaning[9].

The lemmatizer POLEX is a program that performs the morphological analysis of words on the basis of the data contained in the POLEX dictionary. Its main function is to identify the lexeme to which a given form belongs, and also, if it is a different form, to determine its position in the correct inflectional paradigm, which is obviously equivalent to determining the values of the attribute that describe it (case, type, number, etc.). The lemmatizer works on the basis of inflectional-morphological information available in the dictionary, which is the reason for the ambiguity of the output, which is manifested in the fact that the output contains information about all lexemes to which a given form may belong and about all positions of the inflection paradigm on which it appears. Disambiguation requires the use of syntactic-semantic information, which is currently the subject of work on the basis of lexicon-grammar[10] (French *lexique-grammaire*) and has not yet been used in the works reported here (Fig. 1).

```
o_file-proza-all — Notatnik
Plik  Edycja  Format  Widok  Pomoc
podróż={ podróż(N422, 1) podróż(N422, 4) podróż(N423, 1) podróż(N423, 4) }
dookoła={ dookoła(ADV0, 1) dookoła(P, 1) }
malowanego={ malowany(ADJPAP, 4) malowany(ADJPAP, 8) }
pieca={ piec(N320, 2) }
znak niewyświetlany={koniec linii}
najpierw={ najpierw(ADV0, 1) }
napisałam={ napisać(V112, 9) }
do={ do(P, 1) }
felicji={ }
curyłowej={ }
list={ list(N310, 1) list(N310, 4) lista(N411, 9) }
.= {kropka}
chciałam={ chcieć(V021, 9) }
jej={ jej(EXCL, 1) ona(PrsPRO4, 2) ona(PrsPRO4, 3) jej(PosPRO4, 1) jej(PosPRO4, 2) jej(PosPRO4, 3) jej(PosPRO4, 4) jej(PosPRO4, 5
pogratulować={ pogratulować(V104, 1) }
szczęśliwego={ szczęśliwy(ADJ1, 4) szczęśliwy(ADJ1, 8) }
wystąpienia={ wystąpienie(N520, 2) wystąpienie(N520, 8) wystąpienie(N520, 11) wystąpienie(N520, 14) wystąpienie(NV, 2) }
na={ na(P, 1) }
zjeździe={ zjazd(N310, 6) zjazd(N310, 7) }
rady={ rad(N311, 8) rad(N311, 11) rad(N311, 14) rada(N411, 2) rada(N411, 8) rada(N411, 11) rada(N411, 14) }
artystycznej={ artystyczny(ADJ1, 5) artystyczny(ADJ1, 7) artystyczny(ADJ1, 15) }
w={ w(P, 1) }
```

Fig. 1. An example of lemmatization of a fragment of the body of Julia Hartwig's prose[11]

---

[9] See also: Linguistic Problems in the Theory of Man-Machine Communication in Natural Language (Vetulani et al. 1989: 71), where we define lemmatization as "the operation defined on the corpus, which consists in replacing a textual word with a lexem".

[10] The PolNet-Polish Wordnet system takes on the features of lexical-grammar starting from Version 3 (Vetulani et al. 2016), where syntactic-semantic information is introduced into the synsets for predicative words, enabling syntax-semantic desambiguization to be performed.

[11] The beginning of the reportage *From nearby travels* (Polish title: *Z niedalekich podróży*) (Hartwig 1954), which begins with the following sentences: *Podróż dookoła malowanego pieca. Najpierw napisałam do Felicji Curyłowej list. Chciałam jej pogratulować szczęśliwego wystąpienia na zjeździe rady artystycznej w…* (in English: *A trip around the painted stove. First, I wrote a letter to Felicja Curyłowa. I would like to congratulate her on her happy performance at the artistic council convention in…*).

The texts contained in the corpus are displayed in the form of a list of words, where for each of them, after the equality sign in the braces { } were placed possible grammatical interpretation The list also includes punctuation marks, e.g.. = {dot} (in Polish:. = *{kropka}*) and non-displayable characters, e.g. *undisplayed = {end of line}* (in Polish: *znak niewyświetlany = {koniec linii}*).

3.2   Text SubCorpora Creator (TSCC)

The problem of stopping criterion for corpora creation (subcorpora extraction) was at the origin of the system Text SubCorpora Creator (TSCC).

After the initial fascination by *size* as the main quality measure for corpora the reflection came that "a huge corpus is not necessarily a corpus from which generalization can be made; a huge corpus does not necessarily 'represent' a language or a variety of a language any better than a smaller corpus" (Kennedy 1998).

A stopping criterion for a corpus collection permits us to limit collection of linguistic data beyond necessity, i.e. to stop further research once it becomes evident that new observations will not lead to observation of new phenomena. The method to fix the stopping criterion for a given phenomenon consists in observing the speed of the increase of knowledge about this phenomenon. The stopping criterion is satisfied when the knowledge increase speed becomes sufficiently slow, what means that discovering new facts about the phenomenon becomes very rare. Initial application of the TSCC to lexicographical research was to help deciding when to stop corpus collection. In (Vetulni et al. 2018a, 2018b) we described how to apply the TSCC tool to extract from a large corpus subcorpora representative for a given linguistic phenomena. The number of the observed realizations of the phenomenon may be considred as evaluation of its size.

In addition to the above-mentioned applications, TSCC was applied to the study of the literary output of Julia Hartwig (Witkowska 2021) (see Sect. 5.3 below).

3.3   Named Entity Recognizer and Entity Linker

Named Entity Recognizer (NER) and Entity Linker (EL), are tools developed in order to identify the names of entities (such as persons and places) that appear in text and to connect the names that refer to the same entities.

The Named Entity Recognizer used for analyzing the literary works of Hartwig and Boy-Żeleński is a direct successor of the system used by Kubis (2021) for analyzing 19-th and 20-th century Polish prose. However, in order to improve the recognition accuracy, the CRF-based model was replaced with the one that follows the Transformer architecture (Vaswani et al. 2017) and utilizes a pre-trained language model (Dadas et al. 2019) adapted to the name entity recognition task using the manually annotated 1-million word subcorpus of the NKJP corpus (Przepiórkowski et al. 2012) (Fig. 2).

Kiedy/O Przybyszewski/B-persName przybył/O do/O Polski/B-placeName,/O do/O Krakowa/B-placeName,/O pewien/O księgarz/O nabył/O u/O niego/O szereg/O jego/O utworów

Fig. 2. A sentence from Boy-Żeleński T. (1929b) annotated by NER with named entities indicated by *B-persName (person)* and *B-placeName (place)* tags.[12]

The Entity Linker used for connecting names that refer to the same person or place is an unsupervised variant of the system described in (Kubis 2020). It links every name in the text to names that have the same surface forms or are diminutives or abbreviated forms of it. Furthermore, it connects names that are prefixes or suffixes of other names to the closest names that encompass them unless such a connection leads to an ambiguous link (e.g. two characters that share surname, but have distinct forenames are not being linked). As shown in Table 4, the outcome of this procedure consists of clusters of inter-connected names that represent the entities that appear in text.

Table 4. Examples of entities found by the linker in Boy-Żeleński T. (1929b)

Entity	Example
Paulina Zbyszewska	Zbyszewska, Pauliną, Paulinę, Paulina Zbyszewska, Paulinę Zbyszewską
Kraków	Krakowa, Krakowie
Jan Stanisławski	Jana Stanisławskiego, Stanisławski, Jan Stanisławski
Narcyza Żmichowska	Żmichowska, Narcyza_Żmichowska, Żmichowskiej
Monachium	Monachium
Antoni Wysocki	Antoni Wysocki, Antoniego Wysockiego

Examples of application of NER and EL systems to analyse literary texts are presented in Sect. 5.5 below.

4   Other Tools

4.1   NLTK Libraries

The Natural Language Toolkit (NLTK) (Bird et al. 2009) is a set of libraries for Natural Language Processing (NLP) using both statistical and symbolic methods. These libraries are composed of sets of programs and tools written in Python that can be used from the source code and adapted to one own needs. (In our work, we use them for frequency lists generation.)

---

[12] Original in Polish: *Kiedy Przybyszewski przybył do Polski, do Krakowa, pewien księgarz nabył u niego szereg jego utworów.* (English translation: *When Przybyszewski came to Poland, to Krakow, a bookseller bought him a number of his works.*).

4.2 Korpusomat

Korpusomat (in English Corpusomat) (Kieraś et al. 2018) is a constantly developed set of tools that were developed in the Linguistic Engineering Team of the Institute of Computer Science of the Polish Academy of Sciences, and which were included in the Polish resources of infrastructure for humanists as part of the Clarin-pl project. Corpusomat is by definition a tool intended for users who do not need to have specialized knowledge or skills related to its construction. In particular, this tool allows you to load your own corpus and enter metadata[13].

5 Applications to Study Literary Outcome of an Author

Below are some examples of our use of tools for the lexical analysis of literary texts by Tadeusz Boy-Żeleński and Julia Hartwig.

5.1 Tagging and Lemmatizing Texts

To tag adjectives in Julia Hartwig's corpora we used the updated (2018) version of the POLEX lemmatizer implemented in SWI Prolog v.6.4.0. The program was used to find and tag adjective forms in Julia Hartwig corpusesora, omitting all other grammatical categories. In particular, the tags contain the base forms (lemmas) of the tagged words.

On the basis of the lemmatization of the poetry-all corpus, a list of adjective forms in Julia Hartwig's poetry corpus was generated as a reference list for further research on the collected corpus. Similarly, as a result of lemmatizing the prose-all file, a list of adjectival forms included in the corpus of Julia Hartwig's prose was generated (see Marta Witkowska, PhD Dissertation; Witkowska 2021).

5.2 Use of NLTK Libraries for Obtaining Frequency Lists of Adjectival Forms and Lexemes for the Poetry and Prose of Julia Hartwig

With the NLTK library you can, for example, determine the size of the corpus, count the frequency of words, normalize the text, etc. A frequency list is understood as a numbered list of words along with the frequency of their occurrences, in the decreasing order of the number of occurrences in the text. The discussed frequency lists of adjective forms and the corresponding lexemes (ibid.) were developed with the use of appropriate reference lists (see Sect. 5.1) for the prose-all and poetry-all corpora generated from the lemmatization results.

The analysis of the frequency lists of both corpora may indicate some important features of the author's writing technique. For example, in the case of Julia Hartwig's prose and poetry corpus, we notice a clear presence (a large number of occurrences) of adjectives that we would classify as an affirmative lexical field (by affirmative lexical field we understand a class of positive terms regarding phenomena, situations and people, such as beauty, novelty, goodness).

---

[13] Detailed information on the query language can be found in the *Manual on using the corpus search engine in the Corpusomat* (of August 26, 2020) available on the www.korpusomat.pl.

Julia Hartwig's prose and poetry corpus contains 10 prose titles and 22 volumes of poetry, for which attendance lists have been established. This material makes it possible to observe the dynamics of changes in the quantitative distributions of adjective lexemes observed in various works by the author.

5.3   Quantitative Studies of Adjectives in the Corpora of Tadeusz Boy-Żeleński and Julia Hartwig

Initially, the aim of experiments described in (Vetulani et al. 2019) was to show the usefulness of the TSCC tool to investigate the lexical competence of an author by evaluating his/her virtual lexicon. We examined the life-long literary production of two outstanding Polish authors: Julia Hartwig (1921–2017) and Tadeusz Boy-Żeleński (1784–1941).

Lexical Competence Comparison Experiment.   To test usefulness of the TSCC tool for investigating the literary instrumentarium of an author, we observed occurrences of opinion adjectives in texts of Boy-Żeleński and Hartwig. The TSCC system permitted us to observe and compare distributions of opinion adjectives in both corpora. As reference, we used the list of some 400+ main opinion adjectives collected by Steinemann[14]. TSCC was applied to study frequency of occurrence the opinion adjectives from the reference lists in fragments of the corpus and to draw vocabulary increase diagrams (called also saturation graph in Vetulani et al. (2018a, 2018b)).

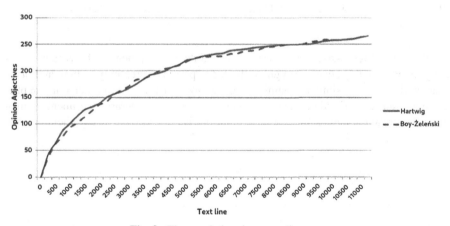

Fig. 3. The vocabulary increase diagram

The vocabulary increase diagram (Fig. 3) was generated for the two investigated corpora. The two almost identical curves show that the speed of finding *new* (i.e. never observed before) opinion adjectives seems change in the same way for both corpora.

---

[14] See more on: https://kathysteinemann.com/Musings/opinion-adjectives/.

This allows to rise the hypothesis that the size of virtual lexicon[15] of both authors is (more or less) the same (Vetulani et al. 2019).

Regression Analysis of Experiment Results. Figure 3 suggests that the number of opinion adjectives used by both authors grows with the square root of the corpus size. In fact, fitting the regression models to the presented data with the square root of the corpus size as an independent variable and the number of opinion adjectives being used as a dependent variable results in the value of the coefficient of determination[16] equal to 0.9682 in the case of Hartwig corpus and to 0.9679 in the case of Boy-Żeleński corpus. The regression coefficients of the fitted models are shown in Table 5.

The vocabulary increase regression model is defined by the following formula:

$$y = a\sqrt{x} + b$$

where x is the corpus size and y is the number of adjectives (see Table 5.)

Table 5. Regression coefficients (calculated in the R statistical software)

Coefficient	Hartwig Corpus (prosa-all)		Boy-Żeleński Corpus	
	Estimate	Standard Error	Estimate	Standard Error
A	2.54479	0.07293	2.61012	0.07517
B	23.43302	5.15712	17.13348	5.31507

The applied regression analysis suggests that the development of the virtual vocabulary in terms of adjectives throughout the entire period of creativity of both authors is characterized by very similar dynamics. An open question remains whether a similar phenomenon can be observed in the case of other elements of the literary instrumentarium of these authors and whether the observed similarity is a specific or regular phenomenon.

---

[15] The concept of virtual lexicon (fr. lexique virtuel or lexique potentiel) was introduced and investigated by Charles Muller. Virtual lexicon of an author is the vocabulary he knows but that was never observed in his/her texts. Muller (1969) used this concept to study the literary instrumentarium of Corneille, Racine and other authors (see e.g. Vetulani 1989). This concept corresponds to the well-known fact that passive language competence is much larger than the active one. The concept of virtual lexicon, both passive and active, is useful to characterize the potential lexical instrumentarium of an author, and thus contributes to the characterization of his/her literary workshop. Evaluation of the size of virtual lexicon of authors brings information about their intellectual potential.

[16] Adjusted R-squared computed using *lm* function from *stats* package of R.

5.4   Study of Saturation of Text Corpora with Adjectival Forms Using TSCC

The incremental method (Vetulani et al. 2018a, 2018b), using the properties of the TSCC system, makes it possible to determine the degree of saturation of the text corpus with adjective forms. For the sake of research, Julia Hartwig's prose corpus (prose-all file) was divided into 40 documents, 280 lines each. The poetry corpus, in turn, was broken down into 50 documents, 410 lines each.

In the conducted research, we were most interested in the frequency lists and in numerical data showing the increments of new lexical units in subsequent parts of the investigated corpus.

Incremental studies of lexical saturation (Vetulani et al. 2018a, 2018b) of Julia Hartwig's prose and poetry corpora did not confirm lexical saturation, both for poetry and for prose (see also Sect. 5.3 above). This observation confirms the otherwise known fact that the author, until the end of her writing activity, significantly expanded the scope of her interests.

We have also observed a similar phenomenon while analyzing the prose of Tadeusz Boy-Żeleński.

In both cases, the comparisons were made using the TSCC incremental method for tranches of the same length (Vetulani et al. 2019). On the other hand, the incremental study for Hartwig's prose and poetry carried out using the diachronic method (subsequent tranches correspond to the original works published in subsequent years) show a linear increase in the number of new uses of adjectives in the last period of the author's life. We hypothesize that this phenomenon is the result of a change in the literary form promoted and used by Hartwig in the last period of her poetry. The literary form developed by her has the form of the so-called "flashes" (pol. "błyski"), which are short, synthetic pieces with a rich message. With such formal assumptions, adjectives carefully selected and adapted to the literary message of the work play a prominent role. This phenomenon was not observed in the author's prose from the same period.

5.5   Use of the Named Entity Recognition and Entity Linking Systems to Study Literary Output

The rationale behind using entity recognition and linking tools for analyzing the literary output is twofold. First, by identifying names of (non-fictional) persons and places in text one can quantify the relationship between a literary work and the external world. Thus, we applied the tools to each literary work of Hartwig and Boy-Żeleński one-by-one and counted the persons and places that were found. The results are shown in Table 6.

Table 6. Numbers of entities in literary works of Hartwig and Boy-Żeleński.

Work	Persons	Places
(Boy-Żeleński 1920)	49	10
(Boy-Żeleński 1922)	5	2
(Boy-Żeleński 1925)	4	5
(Boy-Żeleński 1927)	10	11
(Boy-Żeleński 1929a)	2	0
(Boy-Żeleński 1929b)	29	24
(Boy-Żeleński 1930a)	45	23
(Boy-Żeleński 1930b)	24	9
(Boy-Żeleński 1930c)	6	2
(Boy-Żeleński 1932a)	1	1
(Boy-Żeleński 1932b)	26	16
(Boy-Żeleński 1932c)	12	14
(Boy-Żeleński 1932d)	38	12
(Boy-Żeleński 1934a)	12	10
(Boy-Żeleński 1934b)	8	1
(Boy-Żeleński 1934c)	37	9
(Boy-Żeleński 1958)	20	7
(Hartwig, 1954)	14	54
(Hartwig 1962)	72	17
(Hartwig 1972)	36	13
(Hartwig 1980)	13	30
(Hartwig 2001)	158	55
(Hartwig 2004)	67	20
(Hartwig 2006a)	71	9
(Hartwig 2006b)	32	2
(Hartwig 2011)	160	60
(Hartwig 2014)	191	47

The second reason to use entity recognition and linking tools for analyzing the literary output of Hartwig and Boy-Żeleński is the possibility to asses thematic resemblance among the authors by studying the named entities that repeat across texts. Hence, we report in Table 7 persons that are mentioned by both authors. Numbers of works by Hartwig and Boy-Żeleński that mention the given person are presented in separate columns. One may notice that beside a historical figure of Louis Philippe this group consists solely of other writers with a special attention given to Victor Hugo who appears in three works by Hartwig and six works by Boy-Żeleński.

Table 7. Numbers of entities in literary works of Hartwig and Boy-Żeleński.

Persons	Hartwig Corpus	Boy-Żeleński Corpus
Victor Hugo	3	6
George Sand	3	1
Louis Philippe	2	1
Théophile Gautier	2	1
Walter Scott	1	1
André Gide	1	1
Lechoń	1	1
Jean-Jacques Rousseau	1	1
Jerzy Stempowski	1	1
La Rochefoucauld	1	1
Francis Carco	1	1
Victor Hugo	3	6

We also identified places that appear both in Hartwig and Boy-Żeleński literary works. As shown in Table 8, among the most frequently mentioned cities are Paris (pol. Paryż), Cracow (pol. Kraków) and Warsaw (pol. Warszawa) which are mentioned in at least five works of both authors.

Table 8. Locations common to literary works of Hartwig and Boy-Żeleński.

Location	Hartwig Corpus	Boy-Żeleński Corpus
Paryż	8	9
Kraków	5	11
Warszawa	5	5
Lwów	1	8

(*continued*)

Table 8. (*continued*)

Location	Hartwig Corpus	Boy-Żeleński Corpus
Wiedeń	2	5
Londyn	6	1
Petersburg	1	4
Berlin	1	4
Lublin	4	1
Polska	3	2
Gdańsk	2	2
Bydgoszcz	3	1
Kowno	1	2
Wisła	2	1
Praga	2	1
Łódź	2	1
Mediolan	2	1
Włocławek	1	1
Rennes	1	1
Łowicz	1	1
Turyn	1	1

6 Concluding Remarks

The paper presents the main IT tools that we used to research the literary workshop of two outstanding Polish writers who had a significant impact on Polish literature over the course of one hundred years, starting from the 1920s to the end of the 2010s. They were Tadeusz Boy-Żeleński and Julia Hartwig. We paid particular attention to the lexical aspects of the literary workshop of both writers. Based on the entirety of the works of both authors collected in text corpora (Witkowska 2021), we made a number of observations concerning both authors. In particular, the frequency research confirmed the dominant presence of adjectives related to the affirmative lexical field in Julia Hartwig's works. On the other hand, the analysis of the increase in vocabulary carried out for the entire, more than 60-year-long period of the writer's work made it possible to notice the similarities and differences in the dynamics of the increase in the Hartwig's virtual vocabulary in the corpora of prose and poetry (ibid.). The incremental diachronic comparative analysis of the works of Hartwig and Boy-Żeleński, carried out for adjectives, made it possible to find a similar dynamics of the development of the lexical workshop of both authors, as well as the lack of signs of exhausting their lexical instrumentarium until the end of their literary activity. The application of entity recognition and linking tools reveals similar cultural resources of both authors. It can be a starting point for other comparative interpretations with regard to the intensity of the occurrence of proper names in the poetry and prose of both authors.

As for the future, we plan to analyze the intensity of references to specific cultural circles, especially in the case of Hartwig, who was both a translator and propagator of French and American poetry in Poland.

Appendix

Source texts used in the corpora of literary works by Boy-Żeleński and Julia Hartwig.
Boy-Żeleński Corpus, last accessed 2022/01/29.
Boy-Żeleński T., 1920: Flirt z Melpomeną, Wyd. Biblioteka Polska, Warszawa.

https://wolnelektury.pl/media/book/pdf/flirt-z-melpomena.pdf

Boy-Żeleński T., 1922: Plotka o weselu Wyspiańskiego, Warszawa.

https://wolnelektury.pl/media/book/pdf/plotka-o-weselu-wyspianskiego.pdf

Boy-Żeleński T., 1925: Pani Hańska, Wyd. H. Altenberg, Lwów.

https://wolnelektury.pl/media/book/pdf/pani-hanska.pdf

Boy-Żeleński T., 1927: W Sorbonie i gdzie indziej, Wyd. Ksiegarni F. Hoesicka, Warszawa.

https://wolnelektury.pl/media/book/pdf/w-sorbonie-i-gdzie-indziej.pdf

Boy-Żeleński T., 1929a: Dziewice konsystorskie, Wyd. Księgarnia Robotnicza, Warszawa.

https://wolnelektury.pl/media/book/pdf/boy-dziewice-konsystorskie.pdf

Boy-Żeleński T., 1929b: Ludzie żywi, Wydawnictwo J. Mortkowicza, Towarzystwo Wydawnicze w Warszawie, Warszawa-Kraków.

https://wolnelektury.pl/media/book/pdf/boy-ludzie-zywi.pdf

Boy-Żeleński T., 1930a: Brązownicy, Warszawa.

https://wolnelektury.pl/media/book/pdf/brazownicy.pdf

Boy-Żeleński T., 1930b: Marzeniae i pysk, Towarzystwo Wydawnicze "Rój", Warszawa.
https://wolnelektury.pl/media/book/pdf/boy-marzenie-i-pysk.pdf.
Boy-Żeleński T., 1930c: Piekło kobiet, Warszawa.
https://wolnelektury.pl/media/book/pdf/pieklo-kobiet.pdf.
Boy-Żeleński T., 1932a: Jak skończyć z piekłem kobiet?, Warszawa.
https://wolnelektury.pl/media/book/pdf/jak-skonczyc-z-pieklem-kobiet.pdf.
Boy-Żeleński T., 1932b: Słowa cienkie i grube, Wyd. Biblioteka Boya, Warszawa.

https://wolnelektury.pl/media/book/pdf/slowa-cienkie-i-grube.pdf.

Boy-Żeleński T., 1932c: Zmysły, zmysły, Wyd. Biblioteka Boya, Warszawa.
https://wolnelektury.pl/media/book/pdf/zmysly-zmysly.pdf.

Boy-Żeleński T., 1932d: Znasz-li ten kraj?… Wyd. Biblioteka Boya, Warszawa.
https://wolnelektury.pl/media/book/pdf/boy-znasz-li-ten-kraj.pdf.

Boy-Żeleński T., 1934a: Balzak, Państwowe Wydawnictwo Książek Szkolnych, Lwów.
https://pl.wikisource.org/wiki/Balzak.

Boy-Żeleński T., 1934b: Obiad literacki, Wyd. Biblioteka Boya, Warszawa.
https://wolnelektury.pl/media/book/pdf/obiad-literacki.pdf.

Boy-Żeleński T., 1934c: Obrachunki fredrowskie, nakł. Gebethnera i Wolffa, Warszawa.
https://wolnelektury.pl/media/book/pdf/obrachunki-fredrowskie.pdf.

Boy-Żeleński T., 1958: Proust i jego świat, Państwowy Instytut Wydawniczy, Warszawa.

https://wolnelektury.pl/media/book/pdf/proust-i-jego-swiat.pdf

Julia Hartwig Corpus (prose-all)
Hartwig J., 1954: Z niedalekich podróży, Ludowa Spółdzielnia Wydawnicza, Warszawa.
Hartwig J., 1962 Apollinaire, Państwowy Instytut Wydawniczy, Warszawa.
Hartwig J., 1972: Gérard de Nerval, Państwowy Instytut Wydawniczy, Warszawa.
Hartwig J., 1980: Dziennik amerykański, Zeszyty Literackie, Warszawa.
Hartwig J., 2001: Zawsze powroty. Z dzienników podróży, Wyd. Sic! Warszawa.
Hartwig J., 2004: Pisane przy oknie, Biblioteka Więzi, Warszawa.
Hartwig J., 2006a: Podziękowanie za gościnę. Moja Francja, Słowo/obraz terytoria, Gdańsk.
Hartwig J., 2006b: Wybrańcy losu, Wyd. Sic! Warszawa.
Hartwig J., 2011: Dziennik, Wydawnictwo Literackie, Kraków.
Hartwig J., 2014: Dziennik tom 2, Wydawnictwo Literackie, Kraków.

Julia Hartwig Corpus (poetry-all)
Hartwig J., 1956: Pożegnania, Wyd. Czytelnik, Warszawa.
Hartwig J., 1969: Wolne ręce, Państwowy Instytut Wydawniczy, Warszawa.
Hartwig J., 1971: Dwoistość, Wyd. Czytelnik, Warszawa.
Hartwig J., 1978: Czuwanie, Wyd. Czytelnik, Warszawa.
Hartwig J., 1980: Chwila postoju, Wydawnictwo Literackie, Kraków.
Hartwig J., 1987: Obcowanie, Wyd. Czytelnik, Warszawa.
Hartwig J., 1992: Czułość, Wyd. Znak, Kraków.
Hartwig J., 1995: Nim opatrzy się zieleń. Wybór wierszy, Wyd. Znak, Kraków.
Hartwig J., 1999: Zobaczone Wyd. a5, Kraków.
Hartwig J., 2001: Nie ma odpowiedzi, Wyd. Sic! Warszawa.
Hartwig J., 2002a: Wiersze amerykańskie, Wyd. Sic! Warszawa.
Hartwig J., 2002b: Błyski, Wyd. Sic! Warszawa.
Hartwig J., 2003: Mówiąc nie tylko do siebie. Poematy prozą, Wyd. Sic! Warszawa.

Hartwig J., 2004a: Bez pożegnania, Wyd. Sic! Warszawa.
Hartwig J., 2004b: Zwierzenia i błyski, Wyd. Sic! Warszawa.
Hartwig J., 2007: To wróci, Wyd. Sic! Warszawa.
Hartwig J., 2008: Trzecie błyski, Wyd. Sic! Warszawa.
Hartwig J., 2009: Jasne niejasne, Wyd. a5, Kraków.
Hartwig J., 2011: Gorzkie żale, Wyd. a5, Kraków.
Hartwig J., 2013: Zapisane, Wyd. a5, Kraków.
Hartwig J., 2014: Błyski zebrane, Wyd. Zeszyty Literackie, Warszawa.
Hartwig J., 2016: Spojrzenie, Wyd. a5, Kraków.

References

Bird, S., Klein, E., Loper, E.: Natural Language Processing with Python. O'Reilly Media Inc, Beijing, Cambridge, Farnham, Köln, Sebastopol, Taipei, Tokyo (2009)

Dadas, S., Perełkiewicz, M., Poświata, R.: Pre-training Polish transformer-based language models at scale. In: Rutkowski, L., Scherer, R., Korytkowski, M., Pedrycz, W., Tadeusiewicz, R., Zurada, J.M. (eds.) ICAISC 2020. LNCS (LNAI), vol. 12416, pp. 301–314. Springer, Cham (2020). https://doi.org/10.1007/978-3-030-61534-5_27

Kennedy, G.: An Introduction to Corpus Linguistitics. Addison Weseley Lingman Ltd., Londyn, Nowy Jork (1998)

Kieraś, W., Kobyliński, Ł, Ogrodniczuk, M.: Korpusomat—a tool for creating searchable morphosyntactically tagged corpora. Comput. Methods Sci. Technol. **24**(1), 21–27 (2018)

Kubis, M.: Geometric deep learning models for linking character names in novels. In: Proceedings of the The 4th Joint SIGHUM Workshop on Computational Linguistics for Cultural Heritage, Social Sciences, Humanities and Literature, pp. 127–132. Online, December 2020. International Committee on Computational Linguistics (2020)

Kubis, M.: Quantitative analysis of character networks in Polish 19th- and 20th-century novels. Digit. Scholarsh. Humanit **36**(Supplement_2), ii175-ii181 (2021)

Legeżyńska, A.: Julia Hartwig. Wdzięczność. Wyd. Uniwersytetu Łódzkiego, Łódź (2017)

Przepiórkowski, A., Bańko, M., Górski, R.L., Lewandowska-Tomaszczyk, B. (eds.): Narodowy Korpus Języka Polskiego. Wydawnictwo Naukowe PWN, Warszawa (2012)

Vaswani, A., et al.: Attention is all you need. In: Guyon, I. et al. (eds.) Advances in Neural Information Processing Systems, vol. 30, Curran Associates, Inc. (2017)

Vetulani, Z.: Linguistic Problems in the Theory of Man-Machine Communication in Natural Language. Brockmeyer, Bochum (1989)

Vetulani, Z., Walczak, B., Obrębski, T., Vetulani, G.: Unambiguius coding of the inflection of Polish nouns and its application in electronic dictionaries–format POLEX Wyd. Naukowe UAM, Poznań (1998)

Vetulani, Z., Witkowska, M., Canbolat, U.: TSCC: a new tool to create lexically saturated text subcorpora. In: Diesner, J., Rehm, G., Witt, A. (eds.) Proceedings of the LREC 2018a 1st Workshop on Computational Impact Detection from Text Data, pp. 22–26. ELRA/ELDA, Paris (2018a)

Vetulani, Z., Witkowska, M., Menken, S., Canbolat, U.: Saturation tests in application to validation of opinion corpora: a tool for corpora processing. In: Vetulani, Z., Mariani, J., Kubis, M. (eds.) Human Language Technology. Challenges for Computer Science and Linguistics. LTC 2015. LNCS, vol. 10930. Springer, Cham (2018b). https://doi.org/10.1007/978-3-319-93782-3_27

Vetulani, Z., Witkowska, M., Kubis, M.: TSCC as a tool for lexical structure studies in application to literary output. In: Vetulani Z., Paroubek P. (eds.) Proceedings of the 9th Language and Technology Conference. Human Language Technologies as a Challenge for Computer Science and Linguistics, pp. 22–24. Wyd. Nauka i Innowacje, Poznań (2019)

Witkowska, M.: Analiza warsztatu pisarskiego Julii Hartwig metodami lingwistyki komputerowej (doctoral dissertation in Polish), Adam Mickiewicz University in Poznań (2021). https://hdl.handle.net/10593/26413. Accessed 21 Jan 2022

Neural Nets in Detecting Word Level Metaphors in Polish

Aleksander Wawer[ID], Małgorzata Marciniak[✉][ID],
and Agnieszka Mykowiecka[ID]

Institute of Computer Science PAS, Jana Kazimierza 5, 01-248 Warszawa, Poland
{axw,mm,agn}@ipipan.waw.pl

Abstract. The paper addresses an experiment in detecting metaphorical usage of adjectives and nouns in Polish data. First, we describe the data developed for the experiment. The corpus consists of 1833 excerpts containing adjective-noun phrases which can have both metaphorical and literal senses. Annotators assign literal or metaphorical senses to all adjectives and nouns in the data. Then, we describe two methods for literal/metaphorical sense classification. The first method uses Bi-LSTM neural network architecture and word embeddings of both token- and character-level. We examine the influence of adversarial training and perform analysis by part-of-speech. The second method uses the BERT token-level classifier. On our relatively small data, the LSTM based approach gives significantly better results and achieves an F1 score equal to 0.81.

Keywords: Metaphors · Polish · LSTM · BERT

1 Introduction

Understanding natural language utterances requires addressing very many issues on very different levels. In spite of many attempts to solve Natural Language Processing (NLP) problems as an end-to-end task, there are still many contexts in which we want to understand words, to combine their meanings into larger schemes, and to add context constraints to sentence meaning. At every step, there is a need to resolve ambiguities which are an inherent feature of natural language understanding. Starting at the word level, many of them have several different meanings, like *bat* which can mean either a kind of solid stick or a flying mammal. The process of understanding is even more complicated, and the communication in natural language is at the same time more interesting and challenging, as people "invent" meanings resembling but different from canonical senses, e.g. *blue* means one of the colours but also *sad*. These meanings that have become very popular are listed in language dictionaries. We nevertheless often use a non-literal combination of words whose listing in dictionaries is difficult and not necessary, as the mechanism used to formulate such expressions is both predictable and highly productive. For example, there are many meanings of

© Springer Nature Switzerland AG 2022
Z. Vetulani et al. (Eds.): LTC 2019, LNAI 13212, pp. 277–288, 2022.
https://doi.org/10.1007/978-3-031-05328-3_18

raise noted in the Oxford dictionary, but all of them are somehow connected with changing a position in physical space or in some sorts of lists. And then, we have the phrase *raise a question* which transfers *raise* from concrete space to an abstract one. Such word usages are generally called non-literal, and in this particular case – metaphorical [9].

An efficient application capable of distinguishing literal from non-literal word occurrences can be very useful in many situations as in web search engines, information extraction modules and document clustering. Technically, the task can be treated as a word sense distinguishing one, but as unsupervised methods are still much less efficient than supervised ones, it is treated more as a classification or a sequence labelling task. We adapted a slightly modified approach in our paper – we identify all occurrences of words but only for the nominal and adjectival classes.

2   Related Work

Over the last decade quite a lot of work was done on metaphor detection, see [16]. In these many approaches, the metaphor identification task was defined variously. One group of papers concerned the classification of selected types of phrases (taken in isolation) into those which nearly always have a literal meaning, like *brown pencil* and those which have only figurative usage, e.g. *dark mood*. In this type of task adjective-noun phrases for English [6,19] and Polish [21] were explored as well as verb constructions for English [2]. Phrases which can have different usage can be classify only in the wider context. In this field of research, some papers present experiments with identification of the type of a particular phrase occurrence in text, while in other approaches, all words from a given text are classified into literal or figurative use.

At first, mostly supervised machine learning approaches were used in which apart from features derived directly from the data, many additional data resources have been used. Among others, these features included, concreteness, imageability, WordNet relations, SUMO ontology concepts, sectional preference information, and syntactic patterns. Solutions based on neural nets training were then published. Several new approaches were elaborated and compared due to the shared task on metaphor identification on the VU Amsterdam Metaphor Corpus [17] conducted at the Workshops on Figurative Language Processing (FigLang) which is the successor of Workshops on Metaphor in NLP. The first FigLang workshop was organized at the NAACL 2018 conference [3] while the second one took place at ACL 2020 [7]. Participants were given two tasks: the ALL_POS task, in which they had to repeat annotation at word level of every token in the presented test data, and the Verbs task, in which only verb annotation was taken into account. In 2018, the best performing solution [23] used pretrained word2vec embeddings, embedding clusterings and POS tags as input to CNN and Bi-LSTM layers. In 2020, there were three baselines proposed by the shared task organizers. The best baseline solution used the BERT language model [4] in a standard token classification task. The next one was one of the top-ranked system from FigLang 2018 [18]. It used LSTM BiRNN architecture with fastText word embeddings representing words as input. The last baseline was the best method from 2016 described in [2]. The system used a logistic regression classifier and the following features: lemmatized unigrams, generalized WordNet semantic classes, and differences in concreteness ratings between verbs/adjectives and nouns. The last two baselines were the worst systems in the Figlang 2020 shared task competition on the VUA dataset. It is also worth noting that the BERT based baseline had an F1 score that was more than 10% better than the others. Transformers models dominate among the solutions proposed at the FigLang 2020 shared task.

In the paper, we compare the results of our approach that was reported in the FigLang 2018 shared task [14] (i.e., adversarial training with Bi-LSTM layers) with a solution based on transformers. We tested them on Polish corpus where annotation was carried out on nouns and adjectives.

3   Data Description

Table 1. Most popular AN phrases

Phrase	All	L	M
pełne garście 'handful'	216	52	164
gorzki smak 'bitter taste'	136	68	68
głęboka rana 'deep wound', deeply wounded'	91	65	26
cierpki smak 'sour taste', 'sour grapes'	57	35	22
fałszywa nuta 'false note''deceitfully'	56	23	33
czyste ręce 'clean hands'	33	10	23
kosmiczna katastrofa 'cosmic/huge disaster'	29	29	0
czysta karta 'clean page'	27	9	18
miękkie nogi 'soft legs' or 'strong emotions'	24	24	0
miękkie lądowanie 'soft landing'	22	9	13
słodki owoc 'sweet fruits' or something good	22	13	9
twardy sen 'sound sleep', 'fast asleep'	21	0	21

Table 2. Statistics of M/L annotations in the corpus

	adj		ppas		adj+ppas	subst		ger		subst+ger	total	
	nb	%	nb	%	nb	nb	%	nb	%	nb	nb	%
M	1184	19.1	106	19.0	1290	1306	11.0	81	21.2	1397	2687	14.2
L	5004	80.9	453	81.0	5457	10520	89.0	301	78.8	10821	16278	85.8

The experiment was performed on a corpus consisting of 1833 short pieces of text selected from the NKJP (National Corpus of Polish [15]). The corpus is built from over 45,000 tokens including punctuation marks and excerpt delimiters. Each excerpt consists of one to three sentences and the average length is 24.5 tokens. The part-of-speech annotation is done with the help of the Concraft2 tagger [20].

Each excerpt contains at least one adjective-noun (AN) phrase which could have an (L) literal or a (M) metaphorical meaning depending on the context. The corpus was collected to perform experiments in recognition of M/L senses of 165 different AN phrases. Table 1 shows phrases with the most numerous examples. Quite often, only one element of a metaphorical AN phrase has a metaphorical meaning. For example, in the phrase *gorzka prawda* 'bitter truth', which always has a metaphorical sense, the noun *truth* usually has a literal sense and only *gorzka* 'bitter' has a metaphorical sense. The label L is assigned to an AN phrase if both elements are annotated as literal, while M is assigned if any of two elements (or both) has a metaphorical sense.

In the experiment described in the paper, we decided to annotate all adjectives and nouns in the whole corpus and to detect the meaning of separate adjectives and nouns instead of the whole phrase.

The annotation was done by two researchers specialising in metaphors in Polish: Joanna Marchula and Maciej Rosiński (the corpus is available from: http:// zil.ipipan.waw.pl/CoDeS) The annotators adapted the procedure for recognition of metaphorical usage of individual words developed for the VU Amsterdam Metaphor Corpus [17]. A discussion about difficulties that arise when the method is applied to Polish is given in [11], while a modified procedure for Polish is described in [12]. An inter-annotator agreement was tested on 51 excerpts consisting of 1246 tokens. In this fragment, there are 555 adjectives and nouns which were annotated by two people. 14 words were differently annotated, and the Cohen's kappa was equal to 0.899, so the annotators obtained very good agreement. As the kappa was high and the procedure for annotation was very time-consuming, we divided the corpus into two parts which were annotated separately by one person. The final annotation was reviewed by removing minor inconsistencies and omissions which was done by one of the annotators. 180 decisions were changed, the label M was changed into L in 54 cases, and in the opposite way 126 cases. Table 2 contains information regarding how many adjectives (regular adjectives and past participles fulfilling adjective roles), nouns and gerunds are annotated as having a literal and metaphorical meaning in the final annotation.

4   Experiment Description

4.1   LSTM Based Solution

The basic architecture in our experiment is the BiLSTM-CRF model similar to [10]. In this model, word representation is concatenated from token-level and character-level embeddings. The latter are computed using character-level Bi-LSTMs, by combining final states of each directional network. Thus, generated word embeddings are then used as input to a token-level bidirectional LSTM deep neural network. Finally, inference is carried by a CRF layer instead of traditional softmax. The structure of the model is given in Fig. 1. The number of hidden units in LSTM is set to 150, the initial learning rate to 0.01, and the batch size to 10.

We also experimented with adversarial training, a technique employed in the field of machine learning which, in its original variant, attempts to fool models through malicious inputs. Recent advancements of this technique, introduced in the area of natural language processing, focus on modifying word embeddings in a malicious manner to make the problem more difficult: the worst-case perturbation coefficient η is computed and added to the embeddings. The expected effect is regularisation. This method was found effective in POS tagging as described by [24]. We test this approach using three η values:

- 0 (adversarial component turned off),
- 0.05 (mild adversarial setting),
- 0.1 (aggressive adversarial setting).

In our experiments, we used Wikipedia-trained Polyglot [1] word embeddings for the Polish language.

We divided the data into three classes of tokens: L (literal), M (metaphorical), O (outside, this class covers every other token type).

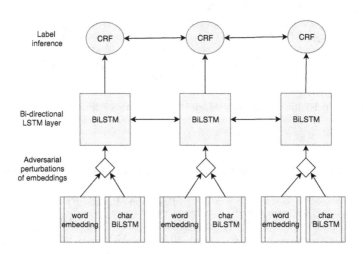

Fig. 1. Diagram of the neural network.

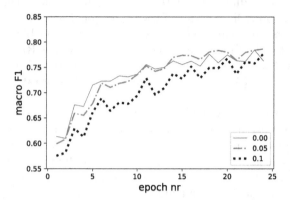

Fig. 2. Macro F1 for various adversarial training rates - initial annotation (LSTM)

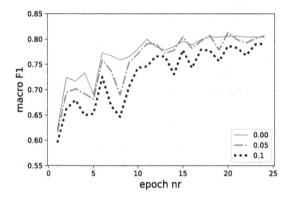

Fig. 3. Macro F1 for various adversarial training rates - final annotation (LSTM)

Table 3. Detailed results of systems on the test data set, 25th epoch (LSTM)

		Precision	Recall	*F1*
Initial data	*M*	0.70	0.58	0.63
	L	0.94	0.94	0.94
	Macro avg	0.82	0.76	0.79
Final data	*M*	0.78	0.61	0.68
	L	0.94	0.92	0.93
	Macro avg	0.86	0.77	0.81

Table 4. Results by POS in numbers (LSTM)

Manual	Auto	adj	ppas	subst	ger	Total
M	M	101	37	92	0	198
L	L	485	5	1070	24	1616
M	L	31	7	50	5	93
L	M	38	1	41	1	81
M	O	1	2	1	4	8
L	O	16	2	40	1	59
Total		672	54	1294	35	2055

Table 5. Evaluation by POS (LSTM)

	adj		ppas		subst		ger	
	M	L	M	L	M	L	M	L
Precision	0.727	0.998	0.833	0.841	0.692	0.955	na	0.828
Recall	0.759	0.900	0.357	0.925	0.643	0.930	na	0.923
F1	0.743	0.946	0.500	0.881	0.667	0.942	na	0.873

4.2 BERT Based Solution

To recognize metaphors at the word level, we attempted to fine-tune two BERT models that had been pre-trained on Polish language data. Namely, PolBERT [8] and HerBERT [13]. The first, PolBERT, was trained on four Polish corpora: Open Subtitles, Paracrawl, Parliamentary Corpus and Wikipedia. The other, HerBERT, was trained on six corpora: CCNet Middle, CCNet Head, National Corpus of Polish, Open Subtitles, and Wikipedia.

We trained and evaluated the PolBERT model in two variants: cased and uncased. In the case of HerBERT, we tested two variants: base and large. We also conducted model training with the use of two values of the learning rate: 2e−5 and 2e−7. The first is within the standard and commonly used range of values mentioned by Google Research for fine tuning BERT models.[1] We chose the second, much smaller value of 2e−7 because of the relatively small size of our data set. Low frequency datasets are usually better handled with more subtle gradient changes.

We used token-level classification API from TensorFlow and softmax as the last neural network layer. We used the Transformers library to train and evaluate BERT models [22], and to perform initial steps such as subword tokenization and padding. For model training and evaluation, we only select the labels of the first subword of each word, as recommended by the Transformers documentation.

---

[1] https://github.com/google-research/bert.

5   Results

In the case of LSTM neural network, we split the data randomly into three partitions: training (80%), development (10%) and test (10%). For each training epoch, results are reported for the test data set, for tokens that are either L or M according to the manual annotation. Figures 2 and 3 contain the macro $F1$-measure computed for various adversarial rates for two versions of the data set, respectively initial annotation and final annotation.

Table 3 contains the results measured at the end of training, 25th epoch, on both versions of the data. We use the mild adversarial setting of η 0.05 as it is the one which provides the best results. Table 4 gives the best results by POS in numbers. The first two columns represent manual and automatic annotation. The next columns gives numbers of annotated adjectives, past participles, nouns and gerunds in the test set. In Table 5, results by POS are given for precision, recall and $F1$-measure. The results of recognition metaphorical meaning of words are much worse as the training data contains almost five times fewer examples. Recognition of adjectives gives better results than nouns despite two times fewer examples in the training data.

In the case of BERT solutions, we split the data randomly into test (10%) and train (90%) sets, preserving the same sets in each experiment for better comparability. We applied the AdamW optimizer from Google Research with the recommended hyperparameter values. We trained the model with the batch size of 1 for 7 epochs.

Table 6 contains the results of the BERT model evaluation on the test set. This table contains three metrics: accuracy, weighted and macro average F1 scores.

Table 6. Evaluation of BERT models

Model	Variant	Learning rate	Accuracy	Macro F1 avg	Weighted F1 avg
HerBERT	Base	2e−5	0.49	0.33	0.47
		2e−7	0.51	0.30	0.45
	Large	2e−5	0.54	0.23	0.38
		2e−7	0.54	0.23	0.38
PolBERT	Cased	2e−5	0.45	0.33	0.45
		2e−7	0.47	0.32	0.45
	Uncased	2e−5	0.48	0.32	0.46
		2e−7	0.45	0.31	0.44

The best results in terms of accuracy were obtained by the HerBERT large. Unfortunately, this can be attributed to the fact that this model did not distinguish between classes and labelled all data as the most frequent class. This is reflected by lower F1 scores. Possibly the best overall compromise is the Her-BERT base model with the learning rate of 2e−7. It managed to outperform PolBERT's accuracy without compromising F1 scores.

Unfortunately, the attempts to train a well-performing model did not yield good results. The most difficult problem is to recognize the M label, as none of the tested BERT models was able to successfully learn and predict the occurrences of this rarest class. This problem is illustrated in Table 7 which contains an example evaluation (precision, recall and F1 scores) of a HerBERT model on the test set. Here, the model was trained for 7 epochs using 2e−5 learning rate. The biggest noticeable issue is the poor recognition of M labels.

Table 7. Example labeling of the test set by the HerBERT model

Class	Precision	Recall	F1
L	0.42	0.33	0.37
M	0.05	0.02	0.03
O	0.55	0.66	0.60

The problem most likely is in insufficient training data size for large models (in terms of the number of parameters) such as BERT. Figure 4 illustrates the loss on training and test sets. It reveals that the loss diminishes on both training and test data, which is an indication of model training process without significant overfitting. The dashed line reflects train set loss while the solid line test set loss.

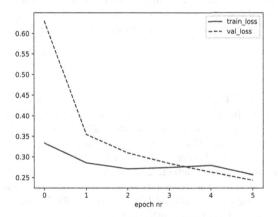

Fig. 4. Train and test set loss for 7 training epochs (BERT)

6 Conclusion

In this paper we tested two architectures for identification of metaphorical use of noun and adjective pairs in Polish texts.

The first of the tested architectures is based on a recurrent neural network with adversarial training. Its application to the metaphor recognition task has proven successful, as the best macro *F1* score achieved 0.81. The best results have been achieved with moderate influence of adversarial training. We tested two variants of the data set and the more coherent version has proven to perform better with the recurrent neural networks. This fact shows that the consistency in annotating the data, in spite of the quite high kappa coefficient on the test fragment, is not perfect. It is an open question whether it comes from the different understanding of the metaphoricity or the annotation task itself. The results obtained with this architecture point to the conclusion that metaphorical senses of adjectives are easier to recognize – the model recognizes them slightly better despite lower frequency.

The second architecture is monolingual Polish BERT [4], pre-trained on large corpora. In our experiments, we attempted to fine-tune the BERT models to recognize metaphors using word (token) level classification. Unfortunately, the attempts did not yield good results. Model training process performed well in terms of decreasing loss, but the biggest problem is the poor recognition of rare metaphorical words. This is likely caused by insufficient frequency of the training set. The conclusion that BERT models need much more training data than recurrent neural network architectures is consistent with the findings reported in [5].

Acknowledgments. This work was supported by the Polish National Science Centre project *Compositional distributional semantic models for identification, discrimination and disambiguation of senses in Polish texts* (2014/15/B/ST6/05186).

References

1. Al-Rfou, R., Perozzi, B., Skiena, S.: Polyglot: distributed word representations for multilingual NLP. In: Proceedings of the Seventeenth Conference on Computational Natural Language Learning, pp. 183–192. Association for Computational Linguistics, Sofia (2013). http://www.aclweb.org/anthology/W13-3520
2. Beigman Klebanov, B., Leong, C.W., Gutierrez, E.D., Shutova, E., Flor, M.: Semantic classifications for detection of verb metaphors. In: Proceedings of the 54th Annual Meeting of the Association for Computational Linguistics, pp. 101–106. Association for Computational Linguistics (2016). https://doi.org/10.18653/v1/P16-2017,http://aclweb.org/anthology/P16-2017
3. Beigman Klebanov, B., Shutova, E., Lichtenstein, P., Muresan, S., Wee, C. (eds.): Proceedings of the Workshop on Figurative Language Processing. Association for Computational Linguistics (2018). http://aclweb.org/anthology/W18-0900
4. Devlin, J., Chang, M., Lee, K., Toutanova, K.: BERT: pre-training of deep bidirectional transformers for language understanding. In: Burstein, J., Doran, C., Solorio, T. (eds.) Proceedings of the 2019 Conference of the North American

Chapter of the Association for Computational Linguistics: Human Language Technologies, NAACL-HLT 2019, pp. 4171–4186. Association for Computational Linguistics (2019). https://doi.org/10.18653/v1/n19-1423

5. Ezen-Can, A.: A comparison of LSTM and BERT for small corpus. arXiv preprint arXiv:2009.05451 (2020)
6. Gutiérrez, E.D., Shutova, E., Marghetis, T., Bergen, B.: Literal and metaphorical senses in compositional distributional semantic models. In: Proceedings of the 54th Annual Meeting of the Association for Computational Linguistics (vol. 1: Long Papers), pp. 183–193. Association for Computational Linguistics, Berlin (2016). https://doi.org/10.18653/v1/P16-1018,https://aclanthology.org/P16-1018
7. Klebanov, B.B., et al. (eds.): Proceedings of the Second Workshop on Figurative Language Processing, Fig-Lang@ACL 2020, Online, 9 July 2020. Association for Computational Linguistics (2020). https://aclanthology.org/volumes/2020.figlang-1/
8. Kłeczek, D.: Polbert: attacking Polish NLP tasks with transformers. In: Ogrodniczuk, M., Łukasz, K. (eds.) Proceedings of the PolEval 2020 Workshop. Institute of Computer Science, Polish Academy of Sciences (2020)
9. Lakoff, G., Johnson, M.: Metaphors We Live by. University of Chicago Press (2008)
10. Lample, G., Ballesteros, M., Subramanian, S., Kawakami, K., Dyer, C.: Neural architectures for named entity recognition. In: Proceedings of the 2016 Conference of the North American Chapter of the Association for Computational Linguistics: Human Language Technologies, pp. 260–270. Association for Computational Linguistics (2016). https://doi.org/10.18653/v1/N16-1030,http://aclweb.org/anthology/N16-1030
11. Marhula, J., Rosiński, M.: Co oferuje MIPVU jako metoda identyfikacji metafory? Polonica **XXXVII**, 37 (2017)
12. Marhula, J., Rosiński, M.: Chapter 9: Linguistic metaphor identification in Polish. In: Metaphor Identification in Multiple Languages: MIPVU Around the World. https://osf.io/phf9q/ (2018)
13. Mroczkowski, R., Rybak, P., Wróblewska, A., Gawlik, I.: HerBERT: efficiently pretrained transformer-based language model for Polish. In: Proceedings of the 8th Workshop on Balto-Slavic Natural Language Processing, pp. 1–10. Association for Computational Linguistics, Kiyv (2021). https://www.aclweb.org/anthology/2021.bsnlp-1.1
14. Mykowiecka, A., Wawer, A., Marciniak, M.: Detecting figurative word occurrences using recurrent neural networks. In: Proceedings of the Workshop on Figurative Language Processing, pp. 124–127. Association for Computational Linguistics, New Orleans (2018). https://doi.org/10.18653/v1/W18-0916
15. Przepiórkowski, A., Bańko, M., Górski, R.L., Lewandowska-Tomaszczyk, B. (eds.): Narodowy Korpus Języka Polskiego. Wydawnictwo Naukowe PWN, Warszawa (2012)
16. Shutova, E.: Design and evaluation of metaphor processing systems. Comput. Linguist. **41**(4), 579–623 (2015)
17. Steen, G.J., Dorst, A.G., Herrmann, J.B., Kaal, A., Krennmayr, T., Pasma, T.: A method for linguistic metaphor identification. From MIP to MIPVU. No. 14 in Converging Evidence in Language and Communication Research, John Benjamins (2010)
18. Stemle, E., Onysko, A.: Using language learner data for metaphor detection. In: Proceedings of the Workshop on Figurative Language Processing, pp. 133–138. Association for Computational Linguistics, New Orleans (2018). https://doi.org/10.18653/v1/W18-0918

19. Tsvetkov, Y., Boytsov, L., Gershman, A., Nyberg, E., Dyer, C.: Metaphor detection with cross-lingual model transfer. In: Proceedings of the 52nd Annual Meeting of the Association for Computational Linguistics, pp. 248–258. Association of Computational Linguistics (2014)
20. Waszczuk, J.: Harnessing the CRF complexity with domain-specific constraints. The case of morphosyntactic tagging of a highly inflected language. In: Proceedings of the 24th International Conference on Computational Linguistics (COLING 2012), pp. 2789–2804 (2012)
21. Wawer, A., Mykowiecka, A.: Detecting metaphorical phrases in the Polish language. In: Proceedings of the International Conference Recent Advances in Natural Language Processing, RANLP 2017, pp. 772–777. INCOMA Ltd., Varna (2017)
22. Wolf, T., et al.: Huggingface's Transformers: State-of-the-Art Natural Language Processing (2020)
23. Wu, C., Wu, F., Chen, Y., Wu, S., Yuan, Z., Huang, Y.: Neural metaphor detecting with CNN-LSTM model. In: Proceedings of the Workshop on Figurative Language Processing (2018)
24. Yasunaga, M., Kasai, J., Radev, D.R.: Robust multilingual part-of-speech tagging via adversarial training. In: Proceedings of NAACL. Association for Computational Linguistics (2018)

Frame-Based Annotation in the Corpus of Synesthetic Metaphors

Magdalena Zawisławska[✉] [iD]

University of Warsaw, Warsaw, Poland
zawisla@uw.edu.pl

Abstract. The research project described in the paper aimed at creating a semantically and grammatically annotated corpus of Polish synesthetic metaphors—Synamet. The texts in the corpus were excerpted from blogs devoted to perfume, wine, beer, cigars, Yerba Mate, tea, or coffee, as well as culinary blogs, music blogs, art blogs, massage, and wellness blogs. Most recent corpus-based studies on metaphors utilize the Conceptual Metaphor Theory by Lakoff and Johnson. Recently, however, a 'domain' has been replaced with the concept of frame. In this project, frames were built up from scratch and were adjusted to the texts. The paper outlines the analytical procedure employed during the corpus compilation, and the main results—statistics of source and target frames and their elements and frame-based models of synesthesia in the corpus.

Keywords: Metaphor · Synesthesia · Frame semantics · Corpus

1 Introduction

This paper outlines the main results of frame-based annotation in Synamet—the Polish Corpus of Synesthetic Metaphors. Synesthetic metaphors [24] were a valuable material for preliminary research on metaphors because their common use in language guarantees that the collected material is rich and varied enough (which should later help to broaden the analysis to include other types of metaphor). Although the majority of recent corpus-based studies utilize the Conceptual Metaphor Theory (CMT), formulated by Lakoff and Johnson [7], there is a trend among recent researchers of metaphor to replace the term domains with frames (e.g., [2, 17]) and to introduce the notion of source and target frames. Arguably, the combination of those two methods aimed at the formalization of metaphor analysis [15, p. 170].

2 Method

The concept of frame is used in various types of study, like anthropology, psychology, and cognitive science [9]. The most recognized theory of frames in linguistic is Fillmore's semantics of understanding (U-semantics) [4]. Fillmore describes frames as a

© Springer Nature Switzerland AG 2022
Z. Vetulani et al. (Eds.): LTC 2019, LNAI 13212, pp. 289–300, 2022.
https://doi.org/10.1007/978-3-031-05328-3_19

complex conceptual system, which underlie the lexical level. The most known is Fill-more's example of time relations—in order to understand the word weekend, one has to activate more broad conceptual systems that include our cultural knowledge about time: that a year has twelve months, one month comprise four weeks, in each week five days are workdays and two are free days. The analytical approach adopted for Synamet draws on both frame semantics [4] and Cognitive Metaphor theory (henceforth CMT), formulated by Lakoff and Johnson [7]. These researchers view metaphor as a primarily conceptual phenomenon consisting of mapping across cognitive domains (from the source domain onto the target domain) e.g., LOVE IS FIRE. Although CMT is the most prominent method of metaphor analysis, it has some disadvantages associated with it—it has not been precisely stated what the term conceptual domain stands for, or how the domain's structure is supposed to be reconstructed. Sullivan [17, pp. 20–21] notices "Conceptual domains are a crucial concept in metaphor theory, yet there is no general agreement on how to define the type of domain used in metaphor".

Moreover, a general schemata *X is Y* results in the metaphors' grammatical properties being ignored. Additionally, CMT does not consider any cultural background in metaphor creation (see [5]), which is very important in the case of verbal synesthesia. Although frames and domains originated on a base of two different theories, several researchers have tried to combine those two tools in order to formalize metaphor analysis [15] since frames are structured entities which contain schematized representations of world knowledge. Sullivan [18] notices that incorporating frames in CMT can help to understand a logic behind the choice of particular lexemes in metaphorical expressions Moreover, frames not only reflect a conceptual level but are also strictly linked to language at both grammatical and lexical levels [15].

2.1 Synamet – A Polish Corpus of Synesthetic Metaphor

Synamet [24] contains texts from blogs devoted to fields where synesthetic metaphors were most likely to be found, e.g., blogs devoted to perfume (SMELL), wine, beer, Yerba Mate, or coffee (TASTE, SMELL, VISION), as well as culinary blogs (TASTE, VISION), music blogs (HEARING), art blogs (VISION), massage and wellness blogs (TOUCH). In total, the Synamet contains:

1. 685,648 tokens,
2. 1,414 annotated texts (entries) from blogs,
3. 2,597 metaphorical topics (i.e., referents of metaphorical expressions),
4. 15,855 activators (words or phrases that activate various frames), and
5. 9,217 grammatically and semantically annotated metaphorical units (MUs).

Synamet consists of 11 categories of texts from thematic blogs (sub-corpuses):

- BEER (beer reviews),
- COFFEE (coffee and café reviews),
- COSMETICS (creams, lotions, masks, shampoos, etc. reviews)
- CUISINE (culinary blogs),
- CULTURE (reviews of theatre, ballet, and operatic performances, exhibitions reviews)

- MASSAGE (blogs written by masseurs),
- MUSIC (reviews of albums, songs, concerts),
- PERFUME (perfumes reviews),
- WELLNESS (blogs concerning physical and mental health)
- WINE (wine reviews),
- YERBA (yerba mate[1] reviews).

In Synamet, the metaphorization process is defined according to Jang *et al.* [6 p. 320] who argue that "a metaphor occurs when a speaker brings one frame into a context governed by another frame, and explicitly relates parts of each, so that the original frame's expectations are extended or enhanced according to the new frame". Therefore, a metaphorical unit (MU) is understood as a single phrase, sentence, or text fragment where two different frames are activated and at least one of them is perceptual. The current project deals solely with linguistic synesthetic metaphors (verbal synesthesia), and not with synesthesia as a psychological phenomenon.

According to Werning et al. [21], a metaphor is synesthetic only when one domain pertains to perception (visual, auditory, olfactory, tactile, or gustatory). If only one of the domains evokes perception, we can talk of *a weak synesthetic metaphor*. If both the source and the target domain evoke perception, it is *a strong synesthetic metaphor*.

In order to collect the most varied material possible, the project adapted the broadest sense of the term *synesthetic metaphor*. In Synamet, both types of metaphors (strong and weak) were annotated.

2.2 Frames in Synamet

The frame semantics has been successfully implemented in FrameNet and MetaNet, both developed at the International Computer Science Institute in Berkeley. Although the FrameNet is a valuable lexical source for English, it wasn't sufficient for the Synamet purposes, as there are too many semantic and grammatical differences between Polish and English. Therefore, the set of frames for the Synamet corpus was constructed from scratch. During annotation, the frames were adjusted to the analyzed texts from the blogs—that is, the project coordinator added new frames or their elements when the annotators signaled that such modifications were needed. Synamet consists of 6 perceptual frames (VISION, HEARING, TOUCH, SMELL, TASTE and MULTIMODAL PERCEPTION—for sensations that activate several senses, e.g., weight or consistency), and 55 non-perceptual frames (e.g. PERSON, ARCHITECTURE, PLANT, SPACE, TIME etc.). For the annotators' convenience, every frame element was associated with one of its typical lexical representations e.g., PERSON/EMOTION (anger), ANIMAL/PART OF ANIMAL (claw). The frame ontology used in the project was hierarchical, meaning that there were very general frames e.g., PERSON, and within them, more specified subframes and elements e.g., the subframe CHARACTER with elements including POSITIVE CHARACTER TRAIT, NEGATIVE CHARACTER TRAIT; the subframe BODY with such elements as BODY PART, GENDER, CORPULENCE, CONDITION, STRENGTH, AGE, etc. Although the frames consisted of subframes and their element,

---

[1] Yerba mate is used to make a tea beverage known as *mate* in Spanish and Portuguese.

the structure was more flat and less complicated than frames in FrameNet. Since a frame in Synamet can be evoked by almost all parts of speech (verbs, nouns, adjectives or even prepositions), one frame consists of all elements essential for an act of perception, that is subject of perception and body part(s) serving as an instrument of perception, object of perception etc. (see Table 1).

Table 1. The TOUCH frame's structure [24].

Subframe	Frame element
SUBJECT OF PERCEPTION	–
BODY PART (*ręce* 'hands', *skóra* 'skin')	–
ACTION OF SUBJECT	PERCEPTION (czuć 'feel') EXAMINATION OF STRUCTURE (naciskać 'press') EXAMINATION OF SURFACE (macać 'feel, grope') CHANGE OF OBJECT'S STATE (zaostrzyć 'strop')
	CHANGE OF STRUCTURE (złamać 'broke') CHANGE OF SURFACE (wygładzić 'polish') CHANGE OF TEMPERATURE (ogrzać 'warm up')
OBJECT OF PERCEPTION (*atłas* 'satin')	–
STATE OF AN OBJECT	STATE (wygładzony 'polished') CHANGE OF STATE (twardnieć 'harden')
SUBJECT'S CONTACT WITH AN OBJECT	CONTACT WITH A WHOLE BODY (otrzeć się 'rub') CONTACT WITH A BODY PART (musnąć 'dab')
SENSATION	–
PATTERN OF SENSATION (*jak jedwab* 'like silk')	–

(continued)

Table 1. (*continued*)

Subframe	Frame element
FEATURE OF SENSATION	TEXTURE (gładki 'smooth') STICKINESS (lepki 'sticky') SUSCEPTIBILITY TO PRESSURE (miękki 'soft') SUSCEPTIBILITY TO STRETCHING (elastyczny 'elastic') SUSCEPTIBILITY TO FLEXION (sztywny 'stiff') SUSCEPTIBILITY TO DESTRUCTION (kruchy 'brittle') TEMPERATURE (ciepły 'warm') MOISTURE CONTENT (mokry 'wet') SHARPNESS (ostry 'sharp')
EVALUATION OF SENSATION (*oślizgły* 'slimy')	–
INSTRUMENT (*gładzik* 'jointer')	–

2.3 Annotation Procedure

MIPVU [14] was employed as a metaphor identification procedure. The procedure constitutes a modified and elaborated version of the MIP as proposed by the Pragglejaz group [11]. Texts excerpted (1,414 entries from blogs) were pre-analyzed by the SPEJD application (Shallow Parsing and Eminently Judicious Disambiguation)[2]—a tool for partial parsing and rule-based morphosyntactic disambiguation. For the distribution of texts among annotators and later—superanotator, the DISTSYS[3] application was used which also stored the annotated data. This tool had already been used in Polish Corpus of Coreference [10] and proved its usability in different types of annotation tasks. Then, a detailed instruction for annotators was prepared, and the analysis procedure was facilitated by a dedicated computer application ATOS (Annotation Tool for Synesthetic Metaphor). The annotation procedure included:

1. Extraction from the text a metaphorical unit, e.g., *tanina gładka* 'tannin smooth'.
2. Correction of the text phrase (if needed), e.g., *tanina jest gładka* 'tannin is smooth'.
3. Defining the referent of the phrase, e.g., taste (of a wine).

---

[2] http://zil.ipipan.waw.pl/Spejd/.
[3] http://zil.ipipan.waw.pl/DistSys.

4. Description of the phrase type: NP.
5. Selection of the metaphor type: strong (because both frames are perceptual).
6. Selection of metaphor category: simple synesthesia.
7. Defining the semantic head of the phrase: tannin.
8. Description of the source frame: TOUCH.
9. Selection of the source frame element: TEXTURE.
10. Description of the source frame evoking word (activator) smooth: ADJ.
11. Description of the target frame: TASTE.
12. Selection of the target frame element: TASTE COMPONENT.
13. Description of the target frame evoking word (activator) tannin: N.

In Synamet, a series annotation procedure was employed instead of a parallel one since the experiment set up for the Polish Coreference Corpus proved that the outcome of such annotation is better [10]. Therefore, texts in Synamet were first analyzed by four annotators and subsequently by three superannotators who amended the initial annotation and checked the corpus coherence. All annotators and superannotators were Masters or PhDs in linguistics with a specialization in semantics. Although the series annotation was chosen as the main method in Synamet, an experiment was performed with a parallel annotation and annotator inter-agreement was tested for a small subset of texts (40 blog entries analyzed independently by two annotators) with the Cohen's κ 'kappa'. Results were as follows: moderate level of agreement for typical metaphors ($\kappa = 0.77$), almost perfect level of agreement for mixed metaphors ($\kappa = 0.95$), a strong level of agreement for entangled metaphors ($\kappa = 0.80$), and almost perfect level of agreement for narrative metaphors ($\kappa = 0.93$). The experiment showed that despite the complex procedure of annotation and problematic data the annotator inter-agreement was high enough to expect that the corpus would be relatively trustworthy.

3   Results

The analysis of frames in Synamet shows interesting and statistically significant results with respect to frequency of perceptual and non-perceptual frames and their elements as well as to frequency and characteristics of activators (i.e., lexical units evoking frames) [see 24].

3.1   Source and Target Frames in Strong Synesthetic Metaphors

The most frequently used source perceptual frames in the analyzed blogs include the **VISION** frame, the **MULTIMODAL PERCEPTION** frame, and the **TOUCH** frame. The ultimate target frame is the **SMELL** frame. Table 2 shows source and target frames in the standardized Pearson residuals ($c^2 = 4594.4$, df $= 5$, p-value $< 0,001$, Cramer's V $= 0.773$).

Table 2. Pearson residual for source (SF) and target (TF) perceptual frames (H[earing], M[ultimodal Perception], S[mell], TA[ste], T[ouch], V[ision]) in strong synesthetic metaphors.

	H	MP	S	TA	T	V
SF	3.21	28.31	-54.04	-14.28	14.28	32.27
TF	-3.21	-28.31	54.043	14.28	-27.77	-32.27

Table 3 presents Pearson standardized residuals of frame pairs in strong synesthetic metaphors (c2 = 870.35, df = 25, p-value < 0.001, Cramer's V = 0.213). The most frequent pairs are HEARING (source) → SMELL (target), VISION (source) → HEARING (target), TASTE (source) → SMELL (target), MULTIMODL PERCEPTION (source) → TASTE (target). Light grey indicate that the result is not statistically significant.

Table 3. Pairs of perceptual source-and target frames in Synamet (S—source, T—target) [24].

Source/ Target	H	MP	S	TA	T	V
H		2.64	4.69	-2.25	1.69	13.75
MP	-1,47		-0.12	-0.87	2.58	0.86
S	19.42	-12.19		11.30	-4.55	-10.84
TA	-4.80	9.62	1.49		2.59	1.13
T	-1.75	2.45	0.14	0.01		0.83
V	-3.31	7.05	2.24	-1.60	3.54	

The strong synesthetic metaphors in Synamet vary depending on a frame element's frequency. Some frame elements are selected more often than others. The target elements apparently arise from the main subjects of the blogs e.g., taste, smell, type of smell, type of taste, or song. More interesting are the elements of the source frames. The broadest set of elements is activated in the VISION frame—52 different elements, from the HEARING frame—39 elements, from the TOUCH—31 elements, from the TASTE frame—21 elements, and from the MULTIMODAL PERCEPTION frame—18 elements.

3.2 Source and Target Frames in Weak Synesthetic Metaphors

In weak synesthetic metaphors, the most frequent pairs of source and target domains include SPORT (source) → TASTE (target), SPACE (source) → HEARING (target), THING (source) → HEARING (target), WEATHER (source) → HEARING (target), ARCHITECTURE (source) → SMELL (target), PLANT (source) → TASTE (target), PERSON (source) → SMELL (target), see Table 4 presenting Pearson standardized residuals (c2 = 526.32, df = 16, p-value < 0.001, Cramer's V = 0.297).

Table 4. Pearson residual for source (S) and target (T) frames (H[earing], TA[ste], S[mell]]) in weak synesthetic metaphors.

Target / Source	H	TA	S
ARCHITECTURE	-2.81	-4.52	**6.26**
ART	-0.43	-2.10	2.08
CLOTHES	1.31	-1.35	-0.15
PERSON	-5.11	0.17	**4.66**
PLANT	-4.64	**6.25**	-0.64
SPACE	**8.06**	-4.22	-4.19
SPORT	-6.69	**16.44**	-6.87
THING	**7.22**	-2.32	-4.92
WEATHER	**6.54**	-4.91	-2.21

The SMELL as a target frame interconnects with the largest and most diverse set of source non-perceptual frames—44 frames (e.g., PERSON, THING, ARCHITECTURE, PLANT, CLOTHES, ART, WILD ANIMAL, SPACE, WEATHER, ELEMENTS, SOCIETY, LANGUAGE, HOME, ARMY, TIME, WEATHER, BASIN, MAGIC, MACHINE, etc.). The next largest target frame, which receives metaphorical transfer from 37 various non-perceptual source frames, is the HEARING frame (e.g., PERSON, THING, SPACE, TRAVEL, VEHICLE, HEALTH SERVICE). A slightly smaller number—27 source frames—appear in synesthetic metaphors with the TASTE as a target frame (e.g., PERSON, SPORT, THING, SOCIETY, ARMY, LANGUAGE). By contrast, the frame VISION as a target frame occurs in metaphors with only 16 different non-perceptual source frames (e.g., PERSON, PLANT, THING). The frames TOUCH and MULTIMODAL PERCEPTION never appear as target frames in weak synesthetic metaphors.

3.3 Frame Activators

All activators (i.e., words that evoke an element of a frame) in Synamet are recorded with respect to two forms: a text form and a base form. Therefore, the base forms of all activators used in synesthetic metaphors form a minidictionary of the Synamet corpus. There are important differences between sets of activators interconnected with frames in the corpus. The perceptual and non-perceptual frames in Synamet differ quite noticeably in regard to being evoked by nouns or verbs. The chi-squared test results are: $c^2 = 95.562$, df = 3, p-value < 0.0001, Cramer's V = 0.187. The non-perceptual frames are evoked more often by verbs, while the perceptual frames exhibit the higher frequency of nouns (see Table 5).

Table 5. Lexemes evoking perceptual (PF) and non-perceptual (NPF) frames in Synamet.

	ADJ	ADV	N	V
PF	1.05	2.21	5.20	-9.54
NPF	-1.05	-2.21	-5.20	9.54

The largest set of lexemes (LUs) evokes the SMELL frame (458) and the HEARING frame (426). The smallest set of lexemes is linked to the MULTIMODAL PERCEPTION frame (142), see Table 6.

Table 6. Lexemes evoking perceptual frames in Synamet.

Frame	ADJ	ADV	N	V	LU
HEARING	86	17	289	34	426
MULT. PERCEPTION	67	19	35	21	142
SMELL	110	3	336	9	458
TASTE	98	13	210	9	330
TOUCH	88	13	35	44	180
VISION	158	23	122	84	387

Statistical analysis shows that the SMELL frame exhibits the dominance of nouns, as well as the HEARING and the TASTE frames. In turn, the TOUCH, the VISION, and the MULTIMODAL PERCEPTION frames are characterized by high frequency of adjectives. The VISION and the TOUCH frames are most often evoked by verbs. The MULTIMODAL PERCEPTION frame exhibits a relatively high frequency of adverbs. The correlation between the number of lexical units and the frequency of parts of speech and how often a frame is source or target was tested by the Pearson correlation coefficient (r). Interestingly, the study revealed that there is no statistically significant correlation between the total number of lexical items and the prevalence of being a source frame, but there is a significant positive correlation between the number of LUs and the prevalence of being a target frame: Pearson's $r = 0.713267$, $t = 2.0353$, $df = 4$, p value $= 0.05577$. There is no correlation between how often adjectives evoke the frame and the prevalence of being a source frame or a target frame. In contrast, tests results show the positive correlation between frequency of adverbs and the prevalence of being a source frame, and the negative correlation between adverbs frequency and the prevalence of being a target frame:

1. ADV and source frame: Pearson's $r = 0.9332103$, $t = 5.1942$, $df = 4$, p-value $= 0.003271$.
2. ADV and target frame: Pearson's $r = -0.8409945$, $t = -3.1088$, $df = 4$, p-value $= 0.01796$.

The test results reveal a significant positive correlation between nouns' frequency and the prevalence of being a target frame—Pearson's r = 0.890914, t = 3.9232, df = 4, p-value = 0.0086; z = 2.2953. In contrast, there is no statistically significant correlation between a noun's frequency and the prevalence of being a source frame. There is also a statistically significant positive correlation between verbs and the prevalence of being a source frame: Pearson's r = 0.7987787, t = 2.6554, df = 4, p-value = 0.02833, while there is no such correlation for verbs' frequency and the prevalence of being a target frame.

4 Discussion

The corpus of synesthetic metaphors shows some interesting properties of synesthetic metaphors. The ultimate "recipient" of metaphorical transfer is the SMELL frame— it can be described by lexemes taken from all other perceptual frames e.g., *korzenno-sandałowy akord* 'spice-sandal chord', *przejrzysty aromat* 'clear aroma', *miękka wanilia* 'soft vanilla', *kwaśno-gorzki zapach* 'sour-bitter smell', *lekkie perfumy* 'light perfume'. The frame that is next most likely a target in metaphorical mapping is the TASTE frame. The HEARING frame equally often serves as a target and as a source frame in synesthetic metaphors. The frames VISION, TOUCH, and MULTIMODAL PERCEPTION are mainly source frames. The widest connectivity with various target frames is typical of the VISION, TOUCH and MULTIMODAL PERCEPTION frames (in both cases— they can be mapped onto five target perceptual frames). The HEARING and TASTE frames have more limited connectivity (with only three target perceptual frames). The SMELL frame is never a source frame in verbal synesthesia. This result calls into question the hypothesis that the most frequent source is the tactile domain, and that the most frequent target is auditory perception [13, 19, 23]. The findings also undermine the attempts to create a universal model of synesthesia in language [19, 20, 22]. Ullmann [20] proposed the following sense hierarchy of synesthesia in language (from lower to higher modalities): touch → heat → taste → smell → hearing → sight. Williams's [22] model was more complex, and the tactile perception was deemed the most typical source of metaphors, while the olfactory perception was the ultimate target. Viberg [20] proposed yet another hierarchy. In his model, the primary source was sight, and the most typical target was likewise smell (sight → hearing → touch → smell/taste). The model of synesthesia in Synamet [24] looks as follows: VISION → MULTIMODAL PERCEPTION → TOUCH → HEARING → TASTE → SMELL.

According to the assumptions of CMT, metaphor operates primarily on the conceptual level. Therefore, analysis of conceptual metaphor typically ignores linguistic forms of metaphorical expressions which appear in texts. Nevertheless, there are researchers who notice that the grammar of verbal metaphor is not accidental and should not be ignored [12, 16, 18]. For example, some analysis shows that verbs are more frequently used metaphorically (that is to evoke metaphoric source domains) than nouns [3, 11, 14, 18]. Sullivan [18, p. 30] proposes an Autonomy Dependence Constraint, which means that "in a metaphorical phrase or clause that can be understood out of context, every source-domain item must be conceptually dependent relative to at least one autonomous target-domain item." Sullivan [18] claims that the metaphoric sentence needs a combination of metaphoric verb and non-metaphoric nouns. Some results of statistical analysis of activators in Synamet support the assumption that verbal metaphor consists of a metaphoric verb and non-metaphoric nouns, and that modifiers dependent of heads (adverbs, adjectives) evoke source frames [1, 3, 11, 14, 18]. There is a significant difference between perceptual and non-perceptual frames with respect to parts of speech—non-perceptual frames exhibit more verbs and fewer nouns than perceptual frames. The statistical odds concerning activators of the perceptual frames also support this assumption. The SMELL frame, which is an ultimate target frame, is evoked by the largest set of nouns, while the MULTIMODAL PERCEPTION, TOUCH, and VISION frames, which serve typically as sources, have fewest evoking nouns. On the other hand, those frames exhibit a large number of modifiers—considerable number of verbs evokes the VISION and the TOUCH frames, while adverbs are characteristic for the MULTIMODAL PERCEPTION frame. All three frames are also quite frequently evoked by adjectives. The Pearson correlation coefficient (r) tests show strong correlation between the number of verbs and the prevalence of being a source frame, while there is no correlation between the number of verbs and the prevalence of being a target frame.

5 Conclusion

The frame-based annotation in Synamet [24] has brought out some important findings, not only about synesthetic metaphors but also about metaphors in general. Deep frame semantic analysis of texts gathered in the corpus proved that there is no universal model of synesthesia in language. Analysis of lexical units evoking frames showed that there is systematicity in linguistic forms of metaphor. Further analysis of frames and frame elements' interconnections could help to establish the extent to which the metaphorization processes are systematic, and whether it is possible to construct a software system that would automatically analyze metaphors in discourse.

Acknowledgments. The paper is funded by the National Science Centre in Poland under the project no. UMO-2014/15/B/HS2/00182 titled: SYNAMET – the Microcorpus of Synaesthetic Metaphors. Towards a Formal Description and Efficient Methods of Analysis of Metaphors in Discourse.

References

1. Cameron, L.: Metaphor in Educational Discourse. Continuum, London & New York (2003)
2. Dancygier, B., Sweetser, E.: Figurative Language. Cambridge Textbooks in Linguistics. Cambridge University Press, Cambridge (2014)
3. Deignan, A.: Metaphor and Corpus Linguistics. John Benjamins Publishing Company, Amsterdam/Philadelphia (2005)
4. Fillmore, Ch.: Frame semantics. In: The Linguistics Society of Korea (eds.) Linguistics in the Morning Calm. pp. 111–137. Hanshin Publishing Co., Seoul (1982)
5. Gibbs, R.W.: Metaphor Wars. Conceptual Metaphors in Human Life. Cambridge Press, Cambridge (2017)

6. Jang, H., Maki, K., Hovy, E., Rose, C.: Finding structure in figurative language: metaphor detection with topic-based frames. In: Jokinen, K., Stede, M., DeVault, D., Louis, A. (eds.) Proceedings of the 18th Annual SIGdial Meeting on Discourse and Dialogue, Saarbrücken, Germany, pp. 320–330. Association for Computational Linguistics (2017)

7. Lakoff, G., Johnson, M.: Metaphors We Live By. University of Chicago Press, Chicago (2008 (1980))

8. Lievers, F.S.: Synaesthesia: a corpus-based study of cross-modal directionality. Funct. Lang. **22**(1), 69–95 (2015)

9. Nerlich, B., Clarke, D.D.: Semantic fields and frames: historical explorations of the interface between language, action, and cognition. J. Pragmat. **32**(2), 125–150 (2000)

10. Ogrodniczuk, M., Głowińska, K., Kopeć, M., Savary, A., Zawisławska, M.: Coreference in Polish: Annotation, Resolution and Evaluation. Walter De Gruyter, Berlin (2015)

11. Pragglejaz Group: MIP: A method for identifying metaphorically used words in discourse. Metaphor Symb. **22**(1), 1–39 (2007)

12. Ronga, I.: Taste synaesthesias: linguistic features and neurophysiological bases. In: Gola, E., Ervas, F. (eds.) Metaphor and Communication, pp. 47–60. Benjamins Publishing, Amsterdam (2016)

13. Shen, Y., Cohen, M.: How come silence is sweet but sweetness is not silent: a cognitive account of directionality in poetic synaesthesia. Lang. Lit. **7**, 123–140 (1998)

14. Steen, G.J., Dorst, A., Herrmann, B., Kaal, A., Krennmayr, T.: A Method for Linguistic Metaphor Identification. From MIP to MIPVU. John Benjamins Publishing Company, Amsterdam (2010)

15. Stickles, E., David, O., Sweetser, E.: Grammatical constructions, frame structure, and metonymy: their contributions to metaphor computation. In: Healey, A., de Souza, R.N., Peškov, P., Allen, M. (eds.) Proceedings of the 11th Meeting of the High Desert Linguistics Society, pp. 317–345. High Desert Linguistics Society, Albuquerque, NM (2016)

16. Strik Lievers, F.S.: Synaesthesia: a corpus-based study of cross-modal directionality. Funct. Lang. **22**(1), 69–95 (2015)

17. Sullivan, K.: Frames and Constructions in Metaphoric Language, vol. 14. John Benjamins Publishing Company, Amsterdam (2013)

18. Sullivan, K.: Integrating constructional semantics and conceptual metaphor. In: Petruck, M.R.L. (ed.) MetaNet, pp. 11–35. John Benjamins Publishing Company, Amsterdam/Philadelphia (2018)

19. Ullmann, S.: The Principles of Semantics. Blackwell, Oxford (1957)

20. Viberg, Å.: The verbs of perception: a typological study. Linguistics **21**(1), 123–162 (1983)

21. Werning, M., Fleischhauer, J., Beşeoğlu, H.: The cognitive accessibility of synaesthetic metaphors. In: Sun, R., Miyake, N. (eds.) Proceedings of the Twenty Eighth Annual Conference of the Cognitive Science Society, pp. 236–570. Lawrence Erlbaum Associates, London (2006)

22. Williams, J.M.: Synesthetic adjectives: a possible law of semantic change. Language **52**(2), 461–478 (1976)

23. Yu, N.: Synesthetic metaphor: a cognitive perspective. J. Lit. Semant. **32**, 19–34 (2003)

24. Zawisławska, M.: Metaphor and Senses. The Synamet Corpus: A Polish Resource for Synesthetic Metaphors. Peter Lang, Bern (2019)

Evaluation

Evaluating Natural Language Processing tools for Polish during PolEval 2019

Łukasz Kobyliński[1]([✉])[iD], Maciej Ogrodniczuk[1][iD], Jan Kocoń[2][iD],
Michał Marcińczuk[2][iD], Aleksander Smywiński-Pohl[3][iD], Krzysztof Wołk[5][iD],
Danijel Koržinek[5][iD], Michal Ptaszynski[4][iD], Agata Pieciukiewicz[5][iD],
and Paweł Dybała[6][iD]

[1] Institute of Computer Science, Polish Academy of Sciences, Warsaw, Poland
lkobylinski@gmail.com
[2] Wrocław University of Science and Technology, Wrocław, Poland
[3] AGH University of Science and Technology, Kraków, Poland
[4] Kitami Institute of Technology, Kitami, Japan
[5] Polish-Japanese Academy of Information Technology, Warsaw, Poland
[6] Jagiellonian University in Kraków, Kraków, Poland

Abstract. PolEval is a SemEval-inspired evaluation campaign for natural language processing tools for Polish. Submitted tools compete against one another within certain tasks selected by organizers, using available data and are evaluated according to pre-established procedures. It is organized since 2017 and each year the winning systems become the state-of-the-art in Polish language processing in the respective tasks. In 2019 we have organized six different tasks, creating an even greater opportunity for NLP researchers to evaluate their systems in an objective manner.

Keywords: Temporal expressions · Lemmatization · Entity linking · Machine translation · Automatic speech recognition · Cyberbullying detection

1 Introduction

PolEval[1] is an initiative started in 2017 by the Linguistic Engineering Group at the Institute of Computer Science, Polish Academy of Sciences, aiming at increasing quality of natural language tools for Polish by organizing a testing ground where interested parties could try their new solutions attempting to beat state-of-the-art. This could be achieved only by setting up formal evaluation procedures according to widely accepted metrics and using newly collected data sets.

---

[1] http://poleval.pl.

© Springer Nature Switzerland AG 2022
Z. Vetulani et al. (Eds.): LTC 2019, LNAI 13212, pp. 303–321, 2022.
https://doi.org/10.1007/978-3-031-05328-3_20

The idea was simple yet it attracted a lot of attention: in first two editions of the contest [10, 22, 35] we received over 40 submissions to 8 tasks and subtasks. In 2019 the number of tasks grew to six, expanding to processing multilingual and multimodal data. Below we describe each of the tasks that have been announced for PolEval 2019[2].

2 Task 1: Recognition and Normalization of Temporal Expressions

2.1 Problem Statement

Temporal expressions (henceforth *timexes*) tell us *when* something happens, *how long* something lasts, or *how often* something occurs. The correct interpretation of a timex often involves knowing the context. Usually, people are aware of their location in time, i.e., they know what day, month and year it is, and whether it is the beginning or the end of week or month. Therefore, they refer to specific dates, using incomplete expressions such as: *12 November, Thursday, the following week, after three days*. The temporal context is often necessary to determine to which specific date and time timexes refer. These examples do not exhaust the complexity of the problem of recognizing timexes.

TimeML [30] is a markup language for describing timexes that has been adapted to many languages. PLIMEX [11] is a specification for the description of Polish timexes. It is based on TIMEX3 used in TimeML. Classes proposed in TimeML are adapted, namely: *date, time, duration, set*.

2.2 Task Description

The aim of this task is to advance research on processing of temporal expressions, which are used in other NLP applications like question answering, summarization, textual entailment, document classification, etc. This task follows on from previous TempEval events organized for evaluating time expressions for English and Spanish like SemEval-2013 [32]. This time we provide corpus of Polish documents fully annotated with temporal expressions. The annotation consists of boundaries, classes and normalized values of temporal expressions. The annotation for Polish texts is based on modified version of original TIMEX3 annotation guidelines[3] at the level of annotating boundaries/types[4] and local/global normalization[5] [11].

---

[2] http://2019.poleval.pl/.

[3] https://catalog.ldc.upenn.edu/docs/LDC2006T08/timeml_annguide_1.2.1.pdf.

[4] http://poleval.pl/task1/plimex_annotation.pdf.

[5] http://poleval.pl/task1/plimex_normalisation.pdf.

2.3   Training Data

The training dataset contains 1500 documents from KPWr corpus. Each document is XML file with the given annotations, e.g.:

```
<DOCID>344245.xml</DOCID>
<DCT><TIMEX3 tid="t0" functionInDocument="CREATION_TIME"
type="DATE" value="2006-12-16"></TIMEX3></DCT>
<TEXT>
<TIMEX3 tid="t1" type="DATE" value="2006-12-16">Dziś
</TIMEX3> Creative Commons obchodzi czwarte urodziny -
przedsięwzięcie ruszyło dokładnie <TIMEX3 tid="t2"
type="DATE" value="2002-12-16">16 grudnia 2002</TIMEX3>
w San Francisco. (...) Z kolei w <TIMEX3 tid="t4"
type="DATE" value="2006-12-18">poniedziałek</TIMEX3>
ogłoszone zostaną wyniki głosowanie na najlepsze blogi.
W ciągu <TIMEX3 tid="t5" type="DURATION" value="P8D">8 dni
</TIMEX3> internauci oddali ponad pół miliona głosów.
Z najnowszego raportu Gartnera wynika, że w <TIMEX3 tid="t6"
type="DATE" value="2007">przyszłym roku</TIMEX3> blogosfera
rozrośnie się do rekordowego rozmiaru 100 milionów blogów.
</TEXT>
```

2.4   Evaluation

We utilize the same evaluation procedure as described in article [32]. We need to evaluate:

1. How many entities are correctly identified,
2. If the extents for the entities are correctly identified,
3. How many entity attributes are correctly identified.

We use classical precision (P), recall (R) and F1-score (F1 – a harmonic mean of P and R) for the recognition.

(1) We evaluate our entities using the entity-based evaluation with the equations below:

$$P = \frac{|\text{Sys}_{\text{entity}} \cap \text{Ref}_{\text{entity}}|}{|\text{Sys}_{\text{entity}}|} \qquad R = \frac{|\text{Sys}_{\text{entity}} \cap \text{Ref}_{\text{entity}}|}{|\text{Ref}_{\text{entity}}|}$$

where, $\text{Sys}_{\text{entity}}$ contains the entities extracted by the system that we want to evaluate, and $\text{Ref}_{\text{entity}}$ contains the entities from the reference annotation that are being compared.

(2) We compare our entities with both strict match and relaxed match. When there is an exact match between the system entity and gold entity then we call it strict match, e.g. *16 grudnia 2002* vs *16 grudnia 2002*. When there is an overlap between the system entity and gold entity then we call it *relaxed match*, e.g. *16 grudnia 2002* vs *2002*. When there is a relaxed match, we compare the attribute values.

(3) We evaluate our entity attributes using the *attribute F1-score*, which captures how well the system identified both the entity and attribute together:

$$\text{attrP} = \frac{\left|\forall x \mid x \in (\text{Sys}_{\text{entity}} \cap \text{Ref}_{\text{entity}}) \wedge \text{Sys}_{\text{attr}}(x) = \text{Ref}_{\text{attr}}(x)\right|}{\left|\text{Sys}_{\text{entity}}\right|}$$

$$\text{attrR} = \frac{\left|\forall x \mid x \in (\text{Sys}_{\text{entity}} \cap \text{Ref}_{\text{entity}}) \wedge \text{Sys}_{\text{attr}}(x) = \text{Ref}_{\text{attr}}(x)\right|}{\left|\text{Ref}_{\text{entity}}\right|}$$

We measure P, R, F1 for both strict and relaxed match and relaxed F1 for value and type attributes. The most important metric is *relaxed F1 value*.

2.5   Results

The best result in the main competition (excluding a baseline system provided by organizers) was achieved by Alium team with its Alium solution. Alium solution is an engine to process texts in natural language and produce results according to rules that define its behaviour. Alium can work either on single words or on triples – *word, lemma, morphosyntactic tag*. Words are additionally masked, so that Alium can work on parts of words as well. More details can be found in [12].

3   Task 2: Lemmatization of Proper Names and Multi-word Phrases

3.1   Problem Statement

Lemmatization relies on generating a dictionary form of a phrase. In our task we focus on lemmatization of proper names and multi-word phrases. For example, the following phrases *radę nadzorczą, radzie nadzorczej, radą nadzorczą* which are inflected forms of *board of directors* should be lemmatized to *rada nadzorcza*. Both, lemmatization of multi-word common noun phrases and named entities are challenging because Polish is a highly inflectional language and a single expression can have several inflected forms.

The difficulty of multi-word phrase lemmatization is due to the fact that the expected lemma is not a simple concatenation of base forms for each word in the phrase [16]. In most cases only the head of the phrase is changed to a nominative form and the remaining word, which are the modifiers of the head, should remain in a specific case. For example in the phrase *piwnicy domu* (Eng. *house basement*) only the first word should be changed to their nominative form while the second word should remain in the genitive form, i.e. *piwnica domu*. A simple concatenation of tokens' base forms would produce a phrase *piwnica dom* which is not correct.

In the case of named entities the following aspects make the lemmatization difficult:

1. Proper names may contain words which are not present in the morphological dictionaries. Thus, dictionary-based methods are insufficient.

2. Some foreign proper names are subject to inflection and some are not.
3. The same text form of a proper name might have different lemmas depending on their semantic category. For example *Słowackiego* (a person last name in genitive or accusative) should be lemmatized to *Słowacki* in case of person name and to *Słowackiego* in case of street name.
4. Capitalization does matter. For example a country name *Polska* (Eng. *Poland*) should be lemmatized to *Polska* but not to *polska*.

3.2 Task Description

The task consists in developing a system for lemmatization of proper names and multi-word phrases. The generated lemmas should follow the KPWr guidelines [24]. The system should generate a lemma for given set of phrases with regards to the context, in which the phrase appears.

3.3 Training Data

The training data consists of 1629 documents from the KPWr corpus [2] with more than 24k annotated and lemmatized phrases. The documents are plain texts with in-line tags indicating the phrases, i.e.
`<phrase id="40465">Madrycie</phrase>`.

All the phrases with their lemmas are listed in a single file, which has the following format:

```
[...]
20250 100619 kampanii wyborczych
kampanie wyborcze
40465   100619 Madrycie             Madryt
40464   100619 Warszawie            Warszawa
40497   100619 Dworcu Centralnym    Dworzec Centralny
40463   100619 Warszawie            Warszawa
[...]
```

3.4 Evaluation

The score of system responses will be calculated using the following formula:

$$Score = 0.2 * Acc_{CS} + 0.8 * Acc_{CI} \qquad (1)$$

Acc refers to the accuracy, i.e. a ratio of the correctly lemmatized phrases to all phrases subjected to lemmatization.

The accuracy will be calculated in two variants: *case sensitive* (Acc_{CS}) and *case insensitive* (Acc_{CI}). In the case insensitive evaluation the lemmas will be converted to lower cases.

4   Task 3: Entity Linking

4.1   Problem Statement

Entity linking [18,29] covers the identification of mentions of entities from a knowledge base (KB) in Polish texts. In this task as the reference KB we use WikiData (WD)[6], an offspring of Wikipedia – a knowledge base, that unifies structured data available in various editions of Wikipedia and links them to external data sources and knowledge bases. Thus making a link from a text to WD allows for reaching a large body of structured facts including: the semantic type of the object, its multilingual labels, dates of birth and death for people, the number of citizens for cities and countries, the number of students for universities and many, many more. The identification of the entities is focused on the disambiguation of a phrase against WD. The scope of the phrase is provided in the test data, so the task boils down to the selection of exactly one entry for each linked phrase.

4.2   Task Description

The following text:

> Zaginieni 11-latkowie w środę rano wyszli z domów do szkoły w **Nowym Targu**, gdzie przebywali do godziny 12:00. Jak informuje "**Tygodnik Podhalański**", 11-letni Ivan już się odnalazł, ale los Mariusza Gajdy wciąż jest nieznany. Source: gazeta.pl

has 2 entity mentions:

1. Nowym Targu[7]
2. Tygodnik Podhalański[8]

Even though there are more mentions that have their corresponding entries in WD (such as "środa", "dom", "12:00", etc.) we restrict the set of entities to a closed group of WD types: names of countries, cities, people, occupations, organisms, tools, constructions, etc. (with important exclusion of times and dates). The full list of entity types is available for download[9]. It should be noted that names such as "Ivan" and "Mariusz Gajda" should not be recognized, since they lack corresponding entries in WD.

The task is similar to Named Entity Recognition (NER), with the important difference that in EL the set of entities is closed. To some extent EL is also similar to Word Sense Disambiguation (WSD), since mentions are ambiguous between competing entities.

In this task we have decided to ignore nested mentions of entities, so names such as "Zespół Szkół Łączności im. Obrońców Poczty Polskiej w Gdańsku, w

---

[6] https://www.wikidata.org.
[7] https://www.wikidata.org/wiki/Q231593.
[8] https://www.wikidata.org/wiki/Q9363509.
[9] http://poleval.pl/task3/entity-types.tsv.

Krakowie", which has an entry in WD, should be treated as an atomic linguistic unit, even though there are many entities that have their corresponding WD entries (such as *Poczta Polska w Gdańsku, Gdańsk, Kraków*). Also the algorithm is required to identify all mentions of the entity in the given document, even if they are exactly same as the previous mentions.

4.3 Training Data

The most common training data used in EL is Wikipedia itself. Even though it wasn't designed as a reference corpus for that task, the structure of internal links serves as a good source for training and testing data, since the number of links inside Wikipedia is counted in millions. The important difference between the Wikipedia links and EL to WD is the fact that the titles of the Wikipedia articles evolve, while the WD identifiers remain constant.

As the training data we have provided a complete text of Wikipedia with morphosyntactic data provided by KRNNT tagger [37], categorization of articles into Wikipedia categories and WD types, Wikipedia redirections and internal links.

4.4 Evaluation

The number of correctly linked mentions divided by the total number of mentions to be identified is used as the evaluation measure. If the system does not provide an answer for a phrase, the result is treated as an invalid link.

5 Task 4: Machine Translation

5.1 Problem Statement

Machine translation is a computer translation of text without human involvement. Machine translation, a pioneer of the 1950s, is also known as machine translation or instant translation.

Currently, there are three most common types of machine translation systems: rule-based, statistical and neural.

- Rule-based systems use a combination of grammar and language rules, as well as a dictionary of common words. Professional dictionaries are created to focus on a particular industry or discipline. Rule-based systems generally provide consistent translations in accurate terms when trained in specialized dictionaries [7].
- The statistical systems do not know the language rules. Instead, they learn to translate by analyzing large amounts of data for each language pair. Statistical systems usually provide smoother but inconsistent translations [13].
- Neural Machine Translation (NMT) is a new approach that uses machines to translate learning through large neural networks. This approach is becoming increasingly popular with MT researchers and developers as trained NMT systems are beginning to show better translation performance in many language pairs compared to phrase-based statistical approaches [36].

310 Ł. Kobyliński et al.

5.2 Task Description

The task was to train the machine translation system as good as possible using any technology with limited text resources. The contest was held in two languages. There were a few languages, more popular English-Polish (Polish direction) and low-resourced Russian-Polish (both directions).

5.3 Training Data

As the training data set, we have prepared a set of bi-lingual corpora aligned at the sentence level. The corpora were saved in UTF-8 encoding as plain text, one language per file. We divided the corpora as in-domain data and out-domain data. Using any other data was not permitted. The in-domain data was rather hard to translate because of its topic diversity. In-domain data were lectures on different topics. As out of domain data we accepted any corpus from http://opus.nlpl.eu project. Any kind of automatic pre- or post- processing was also accepted. The in-domain corpora statistics are given in Table 1.

Table 1. Task 4 corpora statistics.

	No. of segments		No. of unique tokens			
	Test	Train	Test		Train	
			Input	Output	Input	Output
EN to PL	10,000	129,254	9,834	16,978	49,324	100,119
PL to RU	3,000	20,000	6,519	7,249	31,534	32,491
RU to PL	3,000	20,000	6,640	6,385	32,491	31,534

5.4 Evaluation and Results

The participants were asked to translate with their systems test files and submit the results of the translations. The translated files were supposed to be aligned at the sentence level with the input (test) files. Submissions that were not aligned were not accepted. If any pre- or post- processing was needed for the systems, it was supposed to be done automatically with scripts. Any kind of human input into test files was strongly prohibited. The evaluation itself was done with four main automatic metrics widely used in machine translation:

- BLEU [25]
- NIST [4]
- TER [31]
- METEOR [1]

As part of the evaluation preparation we prepared baseline translation systems. For this purpose we used out of the box and state of the art ModernMT machine translation system. We did not do any kind of data pre- or post-processing nor any system adaptation. We simply used our data with default ModernMT settings. Table 2 contains summary of the results limited to BLEU metric for clarity. Full results are available in [21].

Table 2. Task 4: machine translation results

System name	BLUE score		
	EN-PL	PL-RU	RU-PL
Baseline	16.29	12.71	11.45
SRPOL	28.23	n/a	n/a
DeepIf (in-domain)	4.92	5.38	5.51
SIMPLE_SYSTEMS	0.94	0.69	0.57

6   Task 5: Automatic Speech Recognition

6.1   Problem Statement

Automatic speech recognition (ASR) is the problem of converting an audio recording of speech into its textual representation. For the purpose of this evaluation campaign, the transcription is considered simply as a sequence of words conveying the contents of the recorded speech. This task is very common, has many practical uses in both commercial and non-commercial setting and there are many evaluation campaigns associated with it, e.g. [6,9,34]. The significance of this particular competition is the choice of language. To our knowledge, this is the first strictly Polish evaluation campaign of ASR.

$$w* = \arg \max_i P(w_i|O) = \arg \max_i P(O|w_i) \cdot P(w_i) \qquad (2)$$

As shown in formula 2, ASR is usually solved using a probabilistic framework of determining the most likely sequence of words w_i, given a sequence of acoustic observation O of data. This equation is furthermore broken into two essential components by Bayesian inference: the estimation of the acoustic-phonetic realization $P(O|w_i)$, also known as acoustic modeling (AM), and the probability of word sequence realization $P(w_i)$, also known as language modeling (LM):

Each of these steps requires solving a wide range of sub-problems relying on the knowledge of several disciplines, including signal processing, phonetics, natural language processing and machine learning.

A very common framework for solving this problem is the Hidden Markov Model [38]. Currently, this concept was expanded to a more useful implementation based on Weighted Finite-State Transducers [17]. Some of the most recent solutions try to bypass the individual sub-steps by modeling the whole process in a single end-to-end model [8], however knowledge of the mentioned disciplines is still essential to successfully perform the tuning of such a solution.

6.2   Task Description

The task for this evaluation campaign is very simple to define and evaluate: given a set of audio files, create a transcription of each file. For simplicity, only the word sequence is taken into account - capitalization and punctuation is ignored.

Also, the text is evaluated in its normalized form, i.e. numbers and abbreviations need to be presented as individual words.

The domain of the competition is parliamentary proceedings. This domain was chosen for several reasons. The data is publicly available and free for use by any commercial or non-commercial entity. Given the significance of the parliamentary proceedings, there is a wide variety of extra domain material that can be found elsewhere, especially in the media. The task is also not too challenging, compared to some other domains, because of the cleanliness and predictability of the acoustic environment and the speakers.

6.3 Training Data

The competition is organized into two categories: fixed and open. For the fixed competition, a collection of training data is provided as follows:

- Clarin-PL speech corpus [14]
- PELCRA parliamentary corpus [26]
- A collection of 97 h of parliamentary speeches published on the ClarinPL website [15]
- Polish Parliamentary Corpus for language modeling [19, 20, 23]

For those who wish to participate in the competition using a system that was trained on more data, including that which is unavailable to the public, they have to participate as part of the open competition. The only limitation was the ban of use of any data from the Polish Parliament and Polish Senate websites after January 1st 2019.

6.4 Evaluation

Audio is encoded as uncompressed, linearly encoded 16-bit per sample, 16 kHz sampling frequency, mono signals encapsulated in WAV formatted files. The origin of the files is from freely available public streams, so some encoding is present in the data, but the contestants do not have to decompress it on their own. The contestants have a limited time to process these files and provide the transcriptions as separate UTF-8 encoded text documents. The files are evaluated using the standard Word Error Rate metric as computed by the commonly used NIST Sclite package [5].

6.5 Results

The last two entries in Table 3 were the baselines prepared by the competition organizer, with full knowledge of the test data domain. The winner of the competition was the system code named GOLEM with the score of 12.8% WER.

Table 3. Task 5: automatic speech recognition results

System name	WER%	Competition type
GOLEM	12.8	closed
ARM-1	26.4	open
SGMM2	41.3	open
tri2a	41.8	open
clarin-pl/sejm	11.8	closed
clarin-pl/studio	30.9	open

7   Task 6: Automatic Cyberbullying Detection

7.1   Problem Statement

Although the problem of humiliating and slandering people through the Internet existed almost as long as communication via the Internet between people, the appearance of new devices, such as smartphones and tablet computers, which allow using this medium not only at home, work or school but also in motion, has further exacerbated the problem. Especially recent decade, during which Social Networking Services (SNS), such as Facebook and Twitter, rapidly grew in popularity, has brought to light the problem of unethical behaviors in Internet environments, which since then has been greatly impairing public mental health in adults and, for the most, younger users and children. The problem in question, called cyberbullying (CB), is defined as exploitation of open online means of communication, such as Internet forum boards, or SNS, to convey harmful and disturbing information about private individuals, often children and students.

To deal with the problem, researchers around the world have started studying the problem of cyberbullying with a goal to automatically detect Internet entries containing harmful information, and report them to SNS service providers for further analysis and deletion. After ten years of research [28], a sufficient knowledge base on this problem has been collected for languages of well-developed countries, such as the US, or Japan. Unfortunately, still close to nothing in this matter has been done for the Polish language. With this task, we aim at filling this gap.

7.2   Task Description

In this pilot task, the contestants determine whether an Internet entry is classifiable as part of cyberbullying narration or not. The entries contain tweets collected from openly available Twitter discussions. Since much of the problem of automatic cyberbullying detection often relies on feature selection and feature engineering [27], the tweets are provided as such, with minimal preprocessing. The preprocessing, if used, is applied mostly for cases when information about a private person is revealed to the public. In such situations the revealed information is masked not to harm the person in the process.

The goal of the contestants is to classify the tweets into cyberbullying/harmful and non-cyberbullying/non-harmful with the highest possible Precision, Recall, balanced F-score and Accuracy. There are two sub-tasks.

Task 6-1: Harmful vs Non-Harmful: In this task, the participants are to distinguish between normal/non-harmful tweets (class: 0) and tweets that contain any kind of harmful information (class: 1). This includes cyberbullying, hate speech and related phenomena.

Task 6-2: Type of Harmfulness: In this task, the participants shall distinguish between three classes of tweets: 0 (non-harmful), 1 (cyberbullying), 2 (hate-speech). There are various definitions of both cyberbullying and hate-speech, some of them even putting those two phenomena in the same group. The specific conditions on which we based our annotations for both cyberbullying and hate-speech, have been worked out during ten years of research [28]. However, the main and definitive condition to distinguish the two is whether the harmful action is addressed towards a private person(s) (cyberbullying), or a public person/entity/larger group (hate-speech).

7.3 Training Data

To collect the data, we used the Standard Twitter API[10]. The script for data collection was written in Python and was then used to download tweets from 19 Polish Twitter accounts. Those accounts were chosen as the most popular Polish Twitter accounts in the year 2017[11]: @tvn24, @MTVPolska, @lewy_official, @sikorskiradek, @Pontifex_pl, @PR24_pl, @donaldtusk, @BoniekZibi, @NewsweekPolska, @tvp_info, @pisorgpl, @AndrzejDuda, @lis_tomasz, @K_Stanowski, @R_A_Ziemkiewicz, @Platforma_org, @RyszardPetru, @RadioMaryja, @rzeczpospolita.

In addition to tweets from those accounts, we also collected answers to any tweets from the accounts mentioned above from past 7 days. In total, we have received over 101 thousand tweets from 22,687 accounts (as identified by screen_name property in the Twitter API). Using bash random function 10 accounts were randomly selected to become the starting point for further work. Using the same script as before, we downloaded tweets from these 10 accounts and all answers to their tweets that we were able to find using the Twitter Search API Using this procedure we have selected 23,223 tweets from Polish accounts for further analysis.

At first, we randomized the order of tweets in the dataset to get rid of any consecutive tweets from the same account. Next, we got rid of all tweets containing URLs. This was done due to the fact that URLs often take space and limit the contents of the tweets, which in practice often resulted in tweets being cut in

---

[10] https://developer.twitter.com/en/docs/tweets/search/api-reference/get-search-tweets.html.

[11] https://www.sotrender.com/blog/pl/2018/01/twitter-w-polsce-2017-infografika/.

the middle of the sentence or with a large number of *ad hoc* abbreviations. Next, we removed from the data tweets that were perfect duplicates. Tweets consisting only of atmarks(@) or hashtags(#) were also deleted. Finally, we removed tweets with less than five words and those written in languages other than polish. This left us with 11,041 tweets, out of which we used 1,000 tweets as test data and the rest (10,041) as training data.

7.4   Evaluation

The scoring for the first task is done based on standard Precision (P), Recall (R), Balanced F-score (F1) and Accuracy (A), on the basis of the numbers of True Positives (TP), True Negatives (TN), False Positives (FP), and False Negatives (FN), according to the below Eqs. (3–6). In choosing the winners we look primarily at the balanced F-score. However, in the case of equal F-score results for two or more teams, the team with higher Accuracy will be chosen as the winner. Furthermore, in case of the same F-score and Accuracy, a priority will be given to the results as close as possible to BEP (break-even-point of Precision and Recall).

$$Precision = \frac{TP}{TP + FP} \tag{3}$$

$$Recall = \frac{TP}{TP + FN} \tag{4}$$

$$F1 = \frac{2 \cdot P \cdot R}{P + R} \tag{5}$$

$$Accuracy = \frac{TP+TN}{TP + FP + TN + FN} \tag{6}$$

The scoring for the second task is based on two measures, namely, Micro-Average F-score (microF) and Macro-Average F-score (macroF). Micro-Average F-score is calculated similarly as in Eq. (5), but on the basis of Micro-Averaged Precision and Recall, which are calculated according to the below Eqs. (7–8). Macro-Average F-score is calculated on the basis of Macro-Averaged Precision and Recall, which are calculated according to the following Eqs. (9, 10), where TP is True Positive, FP is False Positive, FN is False Negative, and C is class.

In choosing the winners we look primarily at the microF to treat all instances equally since the number of instances is different for each class. Moreover, in the case of equal results for microF, the team with higher macroF will be chosen as the winner. The additional macroF, treating equally not all instances, but rather all classes, is used to provide additional insight into the results.

$$P_{micro} = \frac{\sum_{i=1}^{|C|} TP_i}{\sum_{i=1}^{|C|} TP_i + FP_i} \tag{7}$$

$$R_{micro} = \frac{\sum_{i=1}^{|C|} TP_i}{\sum_{i=1}^{|C|} TP_i + FN_i} \tag{8}$$

$$P_{macro} = \frac{1}{|C|} \sum_{i=1}^{|C|} \frac{TP_i}{TP_i + FP_i}, \tag{9}$$

$$R_{macro} = \frac{1}{|C|} \sum_{i=1}^{|C|} \frac{TP_i}{TP_i + FN_i} \tag{10}$$

7.5   Task 6: Results

Results of Task 6-1: In the first task, out of fourteen submissions, there were nine unique teams: n-waves, Plex, Inc., Warsaw University of Technology, Sigmoidal, CVTimeline, AGH & UJ, IPI PAN, UWr, and one independent. Some teams submitted more than one system proposal, in particular: Sigmoidal (3 submissions), independent (3), CVTimeline (2). Participants used a number of various techniques, usually widely available OpenSource solutions, trained and modified to match the Polish language and the provided dataset when it was required. Some of the methods used applied, e.g., fast.ai/ULMFiT[12], SentencePiece[13], BERT[14], tpot[15], spaCy[16], fasttext[17], Flair[18], neural networks (in particular with GRU) or more traditional SVM. There were also original methods, such as Przetak[19]. The most effective approach was based on recently released ULMFiT/fast.ai, applied for the task by the n-waves team. The originally proposed Przetak, by Plex.inc, was second-best, while third place achieved a combination of ULMFiT/fast.ai, SentencePiece and BranchingAttention model. The results for of all teams participating in Task 6-1 were represented in Table 4.

Results of Task 6-2: In the second task, out of eight submissions, there were five unique submissions. The teams that submitted more than one proposal were: independent (3 submissions) and Sigmoidal (2). Methods that were the most successful for the second task were based on: svm (winning method proposed by independent researcher Maciej Biesek), a combination of ensemble of classifiers from spaCy with tpot and BERT (by Sigmoidal team), and fasttext (by the AGH & UJ team). The results for of all teams participating in Task 6-2 were represented in Table 5. Interestingly, although the participants often applied new techniques, most of them applied only lexical information represented by words

---

[12] http://nlp.fast.ai.
[13] https://github.com/google/sentencepiece.
[14] https://github.com/google-research/bert.
[15] https://github.com/EpistasisLab/tpot.
[16] https://spacy.io/api/textcategorizer.
[17] https://fasttext.cc.
[18] https://github.com/zalandoresearch/flair.
[19] https://github.com/mciura/przetak.

Table 4. Results of participants for Task 6–1.

Submission author(s)	Affiliation	Submission name	Precision	Recall	F-score	Accuracy
Piotr Czapla, Marcin Kardas	**n-waves**	**n-waves ULMFiT**	**66.67%**	**52.24%**	**58.58%**	**90.10%**
Marcin Ciura	Plex, Inc.	Przetak	66.35%	51.49%	57.98%	90.00%
Tomasz Pietruszka	Warsaw University of Technology	ULMFiT + SentencePiece + BranchingAttention	52.90%	54.48%	53.68%	87.40%
Sigmoidal Team (Renard Korzeniowski, Przemysław Sadowski, Rafał Rolczynski, Tomasz Korbak, Marcin Możejko, Krystyna Gajczyk)	Sigmoidal	ensamble spacy + tpot + BERT	52.71%	50.75%	51.71%	87.30%
Sigmoidal Team (Renard Korzeniowski, Przemysław Sadowski, Rafał Rolczynski, Tomasz Korbak, Marcin Możejko, Krystyna Gajczyk)	Sigmoidal	ensamble + fastai	52.71%	50.75%	51.71%	87.30%
Sigmoidal Team (Renard Korzeniowski, Przemysław Sadownik, Rafał Rolczyński, Tomasz Korbak, Marcin Możejko, Krystyna Gajczyk)	Sigmoidal	ensemble spacy + tpot	43.09%	58.21%	49.52%	84.10%
Rafał Pronko	CVTimeline	Rafal	41.08%	56.72%	47.65%	83.30%
Rafał Pronko	CVTimeline	Rafal	41.38%	53.73%	46.75%	83.60%
Maciej Biesek		model1-svm	60.49%	36.57%	45.58%	88.30%
Krzysztof Wróbel	AGH, UJ	fasttext	58.11%	32.09%	41.35%	87.80%
Katarzyna Krasnowska, Alina Wróblewska	IPI PAN	SCWAD-CB	51.90%	30.60%	38.50%	86.90%
Maciej Biesek		model2-gru	63.83%	22.39%	33.15%	87.90%
Maciej Biesek		model3-flair	81.82%	13.43%	23.08%	88.00%
Jakub Kuczkowiak	UWr	Task 6: Automatic cyberbullying detection	17.41%	32.09%	22.57%	70.50%

Table 5. Results of participants for Task 6-2.

Submission author(s)	Affiliation	Submission name	F-score Micro-Average	Macro-average
Maciej Biesek		**model1-svm**	**87.60%**	**51.75%**
Sigmoidal Team (Renard Korzeniowski, Przemyslaw Sadowski, Rafal Rolczynski, Tomasz Korbak, Marcin Mozejko, Krystyna Gajczyk)	Sigmoidal	ensamble spacy + tpot + BERT	87.10%	46.45%
Krzysztof Wróbel	AGH, UJ	fasttext	86.80%	47.22%
Maciej Biesek		model3-flair	86.80%	45.05%
Katarzyna Krasnowska, Alina Wróblewska	IPI PAN	SCWAD-CB	83.70%	49.47%
Maciej Biesek		model2-gru	78.80%	49.15%
Jakub Kuczkowiak	UWr	Task 6: Automatic cyberbullying detection	70.40%	37.59%
Sigmoidal Team (Renard Korzeniowski, Przemyslaw Sadowski, Rafal Rolczynski, Tomasz Korbak, Marcin Mozejko, Krystyna Gajczyk)	Sigmoidal	ensamble + fastai	61.60%	39.64%

(words, tokens, word embeddings, etc.), while none of the participants attempted more sophisticated feature engineering and incorporate other features such as parts-of-speech, named entities, or semantic features.

8 Conclusions and Future Plans

The scope of PolEval competition has grown significantly in 2019, both by means of the number of tasks and by including new areas of interest, such as machine translation and speech recognition. We believe that the successful "call for tasks" will be followed by a large number of submissions, as the interest in natural language processing is rising each year and gradually more and more research is devoted specifically to Polish language NLP.

For the next year, we are planning a more open and transparent procedure of collecting ideas for tasks. We will also be focusing on the idea of open data by establishing common licensing terms for all the code submissions, as well as providing a platform to publish and share solutions, models and additional resources produced by participating teams.

Acknowledgements. The work on temporal expression recognition and phrase lemmatization were financed as part of the investment in the CLARIN-PL research infrastructure funded by the Polish Ministry of Science and Higher Education.

The work on Entity Linking was supported by the Polish National Centre for Research and Development – LIDER Program under Grant LIDER/ 27/0164/L-8/16/NCBR/2017 titled "Lemkin - intelligent legal information system" and also supported in part by PLGrid Infrastructure.

References

1. Banerjee, S., Lavie, A.: METEOR: an automatic metric for MT evaluation with improved correlation with human judgments. In: Proceedings of the ACL Workshop on Intrinsic and Extrinsic Evaluation Measures for Machine Translation and/or Summarization, pp. 65–72 (2005)
2. Broda, B., Marcińczuk, M., Maziarz, M., Radziszewski, A., Wardyński, A.: KPWr: towards a free corpus of polish. In: Calzolari et al. [3]
3. Calzolari, N., et al. (eds.): Proceedings of the 8th International Conference on Language Resources and Evaluation (LREC 2012). European Language Resource Association, Istanbul, Turkey (2012)
4. Doddington, G.: Automatic evaluation of machine translation quality using n-gram co-occurrence statistics. In: Proceedings of the 2nd International Conference on Human Language Technology Research, pp. 138–145. Morgan Kaufmann Publishers Inc. (2002)
5. Fiscus, J.: Sclite scoring package version 1.5. US National Institute of Standard Technology (NIST) (1998). http://www.itl.nist.gov/iaui/894.01/tools
6. Fiscus, J.G., Ajot, J., Michel, M., Garofolo, J.S.: The rich transcription 2006 spring meeting recognition evaluation. In: Renals, S., Bengio, S., Fiscus, J.G. (eds.) MLMI 2006. LNCS, vol. 4299, pp. 309–322. Springer, Heidelberg (2006). https://doi.org/10.1007/11965152_28

7. Forcada, M.L., et al.: Apertium: a free/open-source platform for rule-based machine translation. Mach. Transl. **25**(2), 127–144 (2011)

8. Graves, A., Jaitly, N.: Towards end-to-end speech recognition with recurrent neural networks. In: International Conference on Machine Learning, pp. 1764–1772 (2014)

9. Harper, M.: The automatic speech recognition in reverberant environments (ASpIRE) challenge. In: 2015 IEEE Workshop on Automatic Speech Recognition and Understanding (ASRU), pp. 547–554. IEEE (2015)

10. Kobyliński, Ł., Ogrodniczuk, M.: Results of the PolEval 2017 competition: part-of-speech tagging shared task. In: Vetulani and Paroubek [33], pp. 362–366

11. Kocoń, J., Marcińczuk, M., Oleksy, M., Bernaś, T., Wolski, M.: Temporal Expressions in Polish Corpus KPWr. Cognit. Stud. Études Cognitives **15** (2015)

12. Kocoń, J., Oleksy, M., Bernaś, T., Marcińczuk, M.: Results of the PolEval 2019 shared Task 1: recognition and normalization of temporal expressions. In: Proceedings of the PolEval 2019 Workshop (2019)

13. Koehn, P., et al.: Moses: open source toolkit for statistical machine translation. In: Proceedings of the 45th Annual Meeting of the Association for Computational Linguistics. Companion Volume: Proceedings of the Demo and Poster Sessions, pp. 177–180 (2007)

14. Koržinek, D., Marasek, K., Brocki, Ł., Wołk, K.: Polish read speech corpus for speech tools and services. arXiv preprint arXiv:1706.00245 (2017)

15. Marasek, K., Koržinek, D., Brocki, Ł: System for automatic transcription of sessions of the polish senate. Arch. Acoust. **39**(4), 501–509 (2014)

16. Marcińczuk, M.: Lemmatization of multi-word common noun phrases and named entities in polish. In: Mitkov, R., Angelova, G. (eds.) Proceedings of the International Conference on Recent Advances in Natural Language Processing (RANLP 2017), pp. 483–491. INCOMA Ltd. (2017). https://doi.org/10.26615/978-954-452-049-6_064

17. Mohri, M., Pereira, F., Riley, M.: Weighted finite-state transducers in speech recognition. Comput. Speech Lang. **16**(1), 69–88 (2002)

18. Moro, A., Navigli, R.: Semeval-2015 Task 13: multilingual all-words sense disambiguation and entity linking. In: Proceedings of the 9th International Workshop on Semantic Evaluation (SemEval 2015), pp. 288–297 (2015)

19. Ogrodniczuk, M.: The polish sejm corpus. In: Calzolari et al. [3], pp. 2219–2223

20. Ogrodniczuk, M.: Polish parliamentary corpus. In: Fišer, D., Eskevich, M., de Jong, F. (eds.) Proceedings of the LREC 2018 Workshop ParlaCLARIN: Creating and Using Parliamentary Corpora, pp. 15–19. European Language Resources Association (ELRA), Miyazaki, Japan (2018)

21. Ogrodniczuk, M., Łukasz Kobyliński (eds.): Proceedings of the PolEval 2019 Workshop. Institute of Computer Science, Polish Academy of Sciences, Warsaw, Poland (2019). http://2019.poleval.pl/files/poleval2019.pdf

22. Ogrodniczuk, M., Kobyliński, Ł. (eds.): Proceedings of the PolEval 2018 Workshop. Institute of Computer Science, Polish Academy of Sciences, Warsaw (2018)

23. Ogrodniczuk, M., Nitoń, B.: New developments in the polish parliamentary corpus. In: Fišer, D., Eskevich, M., de Jong, F. (eds.) Proceedings of the Second ParlaCLARIN Workshop, pp. 1–4. European Language Resources Association (ELRA), Marseille, France (2020). https://www.aclweb.org/anthology/2020.parlaclarin-1.1

24. Oleksy, M., Radziszewski, A., Wieczorek, J.: KPWr annotation guidelines - phrase lemmatization (2018). http://hdl.handle.net/11321/591. CLARIN-PL digital repository

25. Papineni, K., Roukos, S., Ward, T., Zhu, W.J.: BLEU: a Method for Automatic Evaluation of Machine Translation. In: Proceedings of the 40th Annual Meeting of Association for Computational Linguistics, pp. 311–318. Association for Computational Linguistics (2002)

26. Pęzik, P.: Increasing the accessibility of time-aligned speech corpora with spokes Mix. In: Calzolari, N., (eds.) Proceedings of the 11th International Conference on Language Resources and Evaluation (LREC 2018), pp. 4297–4300. European Languages Resources Association, Miyazaki, Japan (2018). https://www.aclweb.org/anthology/L18-1000

27. Ptaszynski, M., Eronen, J.K.K., Masui, F.: Learning deep on cyberbullying is always better than brute force. In: IJCAI 2017 3rd Workshop on Linguistic and Cognitive Approaches to Dialogue Agents (LaCATODA 2017), Melbourne, Australia, pp. 19–25 (2017)

28. Ptaszynski, M., Masui, F.: Automatic Cyberbullying Detection: Emerging Research and Opportunities, 1st edn. IGI Global Publishing, Pennsylvania (2018)

29. Rosales-Méndez, H., Hogan, A., Poblete, B.: VoxEL: a benchmark dataset for multilingual entity linking. In: Vrandečić, D., Bontcheva, K., Suárez-Figueroa, M.C., Presutti, V., Celino, I., Sabou, M., Kaffee, L.-A., Simperl, E. (eds.) ISWC 2018. LNCS, vol. 11137, pp. 170–186. Springer, Cham (2018). https://doi.org/10.1007/978-3-030-00668-6_11

30. Saurí, R., Littman, J., Gaizauskas, R., Setzer, A., Pustejovsky, J.: TimeML annotation guidelines, version 1.2.1 (2006)

31. Snover, M., Dorr, B., Schwartz, R., Micciulla, L., Makhoul, J.: A study of translation edit rate with targeted human annotation. In: Proceedings of the 7th Conference of the Association for Machine Translation in the Americas: Technical Papers, Cambridge, Massachusetts, USA, pp. 223–231. Association for Machine Translation in the Americas (2006)

32. UzZaman, N., et al.: SemEval-2013 Task 1: TempEval-3: evaluating time expressions, events, and temporal relations. In: 2nd Joint Conference on Lexical and Computational Semantics (*SEM), Volume 2: Proceedings of the 7th International Workshop on Semantic Evaluation (SemEval 2013), vol. 2, pp. 1–9 (2013)

33. Vetulani, Z., Paroubek, P. (eds.): Proceedings of the 8th Language & Technology Conference: Human Language Technologies as a Challenge for Computer Science and Linguistics. Fundacja Uniwersytetu im. Adama Mickiewicza w Poznaniu, Poznań, Poland (2017)

34. Vincent, E., Watanabe, S., Barker, J., Marxer, R.: The 4th CHiME speech separation and recognition challenge (2016). http://spandh.dcs.shef.ac.uk/chime_challenge/CHiME4/. Accessed 21 Sept 2021

35. Wawer, A., Ogrodniczuk, M.: Results of the PolEval 2017 competition: sentiment analysis shared task. In: Vetulani and Paroubek [33], pp. 406–409

36. Wolk, K., Marasek, K.: Survey on neural machine translation into polish. In: Choroś, K., Kopel, M., Kukla, E., Siemiński, A. (eds.) MISSI 2018. AISC, vol. 833, pp. 260–272. Springer, Cham (2019). https://doi.org/10.1007/978-3-319-98678-4_27

37. Wróbel, K.: KRNNT: polish recurrent neural network tagger. In: Vetulani and Paroubek [33]

38. Young, S., et al.: The HTK Book. Cambridge University Engineering Department, vol. 3, p. 175 (2002)

Open Challenge for Correcting Errors of Speech Recognition Systems

Marek Kubis[1] , Zygmunt Vetulani[1] , Mikołaj Wypych[2],
and Tomasz Ziętkiewicz[1,2(✉)]

[1] Adam Mickiewicz University, ul. Uniwersytetu Poznańskiego 4,
61-614 Poznań, Poland
{mkubis,vetulani,tomasz.zietkiewicz}@amu.edu.pl
[2] Samsung Poland R&D Institute, Pl. Europejski 1, Warszawa, Poland
{m.wypych,t.zietkiewic}@samsung.com

Abstract. The paper announces the new long-term challenge for improving the performance of automatic speech recognition systems. The goal of the challenge is to investigate methods of correcting the recognition results on the basis of previously made errors by the speech processing system. The dataset prepared for the task is described, evaluation criteria are presented and baseline solutions of the problem are proposed.

1 Introduction

The rise in the popularity of voice-based virtual assistants such as Apple's Siri, Amazon's Alexa, Google Assistant and Samsung Bixby imposes high expectations on the precision of automatic speech recognition (ASR) systems. Scheduling a meeting at an incorrect time, sending a message to a wrong person or misinterpreting a command for a home automation system can cause severe losses to a user of a virtual assistant. The problem is even more apparent in the case of deep-understanding systems supposed to work in a very difficult audibility condition, and where ASR errors can appear fatal for the end users. This is the case of systems for crisis situation management (e.g. [23]) where low-quality and emotional voice input can generate a real challenge for speech recognition systems. Furthermore, successful integration of ASR solutions with very demanding AI systems will depend on the degree of being able to take into consideration the non-verbal elements of utterances (prosody). Hence, despite significant improvements to the speech recognition technology in recent years, it is now even more important to search for new methods of decreasing the risk of being misunderstood by the system.

One of the methods that can be used to improve the performance of a speech recognition system is to force the system to learn from its own errors. This approach transforms the speech recognition system into a self-evolving, auto-adapting agent. The objective of this challenge is to investigate to what extent this technique can be used to improve the recognition rate of speech processing systems.

© Springer Nature Switzerland AG 2022
Z. Vetulani et al. (Eds.): LTC 2019, LNAI 13212, pp. 322–337, 2022.
https://doi.org/10.1007/978-3-031-05328-3_21

In order to make the challenge approachable by participants from outside the speech recognition community and to encourage contestants to use a broad range of machine learning and natural language engineering methods that are not specific to the processing of spoken language, we provide the dataset that consists solely of:

1. *Hypotheses* – textual outputs of the automatic speech recognition system.
2. *References* – transcriptions of sentences being read to the automatic speech recognition system.

Thus, the goal of the contestants is to develop a method that improves the result of speech recognition process on the basis of the (erroneous) output of the ASR system and the correct human-made transcription without access to the speech recordings (Figs. 1 and 2).

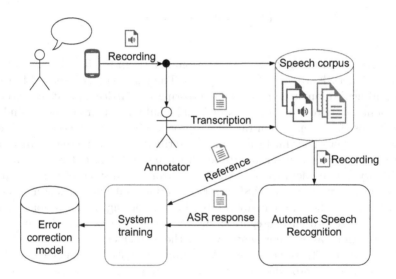

Fig. 1. Error correction model training

2   Related Work

2.1   Shared Tasks

To our best knowledge, our challenge, presented at LTC 2019 [15] was the first one to address the ASR error correction problem. One year later, a similar task was announced as a part of Poleval 2020 competition [21]. The Poleval task introduced datasets with a total of 20 thousands sentences. In addition to providing references and 1-best hypotheses, authors also prepared n-bests list and lattice output for each utterance. Average word error rate (WER) of sentences was

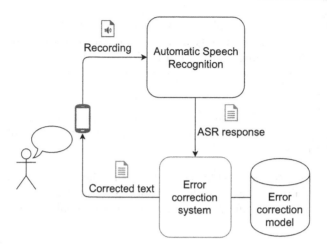

Fig. 2. ASR error correction

much higher than in our data, ranging from 9.49% up to 45.57%, depending on a corpus. The task attracted 8 participants. The winning submission [14] used n-gram and neural language models for rescoring the lattice and lattice extension to account for potential missing words. The submission from the second-best team [27] used edit operation tagging approach, similar to the one described in the Sect. 5 below. The third submission described in the conference proceedings [25] used a machine-translation approach to the problem, but results were not satisfactory (the model increased the word error rate instead of reducing it).

We are not aware of any ASR error correction challenge for languages other than Polish, but there were many competitions targeting similar problems. They can be divided into two categories: speech translation tasks and grammatical error correction tasks. Representatives of the former category are 2 of 3 tasks conducted at the 7th International Workshop on Spoken Language Translation [22]. Tasks 1 and 2 provided sentences in a source language in two forms: ASR recognition results (with errors) and correct recognition results (transcriptions without errors). The goal of the participants was to translate the source text in both forms to the target language. Participants were provided with 3 training corpora composed of 86225 (Task 1), 19972 and 10061 (Task 2) sentence pairs. *CoNLL-2013 Shared Task on Grammatical Error Correction* [20] and *BEA 2019 Shared Task: Grammatical Error Correction* [5] are examples of the second group of tasks. In these competitions participants are given a parallel corpus of texts written by native or non-native English students, containing grammatical, punctuation or spelling errors and their manually corrected versions. The goal of the proposed systems is to correct previously unseen texts. Training corpus in these tasks consisted of 38785 and 57151 pairs of sentences, respectively.

2.2   ASR Error Correction Systems

Errattahi et al. [7] provide review of ASR error detection and correction systems together with a description of ASR evaluation metrics. Cucu et al. [6] propose error correction using SMT (Statistical Machine Translation) model. The SMT model is trained on a relatively small parallel corpus of 2000 ASR transcripts and their manually corrected versions. At an evaluation time, the model is used to "translate" ASR hypothesis into its corrected form. The system achieves 10.5% relative WER[1] improvement by reducing the baseline ASR system's WER from 11.4 to 10.2. Guo et al. [10] describe an ASR error correction model based on LSTM sequence-to-sequence neural network trained on large (40M utterances) speech corpus generated from plain-text data with text to speech (TTS) system. In addition to the spelling correction model, authors experiment with improving the results of an end-to-end ASR system by incorporating an external language model and with combination of the two approaches. The proposed system achieves good results (19% relative WER improvement and 29% relative WER improvement with additional LM re-scoring, with baseline ASR WER of 6.03) but requires a large speech corpus or high-quality TTS system to generate such corpus from a plain text. One of the most recent works [17] presents an error correction model for Mandarin. Authors stress the importance of a low latency of ASR error correction model in production environments. To achieve it, they propose a non-autoregressive transformer model, which is faster than its autoregressive counterpart (i.e. [10]), for the cost of lower error reduction rate. To improve results, the authors propose to train a length prediction model. The model is used to predict the length of reference tokens corresponding to each token from ASR hypothesis. It is then used to adjust the length of a source sentence fed into the decoder. The model is pretrained on a big, artificially created parallel corpus of correct-incorrect sentence pairs, generated by randomly deleting, inserting, and replacing words in a text corpus. Real ASR correction dataset is used to fine-tune the model to a specific ASR system. Relative WER reduction reported by authors on publicly available testset is 13.87, which is slightly worse than the result of autoregressive model (15.53) while introducing over 6 times lower latency.

3   Dataset

We used Polish Wikinews [24] in order to develop the dataset for the task. Wikinews is a collection of news stories developed collaboratively by volunteers around the world in a manner similar to Wikipedia. At the time of writing, the Polish edition of Wikinews contains over 16000 articles and is the sixth largest Wikinews project after Russian, Serbian, Portuguese, French and English. We extracted 9142 sentences from Polish Wikinews stories and asked two native speakers of Polish (male and female) to read them to the speech recognition

---

[1] Word Error Rate, see Sect. 4.

system. Dataset samples consist of transcription of the sentence being read juxtaposed with the textual output captured from the system. Both references and hypotheses are normalized according to the following rules:

– numbers and special characters are replaced by their spoken forms:
   "Początek przemarszu będzie miał miejsce o **10:30** na Targu Rybnym, skąd uczestnicy udadzą się na Targ Drzewny pod pomnik Jana **III** Sobieskiego."
   ⇒ "Początek przemarszu będzie miał miejsce o **dziesiątej trzydzieści** na Targu Rybnym skąd uczestnicy udadzą się na Targ Drzewny pod pomnik Jana **Trzeciego** Sobieskiego."
– all punctuation marks except hyphens are removed:
   "Początek przemarszu będzie miał miejsce o dziesiątej trzydzieści na Targu **Rybnym,** skąd uczestnicy udadzą się na Targ Drzewny pod pomnik Jana Trzeciego **Sobieskiego.**" ⇒ "Początek przemarszu będzie miał miejsce o dziesiątej trzydzieści na Targu **Rybnym** skąd uczestnicy udadzą się na Targ Drzewny pod pomnik Jana Trzeciego **Sobieskiego**"
– all words are uppercased:
   "Początek przemarszu będzie miał miejsce o dziesiątej trzydzieści na Targu Rybnym skąd uczestnicy udadzą się na Targ Drzewny pod pomnik Jana Trzeciego Sobieskiego" ⇒ "POCZĄTEK PRZEMARSZU BĘDZIE MIAŁ MIEJSCE O DZIESIĄTEJ TRZYDZIEŚCI NA TARGU RYBNYM SKĄD UCZESTNICY UDADZĄ SIĘ NA TARG DRZEWNY POD POMNIK JANA TRZECIEGO SOBIESKIEGO"

The dataset is divided into two sets. The training set consists of 8142 utterances randomly sampled from the dataset. The test set contains the rest of the samples.

Simple datasets statistics can be found in Table 1. Average sentence length is about 15 words with few sentences exceeding 50 words length. Histogram Fig. 3 shows length distribution.

The training and test set items consist of:

1. `id`: sample identifier
2. `hyp`: ASR hypothesis - recognition result for the sample voice recording
3. `ref`: reference - human transcription of the sample recording,
4. `source`: copyright, source and author attribution information.

Exemplary dataset items are shown in Table 2. The entire training set and test set (except for reference utterances) are available for download[2] via Gonito online competition platform [8].

Out of all recognition errors in the dataset, 62% are substitutions, 18% deletions and 20% insertions (see Table 1 for details). Average Word Error Rate for both subsets is around 4% (see Fig. 4). Table 3 shows 50 most frequent confusion pairs from the training set, together with their counts (288 words in total) and

---

[2] https://gonito.net/challenge-how-to/asr-corrections.

a manual classification of error types. Most (200) of the mistakes are related to numerals, although they have different causes. Some of them result from normalization issues (such as "II" vs "drugiego"), others from phonetic similarity ("jedną" ⇒ "jedno"). Some of the errors might have originated in the processes of recording (i.e., a different word was read than in the reference text) and normalization of reference sentences (i.e., use of incorrect declination of numerals, "1." becoming "pierwsze" instead of "pierwszy"). There are also misrecognized named entities (24), such as "IBM", "CBOS", which may be out-of vocabulary (OOV) errors. 24 of the errors have phonetic nature - the misrecognized words sounds very similar (but are not homophones). Abbreviation normalization issues account for 18 errors. 4 out of the 50 most frequent confusion pairs, accounting for 10 mistaken words, are orthography mistakes, where ASR returned an unorthographical form (e.g. "drógiego", "członkowstwa"). Most of the substitution errors are caused by replacing or cutting off a suffix or a prefix (e.g. "pierwszy" ⇒ "pierwsze", "sto" ⇒ "to", "czwartym" ⇒ "czwarty"). Insertion and deletion errors are caused mainly by inconsistent normalization of abbreviations across references and hypotheses, and out of vocabulary errors.

Table 1. Dataset statistics.

	Train set	Test set
Number of sentences	8142	1000
Average WER	3.94%	4.01%
Substitutions	2.8%	–
Deletions	0.8%	–
Insertions	0.9%	–
Sentence error rate	25%	25%
Average utterance length (words)	15.40	15.10
Minimum utterance length (words)	2	3
Maximum utterance length (words)	100	48

4 Evaluation

For the purpose of evaluation, the contestants are asked to submit their results via an online challenge set up using the Gonito platform [8] and available at https://gonito.net/challenge/asr-corrections.

The submission consists of a single *out.tsv* file containing the result of running the proposed system on *in.tsv* file, containing ASR system output. Both files contain one sentence per line. The output file should be aligned with the input.

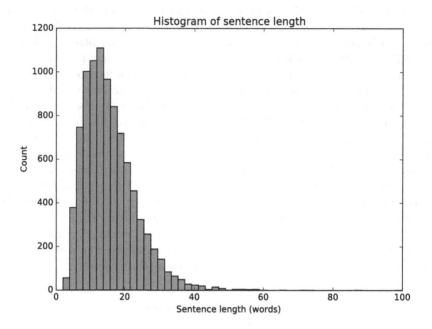

Fig. 3. Histogram of sentence length

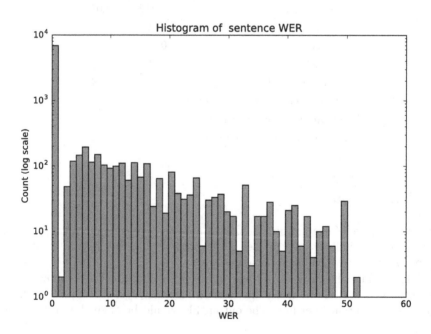

Fig. 4. Histogram of word error rate

The submissions are evaluated using geval tool [9], part of the Gonito platform available also as a standalone tool. Submissions are evaluated using three metrics:

- WER - Word Error Rate of hypothesis corrected by the proposed system, averaged over all tests sentences. WER is defined as follows:

$$WER = \frac{S + D + I}{N = H + S + D}$$

  where: S = number of substitutions, D = number of deletions, I = number of insertions, H - number of hits, N - length of reference sentence. See [18] for in-depth explanation.
- SRR - Sentence Recognition Rate - sentence level accuracy of hypothesis corrected by the proposed system. SRR is defined as ratio of the number of sentences with $WER = 0.0$ (correctly recognized sentences) to the number of all sentences in the corpus.
- CharMatch - $F_{0.5}$ - introduced in [12]. $F_{0.5}$-measure defined in as follows:

$$F_{0.5} = (1 + 0.5^2) \times \frac{P \times R}{0.5^2 P + R}$$

Where: P is precision and R is recall:

$$P = \frac{\sum_i T_i}{\sum_i d_L(h_i, s_i)}, R = \frac{\sum_i T_i}{\sum_i d_L(h_i, r_i)}$$

Where: r_i - i-th reference utterance, h_i - i-th ASR hypothesis, s_i - i-th system output, $d_L(a, b)$ - Levenshtein distance between sequences a and b, T_i - number of correct changes performed by the system, calculated as:

$$T_i = \frac{d_L(h_i, r_i) + d_L(h_i, s_i) - d_L(s_i, r_i)}{2}$$

Table 2. Dataset samples

id	train-1
Hypothesis	DWUDZIESTEGO CZWARTEGO KWIETNIA BIEŻĄCEGO ROKU ROZMAWIALI O WIKIPEDII INTERNECIE WSPÓŁPRACY KLASYFIKOWANIU WIEDZY KSIĄŻKACH I WŁASNOŚCI **LEKTURA LNEJ**
Reference	DWUDZIESTEGO CZWARTEGO KWIETNIA BIEŻĄCEGO ROKU ROZMAWIALI O WIKIPEDII INTERNECIE WSPÓŁPRACY KLASYFIKOWANIU WIEDZY KSIĄŻKACH I WŁASNOŚCI **INTELEKTUALNEJ**
Source	https://pl.wikinews.org/w/index.php?curid=27343&actionaction=history
id	train-2
Hypothesis	**EUROPA POWINNA JĄ** TEŻ ŻE SESJE PE W STRASBURGU SĄ DLA NICH UTRUDNIENIEM BO KOMISJA EUROPEJSKA I RADA UE Z KTÓRYMI PE CIĄGLE WSPÓŁPRACUJE MAJĄ SWOJE STAŁE SIEDZIBY W BRUKSELI
Reference	**EUROPOSŁOWIE PRZYPOMINAJĄ** TEŻ ŻE SESJE PE W STRASBURGU SĄ DLA NICH UTRUDNIENIEM BO KOMISJA EUROPEJSKA I RADA UE Z KTÓRYMI PE CIĄGLE WSPÓŁPRACUJE MAJĄ SWOJE STAŁE SIEDZIBY W BRUKSELI
Source	https://pl.wikinews.org/w/index.php?curid=21290&actionaction=history
id	train-3
Hypothesis	DZIESIĄTEGO WRZEŚNIA DWA TYSIĄCE ÓSMEGO ROKU **LECH MAM BLADES** OGŁOSIŁ WYNIKI FINANSOWE TRZECIEGO KWARTAŁU WYNOSZĄCE TRZY I **DZIEWIĘĆDZIESIĄTYCH** MILIARDA DOLARÓW STRAT
Reference	DZIESIĄTEGO WRZEŚNIA DWA TYSIĄCE ÓSMEGO ROKU **LEHMAN BROTHERS** OGŁOSIŁ WYNIKI FINANSOWE TRZECIEGO KWARTAŁU WYNOSZĄCE TRZY I **DZIEWIĘĆ DZIESIĄTYCH** MILIARDA DOLARÓW STRAT
Source	https://pl.wikinews.org/w/index.php?curid=25282&actionaction=history
id	train-4
Hypothesis	POCHÓD ROZPOCZĄŁ SIĘ NA PLACU SENATORSKIM ALKAMISTA SENAATINTORILLA A **PIERWSZE** W SZEREGU SZŁA SZKOŁA TAŃCA SAMBY SAMBIC TANSSIKOULU
reference	POCHÓD ROZPOCZĄŁ SIĘ NA PLACU SENATORSKIM ALKAMISTA SENAATINTORILLA A **PIERWSZA** W SZEREGU SZŁA SZKOŁA TAŃCA SAMBY SAMBIC TANSSIKOULU
Source	https://pl.wikinews.org/w/index.php?curid=30303&actionaction=history
id	train-5
Hypothesis	DZIESIĄTEGO PAŹDZIERNIKA W KATOWICKIM SPODKU ODBĘDZIE SIĘ DWUDZIESTA DZIEWIĄTA EDYCJA RAWA BLUES FESTIVAL NAJWIĘKSZEJ BLUESOWEJ IMPREZY TYPU INDOOR W EUROPIE
Reference	DZIESIĄTEGO PAŹDZIERNIKA W KATOWICKIM SPODKU ODBĘDZIE SIĘ DWUDZIESTA DZIEWIĄTA EDYCJA RAWA BLUES FESTIVAL NAJWIĘKSZEJ BLUESOWEJ IMPREZY TYPU INDOOR W EUROPIE
Source	https://pl.wikinews.org/w/index.php?curid=25476&actionaction=history
id	train-6
Hypothesis	PRZEPROWADZONE W POŁOWIE GRUDNIA DWUSTRONNE ROZMOWY NIE PRZYNIOSŁY REZULTATU
Reference	PRZEPROWADZONE W POŁOWIE GRUDNIA DWUSTRONNE ROZMOWY NIE PRZYNIOSŁY REZULTATU
Source	https://pl.wikinews.org/w/index.php?curid=5050&actionaction=history

Table 3. Most common confusion pairs

#	Count	Correct word	Incorrect word	Type
1	49	siódmego	siedemset	numerals
2	12	dziewiątego	dziewięćset	numerals
3	11	roku	grog-u	phonetic
4	10	pierwszy	pierwsze	numerals
5	8	roku	mroku	phonetic
6	7	dziesiąte	dziesiąty	numerals
7	7	dziewięć	dziewięćset	numerals
8	7	pierwsze	pierwszy	numerals
9	7	pierwszym	pierwszy	numerals
10	6	ibm	wiem	ne/oov
11	6	tysiąca	tysiące	numerals
12	5	czwartym	czwarte	numerals
14	5	dwie	dwa	numerals
15	5	dziewięćdziesiątym	dziewięćdziesiąty	numerals
16	5	in	innymi	abbreviation
17	5	jedną	jedno	numerals
18	5	m	między	abbreviation
19	5	sto	to	phonetic
20	5	szóstym	szósty	numerals
21	5	trzeciego	trzecie	numerals
22	5	trzeciej	trzeci	numerals
23	5	tysiące	tysiąc	numerals
24	5	w	z	other
25	5	ósmego	ósma	numerals
26	4	czterdziestego	czterdzieste	numerals
27	4	czwartym	czwarty	numerals
28	4	dwadzieścia	dwa	numerals
29	4	dwadzieścia	dzieścia	numerals
30	4	ibm	bije	NE/OOV
31	4	ii	drugiego	numerals
32	4	im	imienia	abbreviation
33	4	mswia	a	NE/OOV
34	4	n	dwa	other
35	4	osiemdziesiątego	osiemdziesiąta	numerals
36	4	osiemdziesiątym	osiemdziesąty	numerals
37	4	piątym	piąty	numerals
38	4	schengen	szengen	orthography
39	4	siódmego	siedmiuset	numerals
40	4	siódmym	siedem	numerals
41	4	szóstym	szóste	numerals
42	4	ulicy	ul	abbreviation
43	4	v	tvn	NE/OOV
44	3	casablance	c	other
45	3	castro	kastro	NE/OOV
46	3	cbos	cbs	NE/OOV
47	3	członkostwa	członkowstwa	orthography
48	3	drugiego	drógiego	orthography
49	3	dwanaście	naście	numerals
50	3	dwudziestego	dwudzieste	numerals

5 Baseline Models

We propose three baseline models targeting the error correction task. All of them are based on the edit operation tagging approach described in [27], but they use different tagging methods.

The edit-operation tagging approach treats the ASR error correction problem as a tagging task. Each token in the ASR hypothesis is tagged either with a label indicating that the token is correct and should be left intact by the error correction system, or with a label indicating how the token should be edited to correct the ASR output. Tokens tagged with an edit operation are then corrected accordingly. Table 4 presents examples of edit operations.

Table 4. Examples of edit operations

Name	Description	Original text	Changed text
`del`	Deletes a token	"z"	""
`add_suffix_{y}`	Appends given suffix to the token	"dystansując"	"dystansujący"
`add_prefix_{s}`	Prepends given prefix to the token	"to"	"sto"
`del_suffix_{1}`	Removes 1 character from the end of the token	"drugie"	"drugi"
`del_prefix_{1}`	Removes 1 character from the beginning of the token	"mroku"	"roku"
`join`	Joins token with previous one	"lgbt i"	"lgbti"
`join_{-}`	Joins token with previous one using given separator	"polsko litewska"	"polsko-litewska"
`replace_suffix_{e}`	Replaces last characters of the token with given string	"pierwszy"	"pierwsze"
`replace_with_{ibm}`	Replaces the token with given string	"wiem"	"ibm"

To prepare training data for the tagger models, pairs of reference-hypothesis sentences need to be aligned, and the differences need to be converted into edit operations tags assigned to tokens in the hypothesis. We use Ratcliff-Obershelp algorithm [13] from difflib[3] library for aligning reference and hypothesis strings. An exemplary reference-hypothesis pair alonh with the tagged hypothesis is shown in Fig. 5.

Once the training data is prepared, a tagging model can be trained. We evaluate 3 different tagging models: rule-based tagger, transformer neural network with linear output layer, and transformer connected with Flair embeddings, LSTM and linear layers. For the sake of training and internal evaluation, we use

---

[3] https://docs.python.org/3/library/difflib.html.

reference
firma ibm znana była tylko z dużych komputerów typu mainframe
hypothesis
firma i wiem znana była tylko z dużych komputerów typu mainframe
tagged hypothesis
firma [$_{del}$ i] [$_{replace\_with\_ibm}$ wiem] znana była tylko z dużych komputerów typu mainframe

Fig. 5. Sample training data

only the 8142 sentence train set. It's further divided into 6513 sentence train set, 814 sentence dev set, and 815 sentence test set. All results are given for the 815 case test subset of the original 8142 train set.

5.1 Brill Tagger

The simplest of the 3 taggers uses Brill tagger [4] - a transformational rule-based tagger. Rules of the tagger are induced automatically using a training data. First, a list of most common tags for each word in a corpus is created. It constitutes a basic lexical tagger, which assigns words with the most likely tag, regardless of their context. This tagger is used to evaluate a separate subset of the training data and to create a list of errors, composed of actual tag-correct tag pairs together with their counts. The list is then used to infer patch rules of a form: "replace tag A with tag B in context C", or "replace tag A with tag B if the current token has property P". The rules are created using patch templates and for each tagging error pair, the template leading to the biggest net reduction of the error number is used. A list of patch rules together with a lexicon of the most likely tags for each word constitute the tagger model. In the experiments, we use Brill tagger implementation from NLTK [2] toolkit[4]. The Brill tagger is very simple, yet easy to implement, train, adjust, and understand. It has the lowest computational requirements among all compared models, both on training and on inference time. We use it as a baseline for other methods - we assume they should perform at least as well as the rule-based tagger.

5.2 Transformer Tagger

This tagger is very similar to the one described in [28]. It uses HerBERT-large [19], a Polish transformer model. A single linear layer has been added at the output, and the whole network was fine-tuned for the tagging task. Specifically, BertForTokenClassification class from Hugging Face Transformers library [26] was used to train the tagger network.

---

[4] https://www.nltk.org/_modules/nltk/tag/brill.html

5.3   Flair Tagger

The third model is similar to the one used in [27]. It uses Sequence Tagger from Flair NLP library [1]. The network is composed of HerBERT-large transformer [19] and Polish Flair word embeddings [3], connected with LSTM [11] and CRF [16] layers. The model can be seen as an extension to the transformer tagger described above, with LSTM and CRF layers instead of linear layer at the output.

5.4   Baseline Results

Results for the proposed baseline systems are shown in Table 5. Not surprisingly, the Brill tagger achieves the lowest scores, but still yields significant 3.2% WER reduction. F1-score of the Brill tagger is 0.94, including the empty class (no edit operation performed). The model was included in the study as a baseline for more sophisticated models, but given the model's simplicity, low time and memory requirements, the result suggests that it can be considered as a real alternative for use in settings with limited computational resources.

Second-best result is obtained with Transformer-based tagger, with 6% WER reduction and micro-average F1-score of 0.971. It's also the second best model when comparing latency of inference (25 ms per sentence). Flair model, which is the most complex one, achieves the highest WER reduction of 7, 8% and micro-average F1-score of 0.972, but it comes for the cost of highest inference time (230 ms per sentence).

Table 5. Results of baseline models. "Relaxed" scores are obtained after performing normalization before comparing reference and hypothesis (lowercasing, removing punctuation characters, trimming trailing spaces, removing double spaces).

	Source	Brill tagger	Flair	Transformer
WER	0.249	0.241	0.229	0.234
WER-relaxed	0.244	0.236	0.225	0.229
WER reduction	–	3.21%	8.03%	6.02%
WER-relaxed reduction	–	3.28%	7.79%	6.15%
Micro-avg F1-score	–	0.940	0.972	0.971
Training time	–	25 s	54 m	3 h
Latency	–	2 ms	230 ms	25 ms

References

1. Akbik, A., et al.: FLAIR: an easy-to-use framework for state-of-the-art NLP. In: Proceedings of the 2019 Conference of the North American Chapter of the Association for Computational Linguistics (Demonstrations), Minneapolis, Minnesota, pp. 54–59. Association for Computational Linguistics, Jun 2019. https://doi.org/10.18653/v1/N19-4010, https://www.aclweb.org/anthology/N19-4010

2. Bird, S., Klein, E., Loper, E.: Natural Language Processing with Python, 1st edn. O'Reilly Media Inc. (2009). https://www.nltk.org/book/

3. Borchmann, Ł., Gretkowski, A., Graliński, F.: Approaching nested named entity recognition with parallel LSTM-CRFs. In: Ogrodniczuk, M., Kobyliński, Ł. (eds.) Proceedings of the PolEval 2018 Workshop, Warszawa, pp. 63–73. Institute of Computer Science, Polish Academy of Science (2018). http://www.borchmann.pl/wp-content/uploads/2018/10/borchmann-{\T1\L}ukasz.pdf

4. Brill, E.: A simple rule-based part of speech tagger. In: Proceedings of the Third Conference on Applied Natural Language Processing, ANLC 1992, USA, pp. 152–155. Association for Computational Linguistics (1992). https://doi.org/10.3115/974499.974526

5. Bryant, C., Felice, M., Andersen, Ø.E., Briscoe, T.: The BEA-2019 shared task on grammatical error correction. In: Proceedings of the Fourteenth Workshop on Innovative Use of NLP for Building Educational Applications, pp. 52–75. Association for Computational Linguistics, Florence (2019). https://doi.org/10.18653/v1/W19-4406

6. Cucu, H., Buzo, A., Besacier, L., Burileanu, C.: Statistical error correction methods for domain-specific ASR systems. In: Dediu, A.-H., Martín-Vide, C., Mitkov, R., Truthe, B. (eds.) SLSP 2013. LNCS (LNAI), vol. 7978, pp. 83–92. Springer, Heidelberg (2013). https://doi.org/10.1007/978-3-642-39593-2_7

7. Errattahi, R., El Hannani, A., Ouahmane, H.: Automatic speech recognition errors detection and correction: a review. Procedia Comput. Sci. **128**, 32–37 (2018). https://doi.org/10.1016/j.procs.2018.03.005

8. Graliński, F., Jaworski, R., Borchmann, Ł., Wierzchoń, P.: Gonito.net - open platform for research competition, cooperation and reproducibility. In: Branco, A., Calzolari, N., Choukri, K. (eds.) Proceedings of the 4REAL Workshop: Workshop on Research Results Reproducibility and Resources Citation in Science and Technology of Language, pp. 13–20 (2016). http://4real.di.fc.ul.pt/wp-content/uploads/2016/04/4REALWorkshopProceedings.pdf

9. Graliński, F., Wróblewska, A., Stanisławek, T., Grabowski, K., Górecki, T.: GEval: tool for debugging NLP datasets and models. In: Proceedings of the 2019 ACL Workshop BlackboxNLP: Analyzing and Interpreting Neural Networks for NLP, pp. 254–262. Association for Computational Linguistics, Florence (2019). https://aclanthology.org/W19-4826/

10. Guo, J., Sainath, T.N., Weiss, R.J.: A spelling correction model for end-to-end speech recognition. In: ICASSP 2019 - 2019 IEEE International Conference on Acoustics, Speech and Signal Processing (ICASSP), pp. 5651–5655 (2019). https://doi.org/10.1109/ICASSP.2019.8683745

11. Hochreiter, S., Schmidhuber, J.: Long short-term memory. Neural Comput. **9**, 1735–80 (1997). https://doi.org/10.1162/neco.1997.9.8.1735

12. Jassem, K., Graliński, F., Obrębski, T.: Pros and cons of normalizing text with Thrax. In: Vetulani, Z., Paroubek, P. (eds.) Proceedings of the 8th Language & Technology Conference, pp. 230–235. Fundacja Uniwersytetu im. Adama Mickiewicza w Poznaniu, Poznań, Poland (2017)

13. Ratcliff, J.W., Metzener, D.E.: Pattern matching: the gestalt approach, July 1988. https://www.drdobbs.com/database/pattern-matching-the-gestalt-approach/184407970

14. Kaczmarek, A., Syposz, T., Martusewicz, M., Chorowski, J., Rychlikowski, P.: t-REX: the rescorer-extender approach to ASR improvement. In: Ogrodniczuk, M., Kobyliński, Ł. (eds.) Proceedings of the PolEval 2020 Workshop, pp. 15–21.

Instytut Podstaw Informatyki Polskiej Akademii Nauk (2020). http://poleval.pl/files/poleval2020.pdf

15. Kubis, M., Vetulani, Z., Wypych, M., Ziętkiewicz, T.: Open challenge for correcting errors of speech recognition systems. In: Vetulani, Z., Paroubek, P. (eds.) Proceedings of the 9th Language and Technology Conference: Human Language Technologies as a Challenge for Computer Science and Linguistics, pp. 219–223. Wydawnictwo Nauka i Innowacje, Poznań, Poland (2019). https://arxiv.org/abs/2001.03041, https://gonito.net/gitlist/asr-corrections.git/

16. Lafferty, J.D., McCallum, A., Pereira, F.C.N.: Conditional random fields: probabilistic models for segmenting and labeling sequence data. In: Proceedings of the Eighteenth International Conference on Machine Learning, ICML 2001, pp. 282–289. Morgan Kaufmann Publishers Inc., San Francisco (2001). http://dl.acm.org/citation.cfm?id=645530.655813

17. Leng, Y., et al.: FastCorrect: fast error correction with edit alignment for automatic speech recognition (2021)

18. Morris, A.C., Maier, V., Green, P.D.: From WER and RIL to MER and WIL: improved evaluation measures for connected speech recognition. In: INTERSPEECH (2004)

19. Mroczkowski, R., Rybak, P., Wr'oblewska, A., Gawlik, I.: HerBERT: efficiently pretrained transformer-based language model for Polish. In: Proceedings of the 8th Workshop on Balto-Slavic Natural Language Processing, Kiyv, Ukraine, pp. 1–10. Association for Computational Linguistics, April 2021. https://www.aclweb.org/anthology/2021.bsnlp-1.1

20. Ng, H.T., Wu, S.M., Wu, Y., Hadiwinoto, C., Tetreault, J.: The CoNLL-2013 shared task on grammatical error correction. In: Proceedings of the Seventeenth Conference on Computational Natural Language Learning: Shared Task, Sofia, Bulgaria, pp. 1–12. Association for Computational Linguistics, August 2013. https://www.aclweb.org/anthology/W13-3601

21. Ogrodniczuk, M., Kobyliński, Ł. (eds.): Proceedings of the PolEval 2020 Workshop. Institute of Computer Science, Polish Academy of Sciences, Warsaw, Poland (2020). http://2020.poleval.pl/files/poleval2020.pdf

22. Paul, M., Federico, M., Stücker, S.: Overview of the IWSLT 2010 evaluation campaign. In: Federico, M., Lane, I., Paul, M., Yvon, F. (eds.) Proceedings of the seventh International Workshop on Spoken Language Translation (IWSLT), pp. 3–27 (2010)

23. Vetulani, Z., et al.: Zasoby językowe i technologie przetwarzania tekstu. POLINT-112-SMS jako przykład aplikacji z zakresu bezpieczeństwa publicznego. Wydawnictwo Naukowe UAM, Poznań, Poland (2010). Language resources and text processing technologies. POLINT-112-SMS as example of homeland security oriented application

24. Wikimedia Foundation: Wikinews. https://pl.wikinews.org (2019). Accessed 01 Mar 2019

25. Wnuk, D., Wołk, K.: Post-editing and rescoring of automatic speech recognition results with openNMT-APE. In: Ogrodniczuk, M., Kobyliński, Ł. (eds.) Proceedings of the PolEval 2020 Workshop, pp. 33–37. Instytut Podstaw Informatyki Polskiej Akademii Nauk, November 2020. http://poleval.pl/files/poleval2020.pdf#page=33

26. Wolf, T., et al.: HuggingFace's transformers: state-of-the-art natural language processing. CoRR abs/1910.03771 (2019). http://arxiv.org/abs/1910.03771

27. Zietkiewicz, T.: Post-editing and rescoring of ASR results with edit operations tagging. In: Proceedings of the PolEval 2020 Workshop, pp. 23–31. Institute of Computer Science, Polish Academy of Sciences, Warsaw, Poland (2020). http://2020.poleval.pl/files/poleval2020.pdf#page=23

28. Zietkiewicz, T.: Punctuation restoration from read text with transformer-based tagger. In: Proceedings of the PolEval 2021 Workshop, pp. 54–60. Institute of Computer Science, Polish Academy of Sciences, Warsaw, Poland (2021). http://poleval.pl/files/poleval2021.pdf#page=55

Evaluation of Basic Modules for Isolated Spelling Error Correction in Polish Texts

Szymon Rutkowski[(✉)](ORCID)

University of Warsaw, Krakowskie Przedmieście 26/28, 00-927 Warsaw, Poland
szymon@szymonrutkowski.pl

Abstract. Spelling error correction is an important problem in natural language processing, as a prerequisite for good performance in downstream tasks as well as an important feature in user-facing applications. For texts in Polish language, there exist works on specific error correction solutions, often developed for dealing with specialized corpora, but not evaluations of many different approaches on big resources of errors. We begin to address this problem by testing some basic and promising methods on PlEWi, a corpus of annotated spelling extracted from Polish Wikipedia. We focus on isolated correction (without context) of non-word errors (ones producing forms that are out-of-vocabulary). The modules may be further combined with appropriate solutions for error detection and context awareness. Following our results, combining edit distance with cosine distance of semantic vectors may be suggested for interpretable systems, while an LSTM network, particularly enhanced by contextualized character embeddings such as ELMo, seems to offer the best raw performance.

Keywords: Spelling correction · Polish language · ELMo

1 Introduction

Spelling error correction is one of fundamental NLP tasks. Most language processing applications benefit greatly from being provided clean texts for their best performance. Human users of computers also often expect competent help in making the spelling of their texts correct.

Because of the lack of tests of multiple common spelling correction methods for Polish, it is useful to establish how they perform in a simple scenario. Here we constrain ourselves to the pure task of isolated correction of non-word errors. It is a kind of error traditionally identified in error correction literature [13]. Non-word errors are defined as incorrect word forms that not only differ from what was intended, but also do not become another, existing word themselves. Thus finding errors of this kind is possible merely with a dictionary of valid words. Much of the initial research on error correction focused on this simple task, tackled without means of taking the context of the nearest words into account.

© Springer Nature Switzerland AG 2022
Z. Vetulani et al. (Eds.): LTC 2019, LNAI 13212, pp. 338–347, 2022.
https://doi.org/10.1007/978-3-031-05328-3_22

It is true that, especially in the case of neural networks, it is often possible and desirable to combine problems of error detection, correction and context awareness into one task trained with a supervised training procedure. In language correction research for English language grammatical and regular spelling errors have been treated uniformly as well, with much success [7]. More recently, fixing the errors has been also formulated as a denoising (sequence to sequence translation) problem [15].

However, when more traditional methods are used, because of their predictability and interpretability for example, one can mix and match various approaches to dealing with the subproblems of detection, correction and context handling (often equivalent to employing some kind of a language model). In this work we call it a modular approach to building spelling error correction systems. This paradigm has been applied, interestingly, to convolutional networks trained separately for various subtasks [4]. In similar setups it is more useful to assess abilities of various solutions in isolation. The exact architecture of a spelling correction system should depend on characteristics of texts it will work on.

Similar considerations eliminated from our focus handcrafted solutions for the whole spelling correction pipeline, primarily the LanguageTool [17]. Its performance in fixing spelling of Polish tweets was already tested [22]. For our purposes it would be given an unfair advantage, since it is a rule-based system making heavy use of words in context of the error.

2   Problems of Spelling Correction for Polish

Polish, being an inflected language, makes it easy for related word forms to transform into another by error. For example, *miska* and *miską* [*bowl*] are different forms despite being distinguished only by the presence of the diacritical mark. This could make using context in detecting and correcting these mistakes more attractive. However, free word order in Polish sentences makes the context harder to interpret without a deeper semantic understanding of text. Models that rely on particular words appearing near each other, or even more so ones depending on a particular word order, are at a disadvantage here.

Published work on language correction for Polish dates back at least to 1970s, when the simplest Levenshtein distance solutions were used for cleaning mainframe inputs [26,27].

Spelling correction tests presented in the literature have tended to focus on one approach applied to a specific corpus. Limited examples include works on spellchecking mammography reports and tweets [5,19,22]. These works emphasized the importance of tailoring correction systems to specific problems of corpora they are applied to. For example, mammography reports suffer from poor typing, which in this case is a repetitive work done in relative hurry. Tweets, on the other hand, tend to contain emoticons and neologisms that can trick solutions based on rules and dictionaries, such as LanguageTool. The latter is, by itself, fairly well suited for Polish texts, since a number of extensions to the structure of this application was inspired by problems with morphology of Polish language [17].

These existing works pointed out more general, potentially useful qualities specific to spelling errors in Polish language texts. It is, primarily, the problem of leaving out diacritical signs, or, more rarely, adding them in wrong places. This phenomenon stems from using a variant of the US keyboard layout, where combinations of `AltGr` with some alphabetic keys produces characters unique to Polish. When the user forgets or neglects to press the AltGr key, typos such as writing *olowek* instead of *ołówek* [*pencil*] appear. In fact, [22] managed to get substantial performance on Twitter corpus by using this "diacritical swapping" alone.

3   Methods

3.1   Baseline Methods

The methods that we evaluated as baselines are the ones we consider to be conceptually and algorithmically basic, and with a moderate potential of yielding particularly good results.

Probably the most straightforward approach to error correction is selecting known words from a dictionary that are within the smallest edit distance from the error. We used the Levenshtein distance metric [14] implemented in Apache Lucene library [1]. It is a version of edit distance that treats deletions, insertions and replacements as adding one unit distance, without giving a special treatment to character swaps.

The SGJP – Grammatical Dictionary of Polish [11] was used as the reference vocabulary. SGJP focuses on providing fully inflected lexemes and all grammatically possible forms, including ones for some from the older Polish vocabulary. It can serve as a source for possible word forms, although is not tailored for any particular context (such as Wikipedia articles or conversational or social media language).

Another simple approach is the aforementioned diacritical swapping, which is a term that we introduce here for referring to a solution inspired by the work of [22]. Given the incorrect form, we try to produce all strings obtainable by either adding or removing diacritical marks from characters. In other words, we treat a as equivalent with $ą$, o with $ó$ etc. In the case of z, $ż$, $ź$, we treat them all as equivalent for our purposes. We then exclude options that are not present in SGJP, and select as the correction the one within the smallest edit distance from the error.

It is possible for the number of such diacritically-swapped options to become very big. For example, the token *Modlin-Zegrze-Pultusk-Różan-Ostrołęka-Lomża-Osowiec* (taken from PlEWi corpus of spelling errors, see below) can yield over $2^{29} = 536,870,912$ states with this method, such as *Módłiń-Żęgrzę-Pułtuśk-Różąń-Óśtrólek ą-Lómzą-Óśówięć*. The actual correction here is just fixing the *ł* in *Pułtusk*. Hence we only try to correct in this way tokens that are shorter than 17 characters.

3.2 Vector Distance

A promising method, adapted from work on correcting texts by English language learners [21], expands on the concept of selecting the correction nearest to the spelling error according to some notion of *distance*. Here, the Levenshtein distance is used in a weighted sum along with cosine distance between semantic word vectors. The word2vec model used as the distributional semantics model carries information about the word forms, but it is static and not based on the context of the actual text being represented. The representation of the words with similar meaning are generally supposed to be more similar (less distant from each other in the multi-dimensional space).

Making use of the semantic vectors for spelling correction is based on the observation that trained vectors models of distributional semantics contain also representations of spelling errors, if they were not pruned. Their representations tend to be similar to those of their correct counterparts. For example, the token *enginir* will appear in similar contexts as *engineer*, and therefore will be assigned a similar vector embedding.

The distance between two tokens a and b is thus defined as

$$D(a, b) = \frac{\mathrm{LD}(a, b) + \mathrm{CD}(\mathrm{V}(a), \mathrm{V}(b))}{2}.$$

Here LD is just Levenshtein distance between strings, and CD – cosine distance between vectors. $\mathrm{V}(a)$ denotes the word vector for a. Both distance metrics are in our case roughly in the range [0,1] thanks to the scaling of edit distance performed automatically by Apache Lucene. We used a pretrained set of word embeddings of Polish [20], obtained with the flavor word2vec procedure using skipgrams and negative sampling [16].

3.3 Recurrent Neural Networks

Another powerful approach, if conceptually simple in linguistic terms, is using a character-based recurrent neural network. Here, we test uni- and bidirectional Long Short-Term Memory networks [10] that are fed characters of the error as their input and are expected to output its correct form, character after character.

This is similar to traditional solutions conceptualizing the spelling error as a chain of characters, which are used as evidence to predict the most likely correct chain of replacements (original characters). This was done with n-gram methods, Markov chains and other probabilistic models [2].

Since nowadays neural networks enjoy a large awareness as an element of software infrastructure, with actively maintained packages readily available, their evaluation seems to be the most practically useful option. Long Short-Term Memory networks (LSTM) are designed to get into account the context of the sequence. The sequence elements can be words, but since we use characters, the constraint forbidding the model from looking at the context of the sentence is preserved. We took advantage of the PyTorch [23] implementation of LSTM in particular.

The bidirectional version [25] of LSTM reads the character chains forward and backwards at the same time. Predictions from networks running in both directions are averaged.

All the LSTM networks, including the ELMo-initialized one described below, have the same basic architecture. The initial word embeddings have 50 dimensions, and are processed by two layers of 512-dimensions LSTM cells. The prediction is computed by the final softmax layer.

In order to provide the network an additional, broad picture peek at the whole error form we also evaluated a setup where the internal state of LSTM cells, instead of being initialized randomly, is computed from an ELMo embedding [24] of the token. The ELMo embedder is capable of integrating linguistic information carried by the whole form (probably often not much in the case of errors), as well as the string as a character chain. The latter is processed with a convolutional neural network. How this representation is constructed is informed by the whole corpus on which the embedder was trained, although in our tests the ELMo embedder is presented only with the token in question, not the whole sentence. The pretrained ELMo model that we used [3] was trained on Wikipedia and Common Crawl corpora of Polish.

The ELMo embedding network outputs three layers as matrices, which are supposed to reflect subsequent compositional layers of language, from phonetic phenomena at the bottom to lexical ones at the top. A weighted sum of these layers is computed, with weights trained along with the LSTM error-correcting network. Then we apply a trained linear transformation, followed by ReLU nonlinearity:

$$\text{ReLU}(x) = \max(0, x)$$

(applied cellwise) in order to obtain the initial setting of parameters for the main LSTM. Our ELMo-augmented LSTM is bidirectional.

Table 1. Test results for all the methods used. The loss measure is cross-entropy.

Method	Accuracy	Perplexity	Loss (train)	Loss (test)
Edit distance	0.3453	–	–	–
Diacritic swapping	0.2279	–	–	–
Vector distance	0.3945	–	–	–
LSTM-1 net	0.4183	**907**	0.3	0.41
LSTM-2 net	0.6634	11182	0.1	**0.37**
LSTM-ELMo net	**0.6818**	706166	**0.07**	0.38

4   Experimental Setup

PlEWi [8] is an early version of WikEd [9] error corpus, containing error type annotations allowing us to select only non-word errors for evaluation. Specifically, PlEWi supplied 550,755 (*error, correction*) pairs, of which 298,715 were unique.

Table 2. Discovered optimal weights for summing layers of ELMo embedding for initializing an error-correcting LSTM. The layers are numbered from the one that directly processes character and word input to the most abstract one.

Layer I	Layer II	Layer III
0.036849	0.08134	0.039395

The corpus contains data extracted from histories of page versions of Polish Wikipedia. An algorithm designed by the corpus author determined where the changes were correcting spelling errors, as opposed to expanding content and disagreements between Wikipedia editors. The annotation is also aware of the difference between word and non-word errors (for our purposes, it is the latter ones that are important).

The corpus features texts that are descriptive rather than conversational, contain relatively many proper names and are more likely to have been at least skimmed by the authors before submitting for online publication. Error cases provided by PlEWi are, therefore, not a balanced representation of spelling errors in written Polish language. But PlEWi does have the advantage of scale in comparison to existing literature, such as [22] operating on a set of only 740 annotated errors in tweets.

All methods were tested on a test subset of 25% of cases, with 70% left for training (where needed) and 5% for development.

The methods that required training – namely recurrent neural networks – had their loss measured as cross-entropy loss measure between correct character labels and predictions. This value was minimized with Adam algorithm [12]. The networks were trained for 35 epochs.

The code for performing the experiments is made public on GitHub[1].

5   Results

The experimental results are presented in Table 1. Diacritic swapping showed a remarkably poor performance, despite promising mentions in existing literature. This might be explained by the aforementioned characteristics of Wikipedia edits, which can be expected to be to some degree self-reviewed before submission. This can very well limit the number of most trivial mistakes.

On the other hand, the vector distance method was able to bring a discernible improvement over pure Levenshtein distance, comparable even with the most basic LSTM. It is possible that assigning more fine-tuned weights to edit distance and semantic distance would make the quality of predictions even higher. The idea of using vector space measurements explicitly can be also expanded if we were to consider the problem of contextualizing corrections. For example, the semantic distance of proposed corrections to the nearest words is likely to carry much information about their appropriateness. Looking from another

---

[1] https://github.com/szmer/spelling_correction_modules.

angle, searching for words that seem semantically off in context may be a good heuristic for detecting errors that are not nonword (that is, they lead to wrong forms appearing in text which are nevertheless in-vocabulary).

The good performance of recurrent network methods is hardly a surprise, given observed effectiveness of neural networks in many NLP tasks in the recent decade. It seems that bidirectional LSTM augmented with ELMo may already hit the limit for correcting Polish spelling errors without contextual information. While it improves accuracy in comparison to LSTM initialized with random noise, it makes the test cross-entropy slightly worse, which hints at overfitting. The perplexity measures actually increase sharply for more sophisticated architectures. Perplexity should show how little probability is assigned by the model to true answers. We measure it as

$$\text{perplexity}(P, x) = 2^{-\frac{1}{N}\sum_{i \leqslant N}\log P(x_i)},$$

where x is a sequence of N characters, forming the correct version of the word, and $P(x_i)$ is the estimated probability of the ith character, given previous predicted characters and the incorrect form. The observed upward tendency of perplexity for increasingly accurate models is most likely due to more refined predicted probability distributions, which go beyond just assigning the bulk of probability to the best answer.

Interesting insights can be gained from weights assigned by optimization to layers of ELMo network, which are taken as the word form embedding (Table 2). The first layer, and the one that is nearest to input of the network, is given relatively the least importance, while the middle one dominates both others taken together. This suggests that in error correction, at least for Polish, the middle level of morphemes and other characteristic character chunks is more important than phenomena that are low-level or tied to some specific words. This observation should be taken into account in further research on practical solutions for spelling correction.

6   Conclusion

Among the methods tested the bidirectional LSTM, especially initialized by ELMo embeddings, offers the best accuracy and raw performance. Adding ELMo to a straightforward PyTorch implementation of LSTM may be easier now than at the time of performing our tests, as since then the authors of ELMoForManyLangs package [3] improved their programmatic interface. Other forms of contextualized embeddings for Polish, such as BERT, are getting traction for Polish [18], and they can also contain morphological information [6].

However, if a more interpretable and explainable output is required, some version of vector distance combined with edit distance may be the best direction. It should be noted that this method produces multiple candidate corrections with their similarity scores, as opposed to only one "best guess" correction that can be obtained from a character-based LSTM. This is important in applications where it is up to humans to the make the final decision, and they are only to be aided by a machine.

It is desirable for further research to expand the corpus material into a wider and more representative set of texts. Nevertheless, the solution for any practical case has to be tailored to its characteristic error patterns. Works on language correction for English show that available corpora can be "boosted" [7], i.e. expanded by generating new errors consistent with a generative model inferred from the data. This may greatly aid in developing models that are dependent on learning from error corpora.

A deliberate omission in this paper are the elements accompanying most real-word error correction solutions. Some fairly obvious approaches to integrating evidence from context include n-grams and Markov chains, although the possibility of using measurements in spaces of semantic vectors was already mentioned in this article. Similarly, non-word errors can be easily detected with comparing tokens against reference vocabulary, but in practice one should have ways of detecting mistakes masquerading as real words and fixing bad segmentation (tokens that are glued together or improperly separated). Testing how performant are various methods for dealing with these problems in Polish language is left for future research.

References

1. Apache Software Foundation: pylucene (2019). https://lucene.apache.org/pylucene/
2. Araki, T., Ikehara, S., Tsukahara, N., Komatsu, Y.: An evaluation to detect and correct erroneous characters wrongly substituted, deleted and inserted in Japanese and English sentences using Markov models. In: Proceedings of the 15th Conference on Computational Linguistics-Volume 1, pp. 187–193. Association for Computational Linguistics (1994)
3. Che, W., Liu, Y., Wang, Y., Zheng, B., Liu, T.: Towards better UD parsing: deep contextualized word embeddings, ensemble, and treebank concatenation. CoRR abs/1807.03121 (2018)
4. Dronen, N.A.: Correcting writing errors with convolutional neural networks. Ph.D. thesis, 200 University of Colorado at Boulder (2016)
5. Dzieciątko, M., Spinczyk, D., Borowik, P.: Correcting polish bigrams and diacritical marks. In: Pietka, E., Badura, P., Kawa, J., Wieclawek, W. (eds.) Information Technology in Biomedicine, pp. 338–348. Springer, Cham (2019). https://doi.org/10.1007/978-3-030-23762-2_30
6. Edmiston, D.: A systematic analysis of morphological content in BERT models for multiple languages. CoRR abs/2004.03032 (2020). https://arxiv.org/abs/2004.03032
7. Ge, T., Wei, F., Zhou, M.: Reaching human-level performance in automatic grammatical error correction: an empirical study. CoRR abs/1807.01270 (2018)
8. Grundkiewicz, R.: Automatic extraction of Polish language errors from text edition history. In: Habernal, I., Matoušek, V. (eds.) Text, Speech, and Dialogue, pp. 129–136. Springer, Heidelberg (2013). https://doi.org/10.1007/978-3-642-40585-3_17
9. Grundkiewicz, R., Junczys-Dowmunt, M.: The wiked error corpus: a corpus of corrective wikipedia edits and its application to grammatical error correction. In: PolTAL (2014)

10. Hochreiter, S., Schmidhuber, J.: Long short-term memory. Neural Comput. **9**(8), 1735–1780 (1997). https://doi.org/10.1162/neco.1997.9.8.1735. http://dx.doi.org/10.1162/neco.1997.9.8.1735

11. Kieraś, W., Woliński, M.: Słownik gramatyczny języka polskiego - wersja internetowa. Język Polski **XCVII**(1), 84–93 (2017)

12. Kingma, D.P., Ba, J.: Adam: a method for stochastic optimization. CoRR abs/1412.6980 (2014). http://arxiv.org/abs/1412.6980

13. Kukich, K.: Techniques for automatically correcting words in text. ACM Comput. Surv. (CSUR) **24**(4), 377–439 (1992)

14. Levenshtein, V.: Binary codes capable of correcting deletions, insertions and reversals. Sov. Phys. Doklady **10**, 707 (1966)

15. Liang, Z., Youssef, A.: Performance benchmarking of automated sentence denoising using deep learning. In: 2020 IEEE International Conference on Big Data (Big Data), pp. 2779–2784 (2020). https://doi.org/10.1109/BigData50022.2020.9377985

16. Mikolov, T., Sutskever, I., Chen, K., Corrado, G.S., Dean, J.: Distributed representations of words and phrases and their compositionality. In: Burges, C.J.C., Bottou, L., Welling, M., Ghahramani, Z., Weinberger, K.Q. (eds.) Advances in Neural Information Processing Systems, vol. 26, pp. 3111–3119. Curran Associates, Inc. (2013). http://papers.nips.cc/paper/5021-distributed-representations-of-words-and-phrases-and-their-compositionality.pdf

17. Miłkowski, M.: Developing an open-source, rule-based proofreading tool. Softw. Pract. Exp. **40**(7), 543–566 (2010). https://doi.org/10.1002/spe.971. https://onlinelibrary.wiley.com/doi/abs/10.1002/spe.971

18. Mroczkowski, R., Rybak, P., Wróblewska, A., Gawlik, I.: HerBERT: efficiently pretrained transformer-based language model for Polish. In: Proceedings of the 8th Workshop on Balto-Slavic Natural Language Processing, pp. 1–10, Kiyv, Ukraine. Association for Computational Linguistics, April 2021. https://aclanthology.org/2021.bsnlp-1.1

19. Mykowiecka, A., Marciniak, M.: Domain-driven automatic spelling correction for mammography reports. In: Advances in Soft Computing, vol. 35, pp. 521–530, April 2007

20. Mykowiecka, A., Marciniak, M., Rychlik, P.: Testing word embeddings for Polish. Cognit. Stud. Études Cognit. **17**, 1–19 (2017). https://doi.org/10.11649/cs.1468. https://ispan.waw.pl/journals/index.php/cs-ec/article/view/cs.1468

21. Nagata, R., Takamura, H., Neubig, G.: Adaptive spelling error correction models for learner English. Proc. Comput. Sci **112**, 474–483 (2017). https://doi.org/10.1016/j.procs.2017.08.065. http://www.sciencedirect.com/science/article/pii/S1877050917314096. Knowledge-Based and Intelligent Information & Engineering Systems: Proceedings of the 21st International Conference, KES-20176-8 September 2017, Marseille, France

22. Ogrodniczuk, M., Kopeć, M.: Lexical correction of Polish Twitter political data. In: Proceedings of the Joint SIGHUM Workshop on Computational Linguistics for Cultural Heritage, Social Sciences, Humanities and Literature, pp. 115–125, Vancouver, Canada. Association for Computational Linguistics (2017). http://www.aclweb.org/anthology/W17-2215

23. Paszke, A., et al.: PyTorch: an imperative style, high-performance deep learning library. In: Wallach, H., Larochelle, H., Beygelzimer, A., d'Alché-Buc, F., Fox, E., Garnett, R. (eds.) Advances in Neural Information Processing Systems, vol. 32, pp. 8024–8035. Curran Associates, Inc. (2019). http://papers.neurips.cc/paper/9015-pytorch-an-imperative-style-high-performance-deep-learning-library.pdf

24. Peters, M.E., et al.: Deep contextualized word representations. CoRR abs/1802.05365 (2018). http://arxiv.org/abs/1802.05365

25. Schuster, M., Paliwal, K.: Bidirectional recurrent neural networks. Trans. Sig. Proc. **45**(11), 2673–2681 (1997). https://doi.org/10.1109/78.650093. http://dx.doi.org/10.1109/78.650093

26. Subieta, K.: Korekcja pojedynczych błędów w wyrazach na podstawie słownika wzorców. Informatyka **11**(2), 15–18 (1976)

27. Subieta, K.: A simple method of data correction. Comput. J. **28**(4), 372–374 (1985)

Assessment of Document Similarity Visualisation Methods

Mateusz Gniewkowski[ID] and Tomasz Walkowiak[(✉)][ID]

Wrocław University of Science and Technology,
Wybrzeże Wyspiańskiego 27, 50-370 Wrocław, Poland
{mateusz.gniewkowski,tomasz.walkowiak}@pwr.edu.pl

Abstract. The article deals with the problem of assessing a visualization of the similarity of documents. A well-known approach for showing the similarity of text documents is a scatter plot generated by projecting text documents into a multidimensional feature space and then reducing the dimensionality to two. The problem stems from the fact that there is a large set of possible document vectorization methods, dimensionality reduction methods and their hyperparameters. Therefore, one can generate many possible charts. To enable a qualitative comparison of different scatter plots, the authors propose a set of metrics that assume that the documents are labeled. Proposed measures quantify how the similarity/dissimilarity of original text documents (described by labels) is maintained within a low-dimensional space. The authors verify the proposed metrics on three corpora, seven different vectorization methods, and three reduction algorithms (PCA, t-SNE, UMAP) with many values of their hyperparameters. The results suggest that t-SNE and fastText trained on the KGR10 dataset is the best solution for visualizing the semantic similarity of text documents in Polish.

Keywords: Similarity visualization · Dimensionality reduction · Document embedding · NLP

1 Introduction

Large text corpora are the basic resource for many researchers in humanities and social science [20]. Therefore, there is a need to automatically categorize documents in terms of subject areas. One solution to this problem is to apply supervised text classification methods. The results reported in the literature [25, 28] are very promising, especially those based on BERT [3] deep neural networks. They show that it is possible to automatically assign text documents to subject categories. However, supervised approaches are very often hard to be applied in real-world scenarios, because in practice, most of the analysed by researcher corpora are lacking a consistent set of labels. Developing such labels is a costly process that also requires annotation rules. One could use an already trained classifier to process a new dataset, but it is highly probable that the documents that had been used for training concerned other areas. Supervised models work

© Springer Nature Switzerland AG 2022
Z. Vetulani et al. (Eds.): LTC 2019, LNAI 13212, pp. 348–363, 2022.
https://doi.org/10.1007/978-3-031-05328-3_23

well only used on texts similar to the training data. Therefore, unsupervised approaches like clustering [4,11,29] or document similarity visualisation in 2-D space [19,30] are essential in practise.

Clustering and document similarity visualisation are quite similar processes. In most cases, they are based on representing documents by multidimensional feature vectors (document vectorization), calculating the similarities or distances between those vectors, and then applying clustering [6] or dimensionality reduction algorithm [1]. Within this paper we will focus on the second problem.

The goal of similarity visualisation is to present documents (represented by feature vectors) as points on a 2D plane. Documents that are more similar should be closer together in the plot than objects that differs. Such charts allow people to easily interpret the corpus and find potential outliers. Often, one can find nonobvious relationships between groups of texts that exhibit subtle similarities hidden to the naked eye but traceable by multidimensional statistical techniques [20]. Similarity visualisation is also a very helpful tool in the process of defining labels for future supervised learning experiments.

The main problem in the application of similarity visualisation methods is the selection of the method parameters. First, there are a large number of possible techniques of representing documents by feature vectors, starting from the bag-of-words technique [26], thorough word embedding [9], to deep neural network models like ELMo [18] or BERT [3]. Next, there are many available pretrained language models. There are also many dimensionality reduction algorithms like PCA [6], t-SNE [10], or UMAP [12], and each of them has many hyperparameters. It raises the question of which combination of the above should be selected for visualization? There is no easy answer to that, because the result of the similarity visualisation is a scatter plot (see Fig. 1) that is interpreted by people. The aim of the paper is to find metrics that are consistent with human perception and allow to automatically compare different approaches of generating plots. Such assessments will not only make it possible to generate better visualizations but also will allow easier selection of any parameter (like the vectorization method) for the dataset that is yet to be labeled.

This work is an extension of the research presented in [30]. We added a new corpus, used new methods of generating document vectors, applied new methods of dimension reduction (originally we only considered t-SNE), and finally proposed a much larger set of evaluation metrics (originally 1, and now 5). The metrics, vectorization and reduction methods were evaluated on labelled (in terms of subject area) corpora in Polish.

The paper is structured as follows. In Sect. 2 we shortly describe the vectorization methods that we used to transform documents into feature space. Next, in Sect. 3 we describe the three dimensionality reduction methods that were used in experiments. Section 4 contains descriptions of the proposed evaluation metrics. In Sect. 5 we discuss the datasets and our results. Conclusions are at the end of the paper.

2 Vectorization Methods

2.1 TF-IDF

The TF-IDF method is based on the bag-of word concept [22], i.e., counting the occurrences of the most common terms (words or their n-grams) in the corpus (term frequencies). Next, these frequencies are weighted by the maximum term frequency in a document and by the inverse document frequency. In the performed experiments, we have used the 1000 most frequent terms (words or bigrams).

2.2 fastText

The big step in the area of text analysis was the introduction of the word2vec method [9]. In this approach, individual words are represented by high-dimensional feature vectors (word embeddings) trained on a large text corpus. The most common solution to generate the document features is to average vector representations of individual words. This approach is known as doc2vec [13].

Due to a large number of word forms in morphological rich languages such as Polish, there are two main approaches: to use lemmas (the text have to be lemmatized) or the word2vec extension [5] from the fastText package. The last one uses the position weights and subword information (character n-grams) that allow to generate embeddings for unseen words.

Doc2vec as well as TF-IDF ignores word order. Therefore, these methods are not aware of word contexts.

2.3 ELMo

The newest approaches in language modeling are inspired by deep learning algorithms and context-aware methods. The first successful is called ELMo [18]. ELMo word embeddings are defined by the internal states of a deep bidirectional LTSM language model (biLSTM), which is trained on a large text corpus. What is important, ELMo looks at the whole sentence before assigning an embedding to each word in it. Therefore, the embeddings are sentence aware and could solve the problem of polysemous words (words with multiple meanings). As the document feature vector, we used the average mean vector of every sentence in it. Generating sentence vectors is built into the model and consists in mean pooling of all contextualized word representations. The main problem with ELMo is its slow performance caused by the bidirectional architecture of LSTM networks.

2.4 BERT

The next step was a usage of the transformer [27] architecture for building language models. The state-of-the-the-art solution is BERT [3]. Due to its bidirectional representation, jointly built on both the left and the right context, BERT looks at the whole sentence before assigning an embedding to each word in it. As the document feature vector, we have used the CLS pooling method, i.e., the embedding of the initial CLS token.

2.5   Method Summary

The above vectorization methods, except TF-IDF, require pretrained language models. The names used in results reporting and sources of the used models are presented in Table 1.

The main drawback of ELMo and partly BERT is the requirement of using GPU even for the vector generation phase. Usage of ELMo on CPU is impractical due too very long processing time. It is slightly better in the case of BERT, but still TF-IDF and doc2vec work much faster on CPU than BERT.

Table 1. Document vectorization methods and sources of language models

Name	Method	Address
kgr10	fastText	hdl.handle.net/11321/606
kgr10-lemma	fastText	hhdl.handle.net/11321/606
fasttext	fastText	https://fasttext.cc/docs/en/crawl-vectors.html
elmo	ELMo	vectors.nlpl.eu/repository/11/167.zip
tfidf	TF-IDF	–
herbert-kgr10	BERT	https://clarin-pl.eu/dspace/handle/11321/851
herbert-base	BERT	https://huggingface.co/allegro/herbert-base-cased

3   Reduction Methods

The aim of the reduction is to present documents in the 2D plane to visualise the distances or dissimilarities between them. The distances between points should reflect similarities in the original multidimensional space of feature vectors (generated as described in Sect. 2).

There are several methods that can be used for 2D visualisation of multidimensional feature vectors. They can be divided in two categories [12]: ones preserving the distance structure within the data such as the PCA [6] or multidimensional scaling [1] and ones that preserve the local distances over the global distance like t-SNE [10], Laplacia eigenmaps, Isomap, and the newest UMAP [12]. Within this work have analysed three methods: PCA, t-SNE, and UMAP.

3.1   PCA

PCA (Principal component analysis) [17] is a traditional and widely used dimensionality reduction technique. It works by identifying the linear correlations with preserving most of the valuable information. PCA algorithm is based on the principal components of the covariance matrix – a set of vectors, the first of which best fits (explain the maximum amount of variance) the data while the rest are being orthogonal to it. To generate low dimensional space, we ignore the less significant principle components by projecting each data point.

3.2   T-SNE

T-SNE, proposed in [10], is a non-linear dimensionality reduction method. It preserves the similarity between points defined as normalised Gaussians. Therefore, it uses Euclidean distance in the original space. The bandwidth of the Gaussian is set by the bisection algorithm, in a way that the resulting perplexity is equal to some predefined value. As a result, the bandwidth, and therefore the similarity, for each point is adapted to the local density of the data. The similarities in low-dimensional space are modeled by a normalised t-Student distribution. The t-SNE method minimises the Kullback-Leibler divergence between the similarities in both spaces with respect to the locations of the points in the low-dimensional space.

3.3   UMAP

Uniform manifold approximation and projection (UMAP) [12] constructs a high-dimensional graph representation of the data and next optimizes a low-dimensional graph to be as structurally similar as possible. It assumes that the data is uniformly distributed on Riemannian manifold which is locally connected [12]. UMAP high-dimensional graph edges represent the likelihood of connection of each pair of data points. UMAP connects (edges) only points for which the local point radius overlaps. Each point local radius is set based on the distance to each point's and number of neighbours. The size of point neighbours is the method hyperparameter. In many papers, it was shown that UMAP outperforms other methods (including PCA and t-SNE) [12,16,24].

4   Evaluation Methods

The main problem addressed in the paper is the measurement of the document visualization quality. Corpus of text is mapped to the multidimensional space by one of the methods described in Sect. 2. Next, this set of high-dimensional vectors (each representing a single document) is projected to a 2D space using one of the methods described in Sect. 3. As a result, we obtain plots like those presented in Fig. 1. The question is: which combination of document feature vector generation methods, reduction methods, and their hyperparameters should be used? Or, in other words, how to quantify individual plots to be able to choose the best one. We need a metric that allows to compare visualisation results automatically, a metric that promotes results with well-separated classes.

This problem does not have a common quality metric. Therefore, in this section, we propose five different methods. All these metrics are based on the assumption that for method comparison purposes we have a set of labels assigned to the documents. We assume that the documents within the same label are similar (at least some of them, a group does not have to be unimodal) and documents assigned to two different labels are different. In other words, the points representing the same label documents should be placed in low-dimensional space close to each other and far away from points representing other classes.

4.1   Closest Match (CM)

In [30], we proposed a simple coherence score defined as an average (over all points) of a number of k-nearest neighbours belonging to the same class, i.e.:

$$\frac{1}{nk}\sum_{p}\sum_{o\in N_k(p)} I(c(p) == c(o)), \tag{1}$$

where n is a number of points (documents), I is the identity function, $N_k(p)$ is the neighbourhood of p defined by the k closest points (using euclidean distance in low dimensional space), and $c(p)$ is a class of point p. The method is parameterized by k - a number of nearest neighbours used in analysis. It measures how many neighbours (in average) of a given point belong to the same label. In our experiments, we used k equaled to 10. In [30] we shown that the value of the metric depends on k in a similar way regardless the used dataset. Therefore, the value of k is not essential (except the extreme values) in the case of comparison.

4.2   KNN

To measure the quality of the reduction, one could also use any classifier that is trained using two-dimensional data. Therefore, we generated ten folds (90% of the data were used for training) using the stratified K-fold strategy and calculated the average accuracy (Exact Match Ratio, MR) as a final score. The formula is as follows:

$$MR = \frac{1}{n}\sum_{i=1}^{n} I(y_i == \hat{y}_i), \tag{2}$$

$$KNN = \frac{1}{K}\sum_{k} MR_k, \tag{3}$$

where K is the number of folds, I is the indicator function, y_i is a true label of a sample i and $\hat{y}_i$ is predicted label of the same sample.

We decided to use a simple KNN classifier (using ten nearest neighbours), which makes the score similar to the one from the previous section. However, almost any classifier could be used here. We have originally started with the multilayer perceptron (MLP) [6]. However, MLP has a much higher computational cost compared to KNN, and within a preliminary experiment gave close to KNN results. Similar approach, i.e., the KNN classifier, was proposed in [12].

4.3   ARI

Instead of using a classification algorithm, it is possible to use any clustering method. If the clusters obtained in a lower space match a ground-truth label, then the clusters should be visually separated. We use the adjusted rand index [7] to calculate the correspondence.

$$n_{ij} = |X_i \bigcap Y_j|, \quad a_i = \sum_{j=1}^{g} n_{ij}, \quad b_j = \sum_{i=1}^{p} n_{ij}, \tag{4}$$

$$ARI = \frac{\sum_{ij} \binom{n_{ij}}{2} - [\sum_i \binom{a_i}{2} \sum_j \binom{b_j}{2}]/\binom{n}{2}}{\frac{1}{2}[\sum_i \binom{a_i}{2} + \sum_j \binom{b_j}{2}] - [\sum_i \binom{a_i}{2} \sum_j \binom{b_j}{2}]/\binom{n}{2}}, \tag{5}$$

where $X = \{X_1, X_2, ..., X_g\}$ defines ground truth labels, $Y = \{Y_1, Y_2, ..., Y_p\}$ defines predicted labels (clusters obtained by the algorithm), g defines number of true labels and p of predicted ones (it is a parameter in the clustering algorithm that we established to be equal g). In the performed experiments we use agglomerative clustering [2] with ward linkage.

4.4   Internal Similarity (INT-SIM)

The next metric we propose to use is the mean distance between samples in the same cluster converted to a similarity measure, i.e.:

$$D(X) = \frac{1}{|C_x|} \sum_{i,j \in C_x, x \in X} d(x_i, x_j), \quad T = \frac{1}{|S|} \sum_{k=1}^{|S|} D(S_k), \tag{6}$$

$$\text{INT-SIM} = \frac{T}{T+1}, \tag{7}$$

where d is a distance between two points, C_x is a set of pairs of points in a cluster X and S is a set of clusters defined by labels. The maximum value of this score is obtained when samples from the same label are gathered in a single coordinate, which can be considered as a defect. There is also a problem with clusters that are made up of subgroups that occur in different places (for example, PRESS data in Fig. 1). The score will be lower in this scenario. To overcome this, we propose to use DBSCAN [23] algorithm for each label in the data to obtain subgroups. In the performed experiments the *eps* parameter of DBSCAN was set to the tenth percentile of a distance distribution between samples in the given group.

4.5   External Dissimilarity (EXT-DIS)

And finally, we propose the external dissimilarity score defined as a normalized (divided by the greatest) mean distance between different groups. We calculate the distance between groups as the average distance between samples in one and the other cluster. The final formula is as follows:

$$D(X, Y) = \frac{1}{|X||Y|} \sum_{x_i \in X} \sum_{y_i \in Y} d(x_i, y_i), \tag{8}$$

$$\text{EXT-DIS} = \frac{\overline{D(X,Y)}}{\max_{i,j \in C}(D(X_i, X_j))}, \tag{9}$$

where X and Y are clusters defined by labels, C is a set of pairs of clusters. The score promotes data reduction with similar distances between groups and might lead to solutions with points from the same group concentrated in one place (similarly to the External Dissimilarity).

5   Experiments

5.1   Datasets

In our experiments, we used three collections of text documents in Polish: *Wiki*, *Press*, and *Qual*. All were labelled in terms of subject area, therefore we can assume that the similarity analysed by the metrics introduced in Sect. 4 is semantic one.

The *Wiki* corpus consists of articles extracted from the Polish language Wikipedia. It was created by merging two publicly available collections [14] and [15]. The original corpus is labeled by 34 subject categories. For clarity of the presented pictures, we have selected a subset of 10 labels, namely: computers, music, aircraft, games, football, cars, chess, coins, shipping, and animation. The resulting corpus consists of 2, 959 elements.

The second corpus, *Press* [31] consists of Polish press news. There are 6, 564 documents in total in this corpus. The texts were assigned by the press agency to 5 subject categories (diplomacy, sport, disasters, economy, business, and transportation). All subject groups are very well separated and each group contains a reasonably large number of members (ca. 1, 300 documents per label) without big differences among label sizes.

The last data set, *Qaul* [11] includes documents containing descriptions of qualifications from a Polish public register of the Integrated Qualifications System and descriptions of degrees from Polish universities. The descriptions mainly consist of so-called learning outcomes statements, which characterize the knowledge, skills, and attitudes required to obtain a given qualification or degree. The data were manually labeled. The labels denote the sectors to which the qualifications belong. Similarly to *WIKI* corpus, we have selected a subset of seven labels, namely: economy, biology, industry, electronics, music, machines, and architecture. The final corpus consists of 1, 419 documents.

5.2   Results

For every corpus and the previously described vectorization and reduction methods (and many different hyperparameters of the last ones), we generated a chart. In total, we obtained almost 3 500 two-dimensional scatter plots and evaluated them using our metrics. To measure the quality and effectiveness of them, we

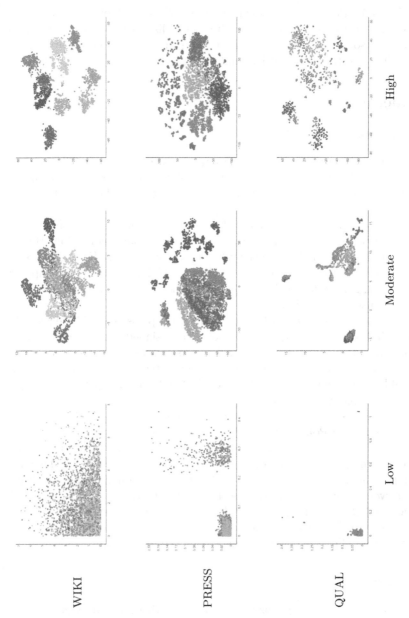

Fig. 1. Exemplary plots related to the highest, lowest, and middle (median) KNN score for every dataset.

Table 2. Scores related to plots in Fig. 1

Dataset	Rating	Scores					Method	Model
		KNN	ARI	CLOSEST MATCH	INT-SIM	EXT-DIS		
WIKI	Low	0.12	0.00	0.20	0.34	0.29	PCA	herbert-base
	Moderate	0.78	0.38	0.76	0.33	0.16	UMAP	tfidf
	High	0.95	0.74	0.94	0.09	0.14	T-SNE	kgr10
PRESS	Low	0.34	0.02	0.36	0.98	0.55	PCA	herbert-base
	Moderate	0.87	0.21	0.85	0.06	0.21	T-SNE	herbert-base
	High	0.94	0.31	0.92	0.03	0.38	T-SNE	kgr10
QUAL	Low	0.37	0.08	0.42	0.97	0.24	PCA	kgr10-lemma
	Moderate	0.76	0.22	0.72	0.55	0.20	UMAP	elmo
	High	0.86	0.32	0.82	0.12	0.16	T-SNE	elmo

Table 3. Statistics of all analysed metrics (scores) for all three corpora.

Method	Score	PRESS			QUAL			WIKI		
		MAX	MIN	AVG	MAX	MIN	AVG	MAX	MIN	AVG
PCA	KNN	0.66	0.34	0.55 ± 0.10	0.61	0.37	0.47 ± 0.08	0.50	0.12	0.29 ± 0.12
	ARI	0.41	0.01	0.14 ± 0.10	0.34	0.05	0.18 ± 0.10	0.25	0.00	0.09 ± 0.08
	CM	0.64	0.36	0.54 ± 0.08	0.59	0.38	0.48 ± 0.06	0.47	0.20	0.31 ± 0.09
	INT-SIM	0.98	0.32	0.85 ± 0.22	0.99	0.43	0.88 ± 0.18	0.95	0.28	0.80 ± 0.23
	EXT-DIS	0.61	0.26	0.43 ± 0.11	0.40	0.13	0.26 ± 0.07	0.50	0.06	0.29 ± 0.13
UMAP	KNN	0.89	0.58	0.75 ± 0.10	0.80	0.53	0.67 ± 0.08	0.88	0.16	0.51 ± 0.22
	ARI	0.39	0.00	0.19 ± 0.08	0.80	0.05	0.31 ± 0.14	0.52	0.01	0.17 ± 0.15
	CM	0.87	0.58	0.73 ± 0.09	0.77	0.52	0.66 ± 0.07	0.85	0.23	0.50 ± 0.20
	INT-SIM	0.55	0.13	0.29 ± 0.08	0.75	0.24	0.44 ± 0.11	0.59	0.15	0.29 ± 0.10
	EXT-DIS	0.71	0.20	0.37 ± 0.08	0.45	0.07	0.22 ± 0.07	0.53	0.10	0.28 ± 0.11
T-SNE	KNN	0.94	0.71	0.88 ± 0.07	0.86	0.62	0.78 ± 0.07	0.95	0.34	0.78 ± 0.20
	ARI	0.59	0.04	0.25 ± 0.10	0.80	0.04	0.34 ± 0.13	0.74	0.01	0.41 ± 0.24
	CM	0.92	0.71	0.86 ± 0.06	0.85	0.61	0.77 ± 0.08	0.94	0.34	0.76 ± 0.20
	INT-SIM	0.14	0.01	0.06 ± 0.02	0.20	0.07	0.13 ± 0.03	0.16	0.00	0.07 ± 0.03
	EXT-DIS	0.62	0.16	0.33 ± 0.10	0.37	0.10	0.20 ± 0.04	0.38	0.07	0.17 ± 0.07

conducted several experiments. The first of them are based on a visual assessment of the correlation between the proposed metrics and the actual plots. In Fig. 1, we present three pictures for each of the corpus that had the lowest, highest, and middle KNN score. In this could be noticed that the top-rated figures contain visually clear and well-separated groups, while the worst-rated ones are rather indistinct. The behavior of KNN measure is following the requirements stated in Sect. 4. Table 2 shows all examined metrics for all plots from Fig. 1. The KNN, ARI, and CLOSEST MATCH (CM) scores act similarly (although the overall promotes different solutions), but the tendency for the remaining scores is the opposite. This means that the best solutions are those where the samples are not too close to each other (INT-SIM) and the distances between pairs of groups are not similar (EXT-DIS). Those two methods should not be used as an out-of-the-box evaluation method.

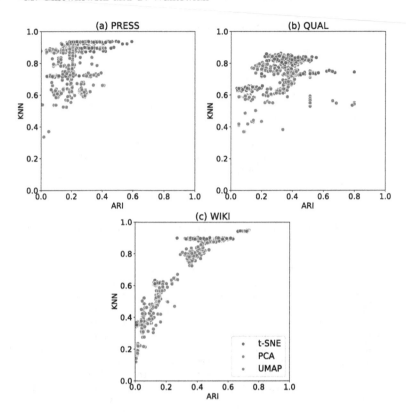

Fig. 2. Values of KNN and ARI metrics (larger values are better) for all methods and data sets. A point represents a single experiment. Experiments differ by data set, document vector generation method and 2-D projection method, and their parameters.

Table 3 shows statistics (average, standard deviation, maximum, and minimum) for each metric and corpus. The statistics are calculated over different metrics, vectorization methods and hyperparameters of reduction algorithms. The Fig. 2 presents the results for each experiment as the relation between KNN and ARI metric for every analysed document in each of the three corpora.

First, it could be noticed that PCA gives the worst results and t-SNE the best. Mind that this statement only applies to the visual aspect of the method. In this paper, we do not address the problem of how well a given method preserves the features of a high-dimensional data. We only focus on the similarity and dissimilarity between documents. Surprisingly, the results for t-SNE outperform UMAP. In the literature, UMAP is considered as a method that outperforms t-SNE [12, 24].

Moreover, we can notice that there is a strong correlation between KNN and ARI for $QUAL$ dataset, but for the other two corpora the relation disappears. Probably, this is due to the existence of different subgroups within the same label (multimodal data within each label). While KNN takes it into consideration, the ARI score is reduced because it is based on a clustering algorithm with a fixed number of classes (that matches the number of labels in the ground truth). The nature of the used algorithm (i.e., agglomerative clustering) might cause wrong assignments in the low dimensional space. The ARI score should rather be used for uni-modal labels (visually one label point should not occur in different areas) or the number of groups in clustering should be at least twice than the number of labels. Since PCA results are not significant, we focus only on t-SNE and UMAP in further analysis. Figure 3 shows correlation between KNN and ARI metric for left reduction methods. It is not clear to determine which of the vectorization methods works the best. It strongly depends on the corpus (the results follow the intuitive statement that document similarity is subjective) and used a reduction method. However, some tendencies can be noticed. First of all, kgr10 (fastText model trained on the KGR10 corpus) is always in the top three. Secondly, we could notice a high position of a simple and old-fashioned TF-IDF method (red circles), especially for KNN metric. It could be explained by the existence of key-words in each label. For example "aircrafts" (label from $WIKI$) can be simply classified by an occurrence of words such as "aircraft" or "plane". Moreover, the results also suggest that the dataset used for training the language models has a big influence on results. It could be noticed comparing the results achieved by models trained on the KGR10 corpus [8] (pink and violet) to the results obtained by default models (orange and blue respectively). KGR10 results outperforms the base models in the case of fastText and BERT models. Moreover, we see that fastText outperforms (i.e., pink) the Bert based vectorization (violet). The fact that BERT based methods are not suitable for similarity/distance calculation is well known in the literature [21]. The surprising results are for elmo (brown). ELMo have a bad performance for $WIKI$ dataset compared to quite good results for $PRESS$ and $QUAL$. The other interesting pattern noticeable in the results is the relatively small dependency of the reduction method hyperparameters (i.e., perplexity, learning rate and number of iterations for t-SNE and number of neighbours and minimal distance for UMAP) on the KNN score. The points in the same color represent results from the same vectorization method but with varying values of the reduction method hyperparameters. It could be noticed that they group and even make vertical lines in case of t-SNE. It is probably due to the fact that the perplexity in the case of t-SNE and k-neighbours in the case of UMAP have a big influence on creating subgroups in each label. And as it was already stated, the KNN is less subjective to this feature than ARI.

Comparing the results of t-SNE and UMAP, we can notice that the achieved plots have some similarities, but they differ in details. It shows that each method focuses on different aspects of multidimensional space.

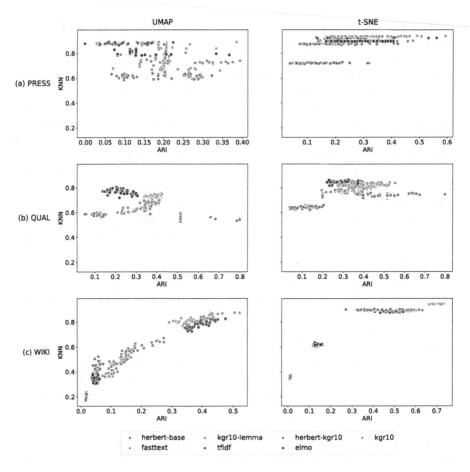

Fig. 3. Values of KNN and ARI metrics (larger values are better) for all three datasets. Points differ by a vector generation method (UMAP and T-SNE) and used hyperparameters.

6 Conclusion

In this work, we proposed a method to assess the visual quality of two-dimensional plots of document similarity obtained using the most popular dimensionality reduction methods. We propose five metrics for quantification of this quality. Based on the testes performed on the three corpora, we conclude that the classifier (KNN), clusterization (ARI) based approaches, or simple score Closest Match can actually determine which of the generated figures better preserves the information of document similarity. This allows for an automatic search of the parameter space to find the optimal ones. We also showed which of the used vectorization methods perform better in the task. The results suggest that fast-Text based approaches outperform the BERT ones and that the language models for Polish trained on KGR10 outperform others in the analysed problem. Even though we focused on texts in Polish, our approach can be used in virtually any problem in the field of data mining.

Although we have shown a convenient way to evaluate the appearance of the plots, there are several aspects that require further research. First, although we believe that the correlation between the human perspective and our scores is true, it is necessary to verify this thesis with a larger number of people using a survey. We hope that having plots evaluated by people, we will be able to suggest a combination of proposed scores as a final method (especially INT-SIM and EXT-DIS which cannot be used alone). Next, we focused on solving the problem with the assumption that ground-truth labels are given. This is not always the case in real-world scenarios, but makes just defining the goal difficult. Building such measures would probably require using also context information (other reductions) and not individual plots.

Acknowledgement. Financed by the European Regional Development Fund as a part of the 2014–2020 Smart Growth Operational Programme, CLARIN - Common Language Resources and Technology Infrastructure, project no. POIR.04.02.00-00C002/19.

References

1. Borg, I., Groenen, P.J., Mair, P.: Applied Multidimensional Scaling and Unfolding, 2nd edn. Springer, Cham (2018). https://doi.org/10.1007/978-3-319-73471-2
2. Day, W.H.E., Edelsbrunner, H.: Efficient algorithms for agglomerative hierarchical clustering methods. J. Classif. **1**(1), 7–24 (1984). https://doi.org/10.1007/BF01890115
3. Devlin, J., Chang, M.W., Lee, K., Toutanova, K.: BERT: pre-training of deep bidirectional transformers for language understanding. arXiv preprint arXiv:1810.04805 (2018)
4. Eder, M., Piasecki, M., Walkowiak, T.: An open stylometric system based on multilevel text analysis. Cognit. Stud. Etudes Cognit. **17** (2017). https://doi.org/10.11649/cs.1430
5. Grave, E., Bojanowski, P., Gupta, P., Joulin, A., Mikolov, T.: Learning word vectors for 157 languages. In: Proceedings of the International Conference on Language Resources and Evaluation (LREC 2018), pp. 3483–3487 (2018)
6. Hastie, T.J., Tibshirani, R.J., Friedman, J.H.: The Elements of Statistical Learning: Data Mining, Inference, and Prediction. Springer Series in Statistics, Springer, New York (2009). Autres impressions : 2011 (corr.), 2013 (7e corr.)
7. Hubert, L., Arabie, P.: Comparing partitions. J. Classif. **2**(1), 193–218 (1985). https://doi.org/10.1007/BF01908075
8. Kocoń, J., Gawor, M.: Evaluating KGR10 Polish word embeddings in the recognition of temporal expressions using BiLSTM-CRF. CoRR abs/1904.04055 (2019). http://arxiv.org/abs/1904.04055
9. Le, Q., Mikolov, T.: Distributed representations of sentences and documents. In: International Conference on Machine Learning, pp. 1188–1196 (2014)
10. van der Maaten, L., Hinton, G.: Visualizing data using t-SNE. J. Mach. Learn. Res. **9**, 2579–2605 (2008). http://www.jmlr.org/papers/v9/vandermaaten08a.html

11. Marcińczuk, M., Gniewkowski, M., Walkowiak, T., Będkowski, M.: Text document clustering: Wordnet vs. TF-IDF vs. word embeddings. In: Proceedings of the 11th Global Wordnet Conference, pp. 207–214. Global Wordnet Association, University of South Africa (UNISA), January 2021. https://www.aclweb.org/anthology/2021.gwc-1.24

12. McInnes, L., Healy, J., Melville, J.: UMAP: uniform manifold approximation and projection for dimension reduction (2020)

13. Mikolov, T., Joulin, A., Grave, E., Bojanowski, P., Mikolov, T.: Bag of tricks for efficient text classification. In: Proceedings of the 15th Conference of the European Chapter of the Association for Computational Linguistics: Volume 2, Short Papers, pp. 427–431. Association for Computational Linguistics (2017). http://aclweb.org/anthology/E17-2068

14. Młynarczyk, K., Piasecki, M.: Wiki test - 34 categories (2015). http://hdl.handle.net/11321/217. CLARIN-PL digital repository

15. Młynarczyk, K., Piasecki, M.: Wiki train - 34 categories (2015). http://hdl.handle.net/11321/222. CLARIN-PL digital repository

16. Parra-Hernández, R.M., Posada-Quintero, J.I., Acevedo-Charry, O., Posada-Quintero, H.F.: Uniform manifold approximation and projection for clustering taxa through vocalizations in a neotropical passerine (rough-legged tyrannulet, phyllomyias burmeisteri). Animals 10(8) (2020). https://doi.org/10.3390/ani10081406. https://www.mdpi.com/2076-2615/10/8/1406

17. Pearson, K.: LIII. on lines and planes of closest fit to systems of points in space. Lond. Edinb. Dublin Philos. Mag. J. Sci. 2(11), 559–572 (1901)

18. Peters, M.E., et al.: Deep contextualized word representations. In: Proceedings of NAACL (2018)

19. Piasecki, M., Walkowiak, T., Eder, M.: Open stylometric system webSty: integrated language processing, analysis and visualisation. CMST 24, 43–58 (2018). https://doi.org/10.12921/cmst.2018.0000007

20. Pol, M., Walkowiak, T., Piasecki, M.: Towards CLARIN-PL LTC digital research platform for: depositing, processing, analyzing and visualizing language data. In: Kabashkin, I., Yatskiv, I., Prentkovskis, O. (eds.) RelStat 2017. LNNS, vol. 36, pp. 485–494. Springer, Cham (2018). https://doi.org/10.1007/978-3-319-74454-4_47

21. Reimers, N., Gurevych, I.: Sentence-BERT: sentence embeddings using Siamese BERT-networks. In: Proceedings of the 2019 Conference on Empirical Methods in Natural Language Processing and the 9th International Joint Conference on Natural Language Processing (EMNLP-IJCNLP), Hong Kong, China, pp. 3982–3992. Association for Computational Linguistics, November 2019. https://doi.org/10.18653/v1/D19-1410. https://aclanthology.org/D19-1410

22. Salton, G., Buckley, C.: Term-weighting approaches in automatic text retrieval. Inf. Process. Manage. 24(5), 513–523 (1988)

23. Schubert, E., Sander, J., Ester, M., Kriegel, H.P., Xu, X.: DBSCAN revisited, revisited: why and how you should (still) use DBSCAN. ACM Trans. Database Syst. 42(3) (2017). https://doi.org/10.1145/3068335. https://doi.org/10.1145/3068335

24. Smets, T., et al.: Evaluation of distance metrics and spatial autocorrelation in uniform manifold approximation and projection applied to mass spectrometry imaging data. Analyt. Chem. 91 (2019). https://doi.org/10.1021/acs.analchem.8b05827

25. Sun, C., Qiu, X., Xu, Y., Huang, X.: How to fine-tune BERT for text classification? In: Sun, M., Huang, X., Ji, H., Liu, Z., Liu, Y. (eds.) Chinese Computational Linguistics, pp. 194–206. Springer, Cham (2019). https://doi.org/10.1007/978-3-030-32381-3_16

26. Torkkola, K.: Discriminative features for textdocument classification. Formal Pattern Anal. Appl. **6**(4), 301–308 (2004). https://doi.org/10.1007/s10044-003-0196-8

27. Vaswani, A., et al.: Attention is all you need. In: Guyon, I., Luxburg, U.V., et al. (eds.) Advances in Neural Information Processing Systems, vol. 30. Curran Associates, Inc. (2017). https://proceedings.neurips.cc/paper/2017/file/3f5ee243547dee91fbd053c1c4a845aa-Paper.pdf

28. Walkowiak, T.: Subject classification of texts in Polish - from TF-IDF to transformers. In: Zamojski, W., Mazurkiewicz, J., Sugier, J., Walkowiak, T., Kacprzyk, J. (eds.) Theory and Engineering of Dependable Computer Systems and Networks, pp. 457–465. Springer, Cham (2021). https://doi.org/10.1007/978-3-030-76773-0_44

29. Walkowiak, T., Gniewkowski, M.: Evaluation of vector embedding models in clustering of text documents. In: Proceedings of the International Conference on Recent Advances in Natural Language Processing (RANLP 2019), pp. 1304–1311. INCOMA Ltd., Varna, September 2019. https://aclanthology.org/R19-1149

30. Walkowiak, T., Gniewkowski, M.: Visualisation of document similarities based on word embedding models for Polish, pp. 148–151. Wydawnictwo Nauka i Innowacje, Poznań (2019)

31. Walkowiak, T., Malak, P.: Polish texts topic classification evaluation. In: Proceedings of the 10th International Conference on Agents and Artificial Intelligence - Volume 2: ICAART, pp. 515–522. INSTICC, SciTePress (2018). https://doi.org/10.5220/0006601605150522

Legal Aspects

Legal Regime of the Language Resources in the Context of the European Language Technology Development

Ilya Ilin$^{(\boxtimes)}$

University of Tartu, Näituse 20, 50409 Tartu, Estonia
ilya.ilin@ut.ee

Abstract. The rapid technological development has created new opportunities for language digitalization and the development of language technology applications. The core element of language technology is language resources, which is in a broad sense, can be considered as a scope of the databases that consists of the myriad of texts both in oral and written forms and used in the machine-learning algorithm. The creation of language resources requires two processes: the first one is language digitalization, meaning the transformation of the speech and texts into the machine-responsible form. The second process refers to text mining, which analyzes data by using a machine-learning algorithm. Adoption of the General Data Protection Regulation (GDPR) and Directive on copyright and related rights in the Digital Single Market (DSM Directive) has been building a renewed legal framework that addresses the demands of the digital economies and unseals challenges, opens prospects for further development. We examine the language resources from two perspectives. Firstly, the language resources are considered a database covered by the protection regulation (the person's rights who created the LR database). Within the second perspective, the legal analysis focuses on the materials used for the language resource creation (data subject's rights, copyright, related rights). The result of the research can be used for further legal investigations and policy design in the field of language technology development.

Keywords: Intellectual property protection · Copyright · Language technology · Text mining

1 Introduction

Nowadays the language technology (LT) is strongly integrated into our life. The examples of the LT applications can be easily found practical in every area (e.g., machine translation systems, spelling and grammar checking applications, chat-bots, apps for reading, etc.). The LT applications are solving the various tasks from ensuring security (e.g., biometric voice systems) to providing health information (e.g., speech analysis systems) and controlling sophisticated machines and mechanisms (e.g., vehicles voice control systems).

© Springer Nature Switzerland AG 2022
Z. Vetulani et al. (Eds.): LTC 2019, LNAI 13212, pp. 367–376, 2022.
https://doi.org/10.1007/978-3-031-05328-3_24

The core element of all LT applications is language resources (LR), which is a broad sense that can be considered as a scope of the databases that consists of the myriad of texts both in oral and written forms (datasets) and used in the machine-learning algorithm [8]. The development of the LT began with the database only of two hundred and fifty words and six grammar rules. This database was used in 1954 in the Georgetown-IBM experiment. Under this experiment, the International Business Machines company (IBM), in collaboration with Georgetown University (U.S.), publicly demonstrated the machine text translation of the approximately sixty Russian sentences from Russian into the English language [3]. Today these databases consist of millions to billions of words and require more complicated algorithms for their analysis.

The LR and their quantitative and quality characteristics have a significant meaning for LT development. Therefore, the question of LT creation should be investigated. From the technical perspective, the creation of LR is done by executing two processes: the first one is language digitalization – collecting and further transformation of the speech and texts into the machine-responsible form. The second process refers to text mining - data analysis by using a machine-learning algorithm [6]. From the legal perspective, the LR can be considered as a database and thus covered by rights corresponding to the database regulation (e.g., the request of LR developer), or the main focus can be made on materials that were used for the LR creation and therefore covered by data subject's rights, copyright, and related rights. This dual legal nature of the LR creates a scope of legal concerns related to the necessity to comply with both tiers of rights.

The present research aims to identify legal challenges appearing within EU jurisdiction, referring to the LR database creation, personal data, and intellectual property protection, and develop practical recommendations on minimizing those risks. We examine the language resources from two perspectives. Within the first perspective, the legal analysis focuses on the materials used for the language resource creation (data subject's rights, copyright, related rights). From the second perspective, the language resources are considered a database covered by the database protection regulation (the person's rights who created the language resources). The structure of the paper is designed with regard to the approaches mentioned above. In the first part of the paper, the main focus is on the materials (data) and process of the LR creation. In the second part, issues related to database protection are investigated.

The result of the research can be used for further legal exploration and policy design in the field of language technology development.

2   Models for Language Resource Creation

The model for LR creation largely depends on the type and characteristics of the materials (raw data) used to create the LR database. Taking into consideration the adoption of the General Data Protection Regulation (GDPR)[1] and Directive on copyright and related

---

[1] Regulation (EU) 2016/679 of the European Parliament and of the Council of 27 April 2016 on the protection of natural persons with regard to the processing of personal data and on the free movement of such data and repealing Directive 95/46/EC (General Data Protection Regulation), dated 27 April 2016, with entry into force on 25 May 2018 <https://eur-lex.europa.eu/eli/reg/2016/679/oj> (accessed 14 October 2021).

rights in the Digital Single Market (DSM Directive)[2], the main legal concerns lie in the field of personal data and intellectual property (IP) protection. The materials used for LR creation can be simultaneously covered by the IP rights and data subject rights. Analysis of the raw data used for the LR creation and defining the tier of rights that covers this data have significant meaning in terms of identifying the model that can be applied for the LR creation. Generally, two models depending on the protected materials: the "contract model" ("consent model") and the "exception model." In further sections, the models are observed from the perspective of the copyright and related rights protection (Sect. 2.1) and the perspective of the data protection regulation (Sect. 2.2).

2.1 The Usage of the LR Materials Protected by Copyright and Related Rights

The language digitalization and further text and data analysis (text mining) presume the usage of the various texts and materials. From the perspective of the copyright protections, the texts can be classified into three groups: non-protected texts (e.g., official documents), "safe" texts that can refer to the non-protected and copyright protected texts depending on their content (e.g., manuals, reports, etc.) and copyright protected texts [12]. To recognize the texts as copyright-protected text or to identify the "safe" text as a copyright-protected, the originality[3] of the text should be evaluated. The originality requirement is harmonized within the EU jurisdiction by the court practice of the Court of Justice of the European Union (CJEU) [10] and covered by the concept of the author's creativity. Most of the EU national jurisdictions define that to receive copyright protection, the work should be created in connection to the author's mind and personality, where the author is defined as a human (e.g., Germany, Spain, France).[4] However, this creates a legal uncertainty to the texts and works created by the computer with minimum human effort (e.g., automatically generated texts).[5] If no human effort was applied for work creation, then the work's author could not be identified, and therefore the work could not be considered copyright-protected work.[6]

---

[2] Directive (EU) 2019/790 of the European Parliament and of the Council of 17 April 2019 on copyright and related rights in the Digital Single Market and amending Directives 96/9/EC and 2001/29/EC (DSM Directive), dated 17 April 2019 <https://eur-lex.europa.eu/legal-content/EN/ALL/?uri=CELEX:32019L0790> (accessed 14 October 2021).

[3] The Berne Convention for the Protection of Literary and Artistic Works, signed at Berne on September 9, 1886 (Berne Convention) states that the work is copyright protected if it fulfils the requirement of the originality.

[4] The "human-centered approach" was also supported by The European Parliament (EP) resolution of 20 October 2020 on intellectual property rights for the development of artificial intelligence technologies <https://www.europarl.europa.eu/doceo/document/TA-9-2020-0277_EN.html> (accessed 14 October 2021). According to the resolution this approach reflects ethical principles and human rights.

[5] SCIgen - An Automatic CS Paper Generator https://pdos.csail.mit.edu/archive/scigen/ (accessed 14 October 2021).

[6] For further discussion of the problem of copyright protection of the works automatically created by the LT application see Ilin, I., & Kelli, A. (2019). The Use of Human Voice and Speech in Language Technologies: The EU and Russian Intellectual Property Law Perspectives. Juridica International, 28, 17–27. https://doi.org/10.12697/ji.2019.28.03.

The LR may also use the materials used to make work available to the public (e.g., using the records and broadcasting audiovisual work). This kind of material is protected by related rights protection. The related rights protection is connected to the beneficiary of these rights (performers, producers, and broadcasting organizations). The process of the identification of the beneficiary of the right causes fewer legal concerns. However, if the beneficiary of the related rights is a performer, then it is required to establish the performer's personality, and therefore the problem of evaluation of the human interaction becomes topical as well as identification of the author of work.

Two models can be applied for the usage of the copyright and related rights-protected materials: the "contract model" and "exception model." The exception model presumes the limitation of the author's exclusive rights. The contract model presumes that before using the copyright-protected text, the appropriate authorization (based on the author's consent) needs to be received [9].

Exception model. The exception model relies on the "Berne three-step test"[7] that allows members of the Berne Convention to set the exceptions and limitations of the copyright rights with the following requirements:

- the formula for provided exception or limitation should be defined in a specific way and does not have a general character;
- provided exception or definition shall not prejudice the normal use of the work;
- provided exception or definition should not unreasonably infringe upon the legitimate interests of the author.

Based on this "Berne three-step test," the following exceptions were implemented to the national legislation of the EU member states: exception for scientific research, temporary copies exception, and exception provided for text and data mining for scientific research (TDM exception).[8]

From the social perspective, the development of the LR based on the exception model is much easier to perform, mainly because there is no need to receive the author's consent and therefore presumes fewer risks [9].

However, the main legal concerns in this model refer to its application to the scientific research that is done for commercial purposes (e.g., cases when LR developed within the company R&D department) or LR after its creation was transferred to the commercial entity (e.g., spin-off company). To identify the possibility of exception model application, the cases of the LR development together with the framework and national legislation should be investigated.

---

[7] Article 9 (2) of the Berne Convention (Note 3).

[8] Article 3, Article 4 DSM Directive (Note 2).

Contract Model. The contract model presumes that the LR developer needs to receive permission from the author to usage copyright-protected materials (e.g., conclude license agreement). From the legal perspective, the contract model is to a large extent, solves the problem of the commercial research and commercial creation of the LR. However, technically, the execution of the contract model has a lot of impediments and drawbacks, mainly related to the IP audit (author's identification, analysis of the work creation process, identification of the exclusive rights holder, etc.) and negotiation process. This process can be quite long and costly processes that could negatively affect LR creation and LT development. However, the application of the contract model become different in the case of developing the LR by Internet giants such as Google, Yandex, etc. For example, in the case of Alisa (Yandex voice assistance), the assistant uses the samples of inputs not only from the assistant app but also from the other Yandex services (e.g., Yandex navigation system, Yandex taxi, Yandex translation, etc.) and insert the relevant provisions in license agreements for these services. At the same time, if the LR developer does not have their materials for LR creation, then applying the contract model is a costly and long-term solution.

Identification of the Legal Risks. The identification of the legal risks from the perspective of the copyright and related rights protection should be done by analyzing specific cases of the LR creation. However, to minimize the risks, the following algorithm of the research could be applied. First, it is necessary to identify the type of material that is used for LR creation. Then one needs to circle the main characteristics of such materials: do they express an idea (e.g., novels, texts, etc.), or should this work be considered as a divertive work (e.g., translation) or used in the process of making the work available to the public (e.g., audiovisual materials)? After that, depending on the work's characteristics, the following steps should be taken:

- in case the work expresses ideas – the originality test should be done;
- in case the work is considered as a divertive work – the authorization of usage of the initial work should be checked;
- in case the voice and speech are used in the process of making the work available to the public: (a) beneficiary of the related rights should be identified, (b) the authorization of usage of the initial work should be checked.

These steps lead to the possibility to outline the following groups of the legal risks systemized in Table 1.

Table 1. Legal risks mapping.

Step	Risk	Solution
Originality test	Risk not to grant the copyright protection	(a) Identify the author (b) Measure the level of human interaction
Divertive work	Risk of the unlawful usage of the initial work	Check the authorization
Bringing work to the public	(a) Risk not to grant the protection of the related rights; (b) Risk of the unlawful usage of the initial work	(a) Identify the author (b) Measure the level of human interaction (c) Check the authorization

2.2 The Usage of the LR Materials that Contain Personal Data

The development of the LR may include the usage of copyright and related rights-protected materials and the materials that contain personal data. The legal ground for the European data protection mainly rests on the GDPR[9], which is directly applicable in all the EU Member States[10] and international regulations such as the European Convention on Human Rights (ECHR)[11] and Convention for the Protection of Individuals with regard to Automatic Processing of Personal Data (Convention 108).[12]

Under the GDPR the personal data is *"any information relating to an identified or identifiable natural person ('data subject'); an identifiable natural person is one who can be identified, directly or indirectly, in particular by reference to an identifier such as a name, an identification number, location data, an online identifier or to one or more factors specific to the physical, physiological, genetic, mental, economic, cultural or social identity of that natural person."*[13]

The GDPR classifies different types of personal data (e.g., general, biometric, genetic, health). However, the legal analysis shows the most crucial differences lie between the general and the special categories of personal data, including biometric, genetic, and health [5].

In accordance with the GDPR, special categories of personal data are defined as: *"data revealing racial or ethnic origin, political opinions, religious or philosophical*

---

[9] General Data Protection Regulation (Note 1).

[10] GDPR territorial scope is not limited by the EU states it is also applying to the members of the European Economic Area (EEA) countries and in certain circumstances to the non-EU/EAA companies.

[11] Art. 8 Convention for the Protection of Human Rights and Fundamental Freedoms (ECHR), reference ETS No.005, Treaty open for signature by the member States of the Council of Europe and for accession by the European Union at Rome 04.11.1950. Entry into force: 03.09.1953.

[12] Convention for the Protection of Individuals with regard to Automatic Processing of Personal Data, reference ETS No.108, Treaty opened for signature by the member States of the Council of Europe and for accession by the European Union at Strasbourg 28.01.1981. Entry into force: 01.10.1985.

[13] Article 4 General Data Protection Regulation (Note 1).

beliefs, or trade union membership, and the processing of genetic data, biometric data for the purpose of uniquely identifying a natural person, data concerning health or data concerning a natural person's sex life or sexual orientation."[14]

Defining personal data as a general or special category of personal data is essential for further processing. For instance, the processing of the special category by general rule is prohibited.[15]

The scope of types of information that could be considered personal data and its forms is very large. Under Article 29 Working Party[16] (WP29), the personal data could be available in different forms (e.g., graphical, photographic, acoustic, alphabetical, and so forth) and kept on different storages (e.g., videotape, paper, computer memory)[17] unless the processing will satisfy the requirements set in the Article 9 (2) GDPR.

The wide scope of the information (its types and formats) leads to the necessity to comply with personal data processing requirements within the LR creation.

Personal Data Processing. The GDPR comprehensively defines data processing so that it covers practically all activities that can be done with personal data. Under the GDPR, data processing involve such operations with data done by automatic or non-automatic means and refers to such activities as collecting, recording, structuring, storing, usage, transmission, and so forth.[18]

The lawful processing of personal data should be based on an appropriate basis. As well as the usage of the copyright and related rights materials, this lawful usage of the personal data can be based on the consent[19] (consent model) or rely on exceptions (e.g., research exception, public interest exception, legitimate interest).[20] The GDPR does not specify the meaning of the research exception. The problem of its application should be analyzed from the perspective of national legislation of the EU member states (for instance, the applicability of the research exception for the research conducted for commercial purposes).

The data processing must be done with strong compliance with the rules. There are three groups of such regulations: rules regarding lawful processing, rules regarding security processing, and rules regarding transparency processing [5].

The problem of the unclear meaning of the research exception and the complexity of the consent concept, and strict rules of data processing make the LR creation quite

---

[14] Article 9 General Data Protection Regulation (Note 1).

[15] Article 9 (1) General Data Protection Regulation (Note 1).

[16] Article 29 Working Party is an advisory committee established with the Data Protection Directive (95/46) (repealed as of 25 May 2018). Its opinions are still relevant since the nature of personal data protection has not changed.

[17] Article 29 Working Party. Opinion 4/2007 on the concept of personal data. Adopted on 20th June, p. 7.

[18] The whole list of the operations that are considered as data processing is set at the Article 4 (2) General Data Protection Regulation (Note 1).

[19] Due to reasons of space and the focus of the papers, the requirements consent requirements are not addressed. For further discussion see WP29. Guidelines on Consent under Regulation 2016/679. Adopted on 28 November 2017.

[20] For the further discussion of the provided exceptions see WP29. Opinion 06/2014 on the notion of legitimate interests of the data controller under Article 7 of Directive 95/46/EC.

challenging. However, in some cases, the problem of compliance with data protection rules can be partly solved by using anonymized data in the LR creation process.

The Anonymized Data. One of the solutions for LR creation is using anonymized data[21]. The usage of non-personal data has fewer legal restrictions. The GDPR states that *"The principles of data protection should therefore not apply to anonymous information, namely information which does not relate to an identified or identifiable natural person or to personal data rendered anonymous in such a manner that the data subject is not or no longer identifiable."*[22]

Data can be non-personal from the day of creation (e.g., machine-generated data that do not contain personal information), or personal data can be anonymized. However, it should be taken into account that based on the definition of the personal data processing provided by the GDPR (provided in a previous chapter), the anonymization process by itself can be considered as personal data processing. Therefore, there is a need to comply with data processing requirements.

3 Criteria and Legal Mechanisms for Language Resource Database Protection

In a broad sense, databases can be considered as an object of intellectual property (intellectual property asset). Investigations of the criteria and legal mechanisms for database protection require first to analyze the legal nature of the database, to distinguish it from other intellectual property objects (e.g., novels, films, etc.) that also consist of different elements, and then investigate the applicable mechanisms of the legal protection.

3.1 Language Resource Database Characteristics

The EU Directive on database protection defines the database as a "collection of independent works, data or other materials arranged in a systematic or methodical way and individually accessible by electronic or other means."[23] Therefore, the following characteristics of the database can be indicated:

- independence of the elements;
- systematic order;
- individual accessibility by electronic means.

The EU Directive on database protection, as well as the CJEU court practice, does not define the meaning of "accessibility by electronic means" [1]. However, the meaning of the other characteristics can be found in the CJEU court practice. The independence

---

[21] There is no need for LR materials to identify individuals.

[22] Recital 26 General Data Protection Regulation (Note 1).

[23] Article 2 (1) Directive 96/9/EC of the European Parliament and of the Council of 11 March 1996 on the legal protection of databases, dated 11 March 1996 < https://eur-lex.europa.eu/legal-content/EN/ALL/?uri=celex:31996L0009 > (accessed 14 October 2021).

of the elements presumes that the database elements can be freely removed without any damage to other elements.[24] The systematic character assumes that each database element is classified and placed in a specific order. The systematic order of the database elements is used to organize the search within the database and separate elements from each other.[25]

3.2 Legal Mechanisms for Database Protection

If the LR database meets the characteristics mentioned above and thus is considered as a database, then it can be protected by law. There are two main options for protecting the LR database as a valuable commercial asset: through copyright protection and through sui generis rights protection [2]. The right choice of the legal frameworks for the protection depends on the level of the database author's creativity and investments that were made to create the database.

The created database should be prepared with a strong connection to the author's intellect and personality to be new, original, unique, and creative to receive copyright protection.[26] In other words, to receive copyright protection, the level of creativity should be evaluated.

The database protection through sui generis rights protection requires that the "qualitatively and quantitatively a substantial investment" was made in regard to the created databases.[27] Therefore, to receive the database base protection through sui generis rights, the level of financial and professional resources (e.g., money, energy, time, effort) should be measured. At the same time, the database protection through sui generis rights does not require the database to be original and creative.

The copyright protection and sui generis rights protection are granted automatically and do not require any registration procedures. However, it means that the database developers should evaluate the database's author's creativity and measure the investment level should be done. Therefore, the authors of the database decide and choose the options for its protection. At the same time, the way of how the database is protected has significant meaning in terms of its further transferring and distribution.

4 Conclusion

In conclusion, it must be said that the development of LT and LR plays an important role in both the social and technology spheres. However, from the legal perspective, the

---

[24] This characteristic separates the database form other works that consists of a scope of elements (e.g. novels).

[25] Judgment of 9 November 2004, Fixtures Marketing Ltd v Organismos prognostikon agonon podosfairou AE (OPAP), C-444/02 2004, EU:C:2004:697; Judgment of 9 November 2004, Fixtures Marketing Ltd v Oy Veikkaus Ab, C-46/02 2004, EU:C:2004:694; Judgment of 9 November 2004, Fixtures Marketing Ltd v Svenska Spel AB, C-338/02 2004, EU:C:2004:696.

[26] The concept of the author's creativity is based on "originality" requirement for copyright protection provided by the Berne Convention (Note 3).

[27] Article 7 Directive on the legal protection of databases (Note 23).

creation and development of the LR and LT raise many concerns, mainly from IP and data protection perspectives.

The problem of LR creation is a starting point in the investigation of LT development. The existing models of LR creation leave much space for discussion (e.g., the problem of applying the exception model to the commercial research). The author argues that due to the specific character of the LT development and LR creation, these processes need to be done within specially developed frameworks. However, as developing the frameworks is a long and costly procedure, the addressed in the paper risk should be taken into consideration and further investigated both from the practical and theoretical perspectives.

References

1. Derclaye, E.: What is a database? J. World Intellect. Prop. **5**(6), 981–1011 (2005). https://doi.org/10.1111/j.1747-1796.2002.tb00189.x
2. Deveci, H.A.: Databases: is Sui generis a stronger bet than copyright? Int. J. Law Inf. Technol. **12**(2), 178–208 (2004). https://doi.org/10.1093/ijlit/12.2.178
3. Hutchins, W.J.: The Georgetown-IBM experiment demonstrated in January 1954. In: Frederking, R.E., Taylor, K.B. (eds.) AMTA 2004. LNCS (LNAI), vol. 3265, pp. 102–114. Springer, Heidelberg (2004). https://doi.org/10.1007/978-3-540-30194-3_12
4. Ilin, I., Kelli, A.: The use of human voice and speech in language technologies: the EU and Russian intellectual property law perspectives. Juridica Int. **28**, 17–27 (2019). https://doi.org/10.12697/ji.2019.28.03
5. Ilin, I., Kelli, A.: The use of human voice and speech for development of language technologies: the EU and Russian data-protection law perspectives. Juridica Int'l. **29**, 71 (2020)
6. Jents, L., Kelli, A.: Legal aspects of processing personal data in development and use of digital language resources: the Estonian perspective. Jurisprudence **21**(1), 164–184 (2014). https://doi.org/10.13165/jur-14-21-1-08
7. Kawashima, N.: The rise of "user creativity" – Web 2.0 and a new challenge for copyright law and cultural policy. Int. J. Cult. Policy **16**(3), 337–353 (2010). https://doi.org/10.1080/10286630903111613
8. Kelli, A., Tavast, A., Pisuke, H.: Copyright and constitutional aspects of digital language resources. Juridica Int'l. **19**, 42 (2012)
9. Kelli, A., Vider, K., Lindén, K.: The regulatory and contractual framework as an integral part of the CLARIN infrastructure. In: Selected Papers from the CLARIN Annual Conference 2015, 14–16 October 2015
10. Rosati, E. (ed.): The Routledge Handbook of EU Copyright Law. Routledge, New York (2021)
11. Street, J., Negus, K., Behr, A.: Copy rights: the politics of copying and creativity. Polit. Stud. **66**(1), 63–80 (2017). https://doi.org/10.1177/0032321717706012
12. Truyens, M., Van Eecke, P.: Legal aspects of text mining. Comput. Law Secur. Rev. **30**(2), 153–170 (2014). https://doi.org/10.1016/j.clsr.2014.01.009

Correction to: ANNPRO: A Desktop Module for Automatic Segmentation and Transcription

Katarzyna Klessa⬤, Danijel Koržinek⬤,
Brygida Sawicka-Stępińska⬤, and Hanna Kasperek⬤

Correction to:
Chapter "ANNPRO: A Desktop Module for Automatic
Segmentation and Transcription" in: Z. Vetulani et al. (Eds.):
Human Language Technology, **LNAI 13212,**
https://doi.org/10.1007/978-3-031-05328-3_5

In an older version of this paper, there were errors in the affiliations of three authors, Katarzyna Klessa, Brygida Sawicka-Stępińska and Hanna Kasperek. Three authors were incorrectly affiliated with the Institute of Romance Studies. This has been corrected to the Faculty of Modern Languages and Literatures.

The updated original version of this chapter can be found at
https://doi.org/10.1007/978-3-031-05328-3_5

Author Index

Printed in the United States
by Baker & Taylor Publisher Services